Frontiers of Broadband,
Electronic and Mobile Commerce

Contributions to Economics

www.springeronline.com/series/1262

Acknowledgments

We owe most to the authors of the 18 chapters that comprise this volume for their intellectual skills, their interest in submitting their papers for consideration for the 2003 International Telecommunications Society Regional Conference or by responding to an invitation to submit a paper for this volume, and their willingness to further refine and revise their papers for this volume. They were a pleasure to work with. As Cooper and Madden co-wrote three chapters, their thanks must come from the other.

We also want to thank Katharina Wetzel-Vandai, the Economics Editor at Physica-Verlag who was enthusiastic about signing on to publish this book. We are pleased to be a part of the first rate list of works they publish in their *Contributions to Economics* series. Physica-Verlag provided all the support we could have hoped for.

Michael Schipp provided superb administrative support from the conference proposal to final publication, support which is critical to the book's timely completion. His exceptional copyediting, desktop publishing and indexing skills are gratefully acknowledged.

Russel Cooper and Gary Madden
February 2004

Contents

PART II: TECHNOLOGY AND PRODUCTIVITY

5 Deciding on Network Architecture for 3G Wireless Services
Hak Ju Kim

6 Measurement of TFP Growth for US Telecommunications
Jeffrey I. Bernstein and Charles J. Zarkadas

PART V: DEVELOPMENT IMPERATIVE

16 North African Information Networks
Andrea L. Kavanaugh

17 OECD Broadband Market Developments
Dimitri Ypsilanti and Sam Paltridge

18 Understanding the Evolving Digital Divide
Russel Cooper and Gary Madden

Introduction

Russel Cooper and Gary Madden

The present volume analyses the frontiers of broadband, electronic and mobile commerce markets. High-capacity and intelligent mobile telecommunication networks have resulted in new services, such as SMS and Internet banking. Growth in mobile Internet network infrastructure and subscription has provided a base for the development of e-commerce. Accordingly, recent research on broadband networks is forward-looking, e.g., forecasting Internet telephony adoption and the structure of future retail markets. The broadband regime brings with it concerns of identifying appropriate standards and delivery for universal service. Regulation and pricing are matters of importance as well as appropriate investment decisions within a market of ongoing innovation.

The volume is divided in five parts: e-commerce business models; network technology and productivity; demand and pricing; market growth, regulation and investment; and issues related to the development imperative. The structure of the volume is guided by the basic themes considered at the International Telecommunications Society's Asia-Australasian Regional Conference "M-business, E-commerce and the Impact of Broadband on regional Development and Business Prospects", which took place in Perth Western Australia on 22-24 June 2003. The volume contains a selection of papers presented at this conference as well as four additional invited papers, commissioned to augment the volume. The invited papers are authored by Jerry Hausman (Chapter 1), Jeffery Bernstein and Charles Zarkadas (Chapter 6), M. Ishaq Nadiri and Banani Nandi (Chapter 8) and Glenn Woroch (Chapter 13).

The Conference was sponsored by: AusAID, Australian Information Industry Association, Curtin Business School and the CBS School of Economics and Finance, the Commonwealth Department of Communications, Information Technology and the Arts, WA Department of Industry and Resources, Edward Elgar Publishers, Elsevier Science Publishers, Ericsson, the Foundation for Development Cooperation, France Telecom, IDATE, Optus, Macquarie Bank, NERA Economic Consultants, Network365, Perth Conference Bureau, Physica-Verlag, The West Australian, Taylor Nelson Sofres, the University of Western Sydney and the UWS School of Economics and Finance, and Wesfarmers Limited. The Conference also received endorsements from the National Office for the Information Economy, the Organisation for Economic Cooperation and Development and the Information for Development Program of the World Bank. The conference was attended by 200 researchers, practitioners and regulators involved in telecommunications market analysis.

Part I: E-Commerce Business Models

The volume begins with four chapters concerned with the 'proper' role of government and of competition in the operation of markets capable of delivering e-commerce activity, and how incumbents and entrants act within these markets. In particular, in Chapter 1, 'Cellular 3G Broadband and WiFi', Jerry Hausman considers competition, and the role of both business and the US Government for this emerging market. Hausman indicates that both 3G cellular and WiFi substitution is occurring, while the broadband Internet market is growing. He argues that policy and regulation need to be designed so as to not retard investment. Also, 3G and WiFi may solve the regulatory problem by offering both voice and broadband data. To succeed, 3G needs compelling applications, spectrum and high-speed access. WiFi and WLL need sufficient spectrum and changes to government regulation. Should 3G and WiFi provide sufficient competition then acceleration in the introduction of new innovation in telecommunications markets would result. The best outcome would be turning the telecommunications sector into one where competition takes over and so eliminates the need for overbearing government regulation. Charles Steinfield (Chapter 2 'Geographic and Socially Embedded B2C and B2B E-Commerce') emphasizes the means through which e-commerce works in concert with established transactions and client practice. An alternative 'situated' e-commerce approach is proposed. This view implies that: physical location matters for both B2C and B2B e-commerce; for firms with a physical retail presence e-commerce is best viewed as complementary to physical market activity; and that—especially for B2B transactions—e-commerce is best employed to strengthen existing business structures as a network-effects only focus is potentially damaging to supply chain relationships. In Chapter 3 ('SME International E-Commerce Activity'), James Tiessen examines SME internationalization and Internet use. In particular, Internet technology assists SME to acquire base foreign market intelligence, establish international networks, and achieve improved economies of scale through inexpensive advertising and the provision of their offerings to niche markets. The model developed provides insight as to how and why SMEs use the Web to export globally. Tiessen argues that it remains unclear whether the Web leads to further SME internationalization. The case study also shows that Internet technology is reducing the volume of personal calling necessary to secure new sales. However, it cannot be assumed that inevitability most firms will become sophisticated Web users. Phil Malone (Chapter 4 'SME Interaction in Supply Chains') concludes Section I by noting that e-business encompasses e-procurement, supply chain management and the transformation of firms' internal functions so as to allow the seamless transfer of information along supply chains. The role of the NOIE, in the development of Australian e-business markets, involves the brokering of business relationships and the playing of a catalytic role in innovative projects, so as to: promote the effective use of e-business that results in productivity growth for the Australian economy; facilitate firm e-business adoption—especially by small business; remove impediments to small business participation in e-business by reducing associated costs, complexity and risk; and ensuring government e-business activity is consistent with broad industry contexts. Accordingly, NOIE provides independent assessment of contemporary developments in Australian e-business markets. The core emerging issues identified by NOIE for Australian e-business growth revolved around necessary cooperation on standards, greater interoperability between e-business frameworks and the effective integration of new and established technology.

Part II: Technology and Productivity

The next four chapters analyze traditional areas of concern to firms and regulators in telecommunications markets, i.e., technology choice and its efficient operation. Hak Ju Kim (Chapter 5, 'Deciding on Network Architecture for 3G Wireless Services') considers that evolution from existing architecture to emerging 3G wireless networks will increase both technological complexity and uncertainty facing network operators. As such, optimal technology choice requires that carrier management examine technology scenarios as strategic decisions. The principal intention is to develop a conceptual framework through which wireless network operators can support their strategic decisions concerning next generation network architecture. Typical network architecture migration paths, such as GSM and CDMA based network scenarios, are examined using innovation theory. Strategic options for network migration based on the real options are also discussed. In Chapter 6, 'Measurement of Total Factor Productivity Growth for US Telecommunications', Jeffrey Bernstein and Charles Zarkadas develop two sets of total factor productivity indexes. First, an FCC TFP study used to calculate X-factors for ILEC interstate services is updated. This method is based on the ILECs regulated books of account, which includes the output categories: local, intrastate, and interstate services. A modified TFP index disaggregates intrastate services into switched access and toll services. The revised measure of intrastate output accounts for toll and access price changes, which are regulated by state regulatory agencies and the changing pattern of respective volumes over time. Another modification involves the introduction of a broadband service into the ILECs supply high bandwidth transmission services to carriers who offer broadband DSL or high-speed Internet services. Chapter 7 by Russel Cooper, Gary Madden and Grant Coble-Neal ('Measuring TFP for an Expanding Telecommunications Network') utilizes a generalization of the translog cost specification that allows for structural breaks across long time series to decompose TFP into continuous and discontinuous effects. This represents an alternative dichotomy to standard short-run (disequilibrium) versus long-run (equilibrium) classifications. While a dynamic model may be necessary to investigate network disequilibrium associated with ongoing network geographic expansion for much of the 20th century it is also arguable that predictable or continuous aspects of this expansion could have been fully factored into carrier provider plans, so that massive technological change over this period is more suggestive of an overlay of steady technological progress with periodic discontinuities than it is of disequilibrium adjustment. M. Ishaq Nadiri and Banani Nandi (Chapter 8, 'Dynamic Aspects of US Telecommunications Productivity Measurement') concludes this section with a discussion of the contribution of the telecommunications industry to the US economy by addressing the trend in TFP growth in the US telecommunications since the mid-1930s, and the changing sources of TFP growth. Dynamic aspects of the telecommunications sector and its impact on the production structure of other industries, i.e., changing factor ratio in different industries, and spillover or network effects on the output and productivity growth of other sectors and industry are also examined. The spillover or network effects of telecommunications infrastructure are then explored. Total benefit derived from telecommunications infrastructure by US industry is presented.

Part III: Demand and Pricing

The following three complementary chapters concern wireless demand and the pricing of information goods. Sang-Kyu Byun, Jongsu Lee, Jeong-Dong Lee and Jiwoon Ahn in Chapter 9 ('Korean Wireless Data Communication Markets and Consumer Technology Choice') forecast wireless data communication services demand patterns. To do so, the characteristics of wireless data communication (technology and services) are used to estimate consumer attribute preferences. Conjoint designs are employed to construct choice sets containing these characteristics. From these data the model is estimated and predictions made about communication service market evolution. In particular, the analysis considers mobile Internet and wireless LAN services. Common and competing technology used by the services is also discussed. Paul Rappoport, Lester Taylor and James Alleman (Chapter 10, 'WTP Analysis of Mobile Internet Demand') analyze US wireless Internet access demand with household WTP survey data. The theoretical framework identifies WTP for consumer surplus obtained from network use. Their results suggest that demand is elastic for prices currently charged by wireless service providers. Determinants of demand are consistent with those for Internet and broadband access. Age of the household head and household income are important determinants in the WTP function. Indicators of Internet use include the presence of a PCS telephone and personal computer within the household. Additionally, WTP is strongly related to whether the household has recently relocated. A WTP frequency distribution indicates that many respondents are willing to pay US$ 50 for wireless access. Finally, concentrations of WTP values at US$ 20, US$ 25 and US$ 30 suggest wireless providers should vary their plans to better meet customer needs. In Chapter 11 ('Asymmetry in Pricing Information Goods'), Yong-Yeop Sohn investigates the pricing of an information good, and analyzes factors that determine the optimal pricing strategy. He shows that a firm producing an information good (characterized by a network externality) should adopt an introductory pricing strategy to secure an installed base of customers. When the product is an upgrade of an existing version, the firm is better off charging a profit-maximizing price initially to the installed base of established customers, and then lower price to help secure the customer base for the next version. Another finding is that an information good with a network externality is likely to provide less profit than non-information goods, since information good producers must ensure they retain a substantial customer base.

Part IV: Market Growth, Regulation and Investment

The four chapters that appear next concern the importance of network effects for system growth, and the role of appropriate regulation to provide a suitable backdrop to enable this growth to occur. In Chapter 12 ('Measuring Telecommunication System Network Effects'), Gary Madden, Aniruddha Banerjee and Grant Coble-Neal analyze consumer demand for network services. A subscription model is specified that allows interaction among the network effects for competing networks. The model also examines whether consumers consider the combined network size of compatible fixed-line and mobile telephone systems when deciding to subscribe. Estimated network effect magnitudes for fixed-line telephony, mobile telephony and the Internet are derived, and interactions among network effects are reported. In Chapter 13 ('Open Access Rules and Equilibrium Broadband Deployment') Glenn Woroch examines the impact of alternative open access rules using an equilibrium model of a broadband deployment race. The rules alter deployment timing by contestants, typically resulting in delay. Delays are traced to a reduced investment incentive in broadband facilities relative to service-based alternatives, or relative to no investment. Glenn also finds that asymmetric treatment of carriers has substantial effects on the pattern of deployment. The analysis is limited to assessing open access rules in terms of their impact on timing of broadband deployment. Johannes Bauer (Chapter 14, 'Spectrum Management and Mobile Telephone Service Markets') examines the implications of policy for innovation within the mobile services industry. The analysis indicates a framework of spectrum use markets fits broadly with innovation scenarios. Proponents of free spectrum access argue that ubiquitous underlay rights are sufficient for emerging mobile technology. However, conflict of interest exists between primary and underlay users. An alternative is to create regimes for designated bands however this requires a metric for comparison. Further, spectrum policy is only a component shaping innovation processes. Other factors include policy governing access to a communications gateway or business strategy that interacts with spectrum policy. Also, the complexity of the value chain makes open access regimes neither necessary nor sufficient for vibrant upstream industry. A system of exclusive rights in spectrum is conducive to innovation. As a best regime does not exist, policy makers are advised to allow competition between institutional regimes. In Chapter 15 ('Rational Explanations of ICT Investment'), Russel Cooper and Gary Madden construct a model of investment behavior by a typical ICT firm. As investment decisions are made in an uncertain environment, occasionally 'bad news' dominates. To rationally explain the recent ICT sector investment experience, components of investment are decomposed into factors that influence optimal investment decision making, that is, profitable production; optimal portfolio choice; strategic merger and acquisition; shareholder satiation; and futures preparation. Strands of the received investment literature are associated with these factors; however it is the combination of influences that leads to investment outcomes. Finally, interaction among these factors may lead to complex and possibly undesirable outcomes that can be characterized as the result of rational decision making.

Part V: Development Imperative

The final three chapters concern the digital divide and regional information net-works. Andrea Kavanaugh (Chapter 16, 'North African Information Networks') argues that mobile telephony, especially in developing countries, is an appropriate transition technology for teaching computer networking skills to mostly illiterate populations, as technical support is provided through social networks. North African data show poor quality fixed-line telephone service, rapid mobile telephone and Internet subscription growth when service is provided via competitive markets. To effectively use the Internet requires reading and computer literacy skills, and an on-line social network. Social networks are important in motivating individuals to access the Internet. Simply, the Internet will diffuse more broadly in North Africa and developing country populations, more generally, via mobile communication, and not through desktop computers attached to fixed-line telephone network connections. Policy should reflect this reality so that the economic benefits of information networks can be more widely appropriated by developing country populations. Dimitri Ypsilanti and Sam Paltridge (Chapter 17, 'OECD Broadband Market Developments') argue that mapping broadband availability and allowing firms to define their threshold for the providing service is integral to the process. To ensure transparency this information should be made publicly available. Further, government should strive to strengthen competition in local access markets. In rural areas, entrants may provide innovative solutions for areas that are not profitable to serve via DSL or cable modem. A corollary proposal is to support competing platforms and open existing platforms to competition. Initial evidence indicates that unbundling has some entrants setting lower exchange upgrade thresholds than incumbents. Government should review spectrum allocations to facilitate entry and experimentation. Further, government can act to aggregate broadband demand so that it is profitable for efficient firms to upgrade their net-works—both access and backhaul. Finally, government should deliver services via broadband networks to stimulate use, provide more efficient delivery and increase rural access. The final chapter in this volume (Chapter 18, 'Understanding the Evolving Digital Divide') by Russel Cooper and Gary Madden posit the existence of representative North (developed country) and South (developing country) firms that are differentiated by their productivity. Also South downstream application sector firms are allowed to invest in North ICT. Asymmetry in the impact of un-certainty specifies different responses to 'good' and 'bad' news. Implications of these features on the digital divide are analyzed. Further, the structure of capital stock transition equations obtained from North-South firm investment decisions made in an uncertain environment are derived. The model uses isoelastic con-sumer preferences and simplified technology to illustrate that optimal investment results highlight the complexity of North and South firm investment and growth relationships, and suggests that hoped for technology transfer 'solutions' to neces-sitate growth and convergence are too simplistic. In particular, study findings show that asymmetric uncertainty can exacerbate the digital divide.

Part I: E-Commerce Business Models

1 Cellular 3G Broadband and WiFi

Jerry Hausman

Introduction

This chapter focuses on a potentially important emerging telecommunications market, cellular 3G broadband and WiFi. This study also concerns competition, and the role of both business and the Government. This topic is particularly interesting, because through to 2000 the industry remained buoyant with stock prices appreciating. Until then, telecommunications companies had invested in new technology, wireless spectrum licenses and acquisitions at an extremely high rate. From the enactment of the US Telecommunications Act of 1996, through 2000, the telecommunications industry raised approximately US$ 1.3 trillion in new equity capital from Wall Street. An even greater debt accumulated. According to the *Wall Street Journal*, telecommunications firms borrowed about US$ 1.5 trillion in debt from banks, and issued US$ 630 billion in corporate bonds from 1996 through 2001. By 1999-2000, new telecommunications-related equipment accounted for 12% to 15% of capital investment among the Standard & Poor 500 firms. As this equipment was being installed demand for new telecommunications services grew at up to 100% per quarter. Demand for additional services continues to grow, but at a tenth of that rate. Much of this optimism, and related aggressive investment activity, was based on the premise that high demand for broadband access to voice, video and data would provide room for many competitors backed by their infrastructure. Financial markets eventually realized the true state of the industry. About US$ 2 trillion of telecommunications industry value was lost through plummeting stock values and defunct corporate bonds. Of course, now things are in the basement, the important question is, where to next?

Consumer Welfare and Cellular Telephony

Cellular telephony was introduced in 1983, and with the Internet, is among the most beneficial new products of the last fifty years. It is quite remarkable that—in a field like telecommunications—two new products could have had such a profound affect. A salient feature of cellular telephony markets is that penetration is very high, with Australian penetration approximately 65%, for the US slightly lower at 50%, and 80% for Scandinavian countries. However, Asian nation penetration rates are even higher with, e.g., Korea at 90% and Taiwan 104%.[1] Finally,

---

[1] Taiwanese government statistics report cellular telephone handset numbers exceed the population. During a visit in 2002, I was assured their government statistics are accurate.

The People's Republic of China (China) has a cellular telephone penetration rate of 16%. That is, 16% of 1.3 billion persons translate to 208 million cellular subscribers, so the future of the cellular industry will be deeply affected by Chinese growth toward maturity. For many countries, cellular penetration exceeds that for fixed-line telephony. This is the case for Australia, Scandinavia and several Asian countries. Such acceptance has led to substantial improvement in consumers' welfare.

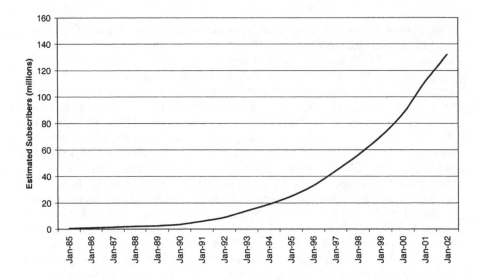

Fig. 1.1. US Cellular Penetration. Source: CTIA Semi-Annual Wireless Industry Survey

Cellular telephones were introduced to the US telecommunications market in 1984 in time for the Los Angeles Olympics. US market penetration is graphed from in January 1985 in Fig.1.1 through to mid-2002. Through to the late-1980s adoption remained sluggish. At that time, PacTel Cellular (later AirTouch) received advice that cellular adoption was poor because consumers placed too much weight on equipment purchase costs and underweighted subsequent service prices (Hausman 1999a). A selective review of the market research literature suggests that this behavior holds for consumers quite pervasively. The cellular telephones then available are very different from those in current supply. Cellular telephones were practically found only in cars, and priced at about US$ 2,000. In response to advice received, PacTel subsidized the price of the handsets provided to network subscribers. The handset supply cost was amortized and recovered through minimum service contracts with fixed service charges. However, in doing so PacTel entered into fierce price competition with Californian cellular telephone competi-

---

However, officials showed some surprise when I indicated, based on these data, that the penetration rate was 104%.

tors. This competition caused cellular handset prices to decline rapidly, until by 1993-1994, the price of handsets was effectively zero. This competitive strategy led to substantial increases in cellular adoption. This pricing strategy is now prevalent globally.

Consider next the increase in consumer welfare caused through the introduction of cellular telephony. Economists use the expenditure function to measure consumer welfare effects.[2] The idea is to find out how much income a representative consumer would relinquish and be as well off as without cellular telephony. Alternatively, if cellular telephone was taken away from the representative consumer, how much monetary compensation is required to make this individual as well off as with the telephone? For consumer surplus (CS), the expenditure function

$$CS = e(p_1, p_n^*, u^1) - e(p_1, p_n, u^1) \tag{1.1}$$

is defined as a function of p_1, the prices of all other goods, i.e., prices of cars, clothes, houses, gasoline, whatever. p_n^* is a virtual price that sets demand equal to zero, u^1 is the utility level and p_n is the price paid for cellular telephony and keeps the same utility level.[3] The expenditure function indicates the minimum income that a consumer needs at prices p to reach utility level u^1. Thus, from the virtual and actual prices of cellular telephony, the consumer welfare effect can be estimated.

Hausman (1997) estimates the virtual price for cellular at US$ 156 per month. This estimate indicates the upper limit price that an individual would typically pay to have cellular telephony. To obtain this estimate requires the use of panel data econometric techniques. Post estimation, it is always sensible to ask whether the estimate obtained is reasonable. The Hausman estimate is certainly a realistic number, because by examining the historical prices paid for cellular telephony service, it was not uncommon at the beginning, for more than US$ 156 a month to be paid for cellular service. The price has fallen a great deal since, but, this estimate is certainly in a range of what was paid at that time. The current price is about US$ 50 per month in the US. These data are obtained from the Cellular Telecommunications and Internet Association (CTIA) Web site.[4] With approximately 135 million US subscribers, the consumer surplus is estimated to be approximately US$ 150 billion p.a. The lower bound estimate, discussed in Hausman, is US$ 80 billion p.a. The estimates find consumer welfare from cellular telephony is between US$ 80 and US$ 150 billion, and this amount is from 0.8% to 1.5% of US GDP of approximately US$ 10 trillion.

---

[2] An older literature uses consumer surplus.

[3] The concept of virtual price, although he did not use the name, is first employed by Hicks (1940). Hausman (1996) resuscitated the concept when calculating how much utility is gained from the introduction of new cereals.

[4] http://www.wow-com.com/

Government policy can have a major effect on the introduction of new technology adoption and associated consumer welfare. In the US, the introduction of cellular telephony was delayed for approximately ten years, and this outcome is mainly because the government could not decide exactly who should be allowed to provide service (see Hausman 1997). Cellular technology was invented in the 1960's at Bell Labs.[5] The question the US Federal Communications Commission (FCC) was deciding on was who should be allowed to provide cellular services, because the FCC controlled the radio spectrum. Through the FCC's control of the radio spectrum, a company had to obtain permission to use the radio spectrum before providing cellular service. In the US the actual introduction of cellular service took place in 1983-1984. By that time cellular telephony had already been introduced in Japan and Scandinavia. Partly, pressure on the FCC to introduce cellular service came from returning US visitors to these countries.

Hausman (1997) estimates that this ten year delay in cellular telephony introduction in the US led to a loss of consumer welfare of about US\$ 50 billion p.a.. This estimate has never been challenged. In coming to terms with the magnitude of this estimate it is important to think about new technology differently than for most economic goods. For instance, Western Australia is a mineral rich part of the country, so delay in opening an oil well for ten years means no oil removed from the ground during that time. Still, ten years later the oil is still there, the time value of money is lost, but you can still remove the oil later. However, for new technology if the government had allowed cellular technology entry ten years earlier, it is not as if the resource is being depleted. So, government delay in this case is really deleterious to consumers.

Next consider China. In China, there are approximately 200 million cellular subscribers, more than any other country. While market penetration is only 16%, it now has more subscribers than the US. In considering how government policy can harm consumers, compare this outcome to that for India. In India, there are relatively few cellular telephone subscribers. The level of Indian corruption is such that they cannot seem to provide new technology. In India, cellular auctions are held and won regularly, but somehow the cellular industry never actually commences operation.

The Chinese wire-line market is experiencing 30% growth p.a., however by comparison the cellular market is growing at 80% p.a. In Shanghai or Beijing, which are relatively affluent, you observe more cellular handsets than you would in Boston. In China, the gain in consumer welfare is 3% of the GDP. Cellular and wire-line market growth are reported in Fig. 1.2. It was in 2000 that the market penetration lines crossed, and mobile is now surpassing wire-line penetration. There has been much recent debate that mobile telephony is less expensive than wire-line, and China seems to be the first large country where mobile growth may severely constrain wire-line penetration.

---

[5] Anecdotal evidence of the existence of latent demand for cellular telephony is found from US movies of the time. Car telephones were available, and there was a ten year wait list for car telephones.

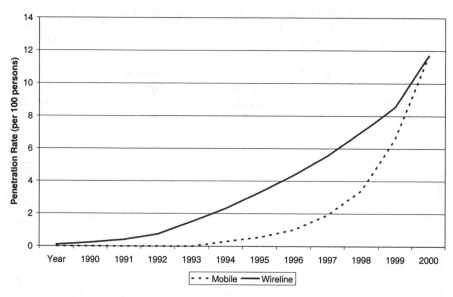

Fig. 1.2. Chinese Cellular and Wire-line Market Penetration

Cellular Broadband Internet Access

First-generation (1G) cellular telephones are analogue telephones, and second-generation (2G) handsets are digital telephones. In Australia, Telstra the former government company which is now part-privatized provides 2G. Competitors, Optus and Vodafone also provide 2G service. Vodafone is the largest cellular telephone provider in the world and third largest in Australia. Younger people increasingly use cellular service, among the reasons for this is that they do not have to pay a network connection fee when they move as with wire-line, also with cellular access the subscriber can always be reached. Cellular calling prices remain somewhat higher than those for wire-line service however, this difference is decreasing. The US cellular market is competitive, given the presence of 'bucket plans' for long-distance calling, when compared to wire-line long distance calling. Another lesson from US telecommunications markets is that subscribers purchase bucket plans, in which for US$ 30 a month, say, subscribers receive 120 calling minutes during peak hours, and basically unlimited off peak usage. What is interesting about bucket plans is that subscribers consistently purchase bucket plans that are too large based on their typical use. When usage is examined, subscribers almost never exhaust their minutes, which they should do occasionally if they are purchasing a plan that yields minimum cost over time. Currently, in the US about 3% of customers use only cellular telephony, viz., 3% of the population does not have a fixed-line telephone. Cellular telephony is also widely used for short message service. The main disadvantage of cellular service is lack of convenient use of the Internet. An earlier approach, wireless application protocol (WAP) was entirely unsuccessful, as it did not work at all well. WAP was too slow, with a speed of about 9.6 kbps, which is six generations ago for wire-line Internet access from modems.

Consider Korea which has the highest global broadband penetration. Korean broadband penetration approximately doubles annually, from 4.3 to 9.0 to 19.2 connections per 100 inhabitants. Canada is next, and the US interestingly enough, once third is now in sixth place, with 5.8% penetration reported in Fig. 1.3.

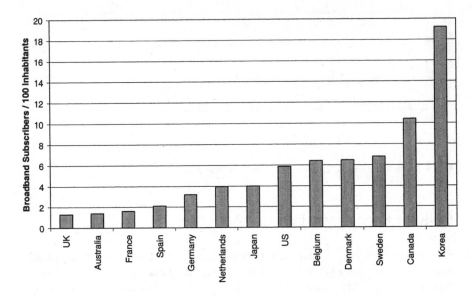

Fig. 1.3. Broadband Connections per 100 Inhabitants. Source: OECD

An interesting question is whether this outcome is a failure of industry, government or something that is not a matter of concern. In the US, there has been controversy about broadband penetration and whether government policy is slowing down the spread of broadband (Crandall and Alleman 2002). Hausman (2002) estimated an econometric model of the determinants of broadband penetration. Per capita income is isolated as an important determinant of penetration. High per capita income causes higher demand for computers, and also higher demand for broadband. However, another important factor is the relative price of narrowband. Narrowband is when broadband access is obtained from a telephone hook up. The US and New Zealand are the only countries in which there is no charge for local telephone calling, so unlimited narrowband is available. Narrowband Internet access is slow, but pricing is not time sensitive. In Australia, the price for local calling is about 18 US cents, so this is the cost for narrowband Internet access. This relatively small amount is among the reasons that the US and Australia have low broadband penetration. Almost everywhere else, including Korea, narrow band Internet access bears per minute charge. But, in the US, narrowband access is 'free', once the telephone line rental is paid. In Australia 18 US cents is paid and customers can remain on-line for as long as is desired. However there are other differences, in the US there is no charge for broadband beyond a monthly fee. In Australia both Optus and Telstra, which are the main broadband providers, charge depending on the volume of data that is transmitted. A final factor is government regulation. Korea, as it turns out, was completely unregulated for wire-line broadband until very recently. The US and Australia are both regulated. Both countries have unbundled network regulation. Clearly, this government regulation can actually have a retarding effect on the spread of broadband. The Australian Competition and Consumer Commission (ACCC) is the telecommunication sector regulator in Australia, and along with the FCC, may have retarded broadband adoption in Australia.

Effect of Regulation

When a network is unbundled incumbent telephone providers are required by the regulator to rent their network elements to competitors at regulated prices. The goal of such regulation should not be a competitor welfare goal, as regulators often seem to believe, but a consumer welfare goal. Indeed, the ACCC has adopted a consumer welfare goal. However, the FCC often seems more concerned with the welfare of competitors rather than those of consumers. This approach often leads to regulatory mistakes, and the courts often reverse FCC regulatory policy. Returning to the Korean case, the technical means to obtain broadband service now, and cellular 3G and with WiFi in the future, is through digital subscriber lines (DSL), i.e., over a telephone wire. So, the copper wires that carry voice also carry broadband. In Korea, Fig.1.4 shows that approximately 70% of broadband service is delivered by DSL. Another way to subscribe to broadband service is via cable modem. Cable modems provide broadband access for the remaining 30% of Korean subscribers. Regulators are often concerned with an incumbent having a dominant market share. The incumbent, Korea Telecom, has only 45% of the market, viz.; competitors have 55% of the market. While 45% is still a large share, Korea Telecom initially had 100%. So, Korea's case is a success story. With no or minimal regulation, Korea has achieved the highest national broadband penetration, and the incumbent retains less than a 50% market. This situation is unique to Korea. In the US, by contrast, DSL has 30% market share, and cable which is unregulated, has a 68% to 70% share, viz., essentially the reverse market shares to those of Korea (see Fig. 1.5).[6] The FCC changed the regulatory rules in February 2003, and removed the incumbent's obligation to provide DSL to new competitors. A Federal Appeals Court decision told the FCC that it had to change this policy. In the US, cable passes 96% of homes. DSL is employed for about 67% because it is distance limited, and cable retains 70% market share. So perversely, in

---

[6] Students of statistics or econometrics know not to place too much weight on a result from a sample size of two observations. Nevertheless, it is interesting what has happened in the two countries.

the US a technology with 30% market share was regulated, while a technology with 70% share was not. Clearly, regulation should not lead to this type of outcome (Hausman 2002).

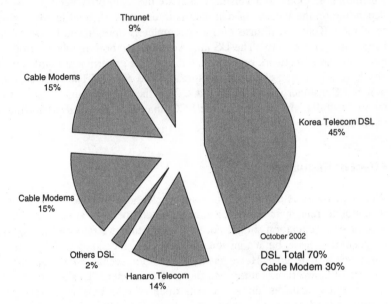

Fig. 1.4. Korean Broadband Internet Access

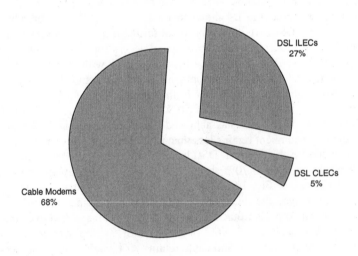

Fig. 1.5. US Broadband Market Shares

Sunk Investment and Regulation

An important question is what is the effect of this type of regulation has had on competition. Another concern is whether such regulation has had a negative effect on telecommunications sector investment. Lastly, have any potentially large negative effects on consumer welfare occurred. The outcome of such asymmetric regulation is the prevention of DSL being an effective competitor, and decreased investment by incumbents. To consider the position of a firm undertaking new investments in telecommunications, I recall that Sir John Hicks said, when I was a graduate student at Oxford, "When you make an investment, you are giving a hostage to fortune". So, what did he mean by that? Well, most investment in telecommunications is a sunk cost or investment. A firm installs wires under the ground. Should the new service fail, the investment is gone forever, because the firm is unable to remove the wires and reuse them for an alternative purpose. It is economically infeasible to do so. So the sunk cost nature of the investment makes the investment much more risky when compared to many investments in other industries.

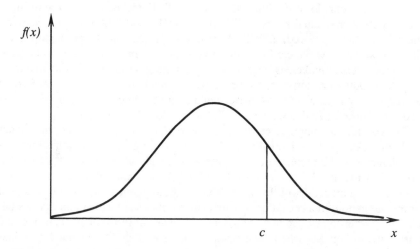

Fig. 1.6. Effect of Regulation on Expected Returns

Next, consider what happens when the industry is regulated. Consider the frequency curve of possible expected returns in Fig.1.6. With network unbundling regulation in place, the regulator sets some price c and requires an incumbent to sell product to competitors at that price. Thus, regulation truncates the distribution of possible returns. Should the investment fail the firm's shareholders are made worse off. Conversely when the investment is successful, the regulator requires the incumbent to sell the innovation to competitors at the regulated price. This type of regulation can profoundly affect telecommunications market outcomes. In particular, by truncating the right tail of the distribution, the regulator exogenously decreases the expected return. Consider a product with return y, and using a normal distribution with standard deviation σ,

$$E(y \mid y < c) = \mu - \sigma M(c) \tag{1.2}$$

where the formula for the expected value of y, given that it has to be less than c, is equal to the mean μ, minus the standard deviation times the function $M(c)$ or Mills ratio. After truncation the mean is smaller than before truncation. The riskier the investment, the greater is the magnitude of standard deviation, so the greater is the impact of truncation. To sum, the more risky is the investment, the more expected value the regulator eliminates. In telecommunications, because of positive network effects relating to product adoption, technology is often quite risky, and so the effect on innovation can be substantial. The tighter is the cost standard, the lower the economic incentive to innovate. As returns to innovation become more uncertain, the more expected returns decrease, and the greater the disincentive to invest in new technology. To illustrate, consider WiFi technology, a type of high-speed wireless broadband access. Will this 80211B or 80211A standard succeed? When WiFi first appeared, a PCIMA card was placed into users' laptop computers, however, now the card is built into computers, and Intel, the world's largest microprocessor maker, is aggressively promoting WiFi. WiFi is available in many 'hot spots', i.e., airports, coffee shops and hotels in Asia, Australia, Europe and the US. WiFi could be a superior technology to ADSL, because ADSL depends on copper wires that were not built for broadband but for voice.

Now consider the importance of options in regulation. What is an option under regulation? An option is a legal right, but not the obligation, to purchase unbundled elements at the regulated price. Another example of an option is a financial option. Consider a call option to purchase Intel stock, should the Intel stock price rise to 20, then the holder of the option has a right to purchase this stock at 20, but not the obligation. In a regulatory context, such options are called real options because they refer to real products or assets, not financial instruments. Consider the distribution in Fig. 1.6, when the economic outcome is 'good', the entrant utilizes this option. The entrant approaches the regulator and resells the innovative product, and the regulator requires the incumbent to provide the product at the regulated price. However, should the outcome be 'bad', then the entrant does not exercise the option. It is as if Intel stock falls to 10, and option is not exercised. However, the Intel options investor must purchase the option. Under regulation the entrant receives a free option from the regulator. Hausman (1997, 1999b, c), demonstrates that the value of options with sunk investment and rapidly changing technology is very high. Thus, the regulatory grant of free options distorts investment decisions. If the regulator truncates the distribution and gives away these real options for free, the regulated firm has a substantially reduced incentive to invest. The outcome is the same for Intel. If the regulator started to give away options on Intel at 20 for free, Intel would have less reason to invest in new generations of microprocessors. If Intel went above 20, individuals would exercise their free options limiting the upside movement of Intel stock.

What is the source of a free option? Consider a new investment by an incumbent telecommunications provider and a competitor wishes to rent the unbundled element. The incumbent can offer the entrant competitor a contract for the economic life of the investment. In this case, the competitor would become similar to a joint venture partner with the incumbent. That outcome does not create any economic distortion. If demand is not realized or price falls, then the entrant bears this economic risk. However, in the US, Australia and several other countries, the regulator has not required entrants to sign a contract for the life of the investment. Instead, regulators allow the entrant to rent the unbundled element on a month-by-month basis at the regulated price, set at long-run incremental cost (LRIC). When economic events work out well, the new competitor continues to rent. However, if economic events are not successful then the competitor is able to not rent the unbundled element the following month. So, the incumbent is required by regulation to provide a free option to the entrant. The monthly price of the unbundled element should be significantly higher than c because of this risk of investment with sunk costs. To see this, consider the fundamental equation for investment with economic depreciation, but no sunk costs:

$$V(P) = \int_0^\infty \lambda \exp(-\lambda t)\, P\, \frac{1 - \exp(-\delta t)}{\delta}\, dt = P/(\lambda + \delta)\,. \tag{1.3}$$

To obtain the value of an investment, integrate from zero to infinity, where δ is the economic depreciation rate, λ is the combined cost of capital, the discount rate plus the expected change in the product price. For simplicity assume constant exponential rates of change, so the integration is in closed form. The value of Eq. 1.3 is equal to the price the company sets (P) divided by $(\lambda + \delta)$. To invest this value must exceed the cost of investment (I). Next, consider what happens when the investment cost is sunk. The last term in Eq. 1.3, λ, which is added to the expression to account for the decreasing price of capital goods, is economic depreciation. It is important to note, that in the telecommunications industry prices of new equipment decrease rapidly through time. As the prices of PC's have fallen so too have prices for telecommunications equipment. This downward price movement is among the reasons that so many telecommunications companies have collapsed, viz., telecommunications equipment prices decreased much faster than expected. To see the impact of a faster than expected fall in network equipment prices, assume a company called New Telecom enters the Internet access market. To do so the company purchases a switch or router from Cisco for US$ 10,000. The expectation is that the company will serve a customer base of a hundred subscribers at a variable cost of US$ 500 p.a. So the average variable cost is US$ 5 per customer and with a 10% cost of capital. Further, the company assumes that routers last for five years, at which time they are scrapped with no residual value. The cost of the project is US$ 11,895, an amount that has to be earned to break even, or on a per customer basis, US\$ 118.95. If the price is set at US\$ 31.38 p.a., the company breaks even. However, this calculation does not allow for economic depreciation, and at this price the company loses money. Assuming that prices obey Moore's law, viz., equipment price falls monotonically through time. In particular, assume that router prices decline by 10% p.a., and other costs remain the same. This assumption of an annual 10% price decrease is not unreasonable by telecommunications sector standards. Following this simulation through, assume in year-2 a new competitor enters the market and pays US\$ 1,000 less for a router, because router prices have fallen, and so, their discounted cost is US\$ 10,895— exactly US\$ 1,000 less. The early entrant New Telecom has to reduce its service price by US\$ 2.64 to compete with the late entrant, or it incurs a loss. This annual loss increases as competition continues, in the following year the router price falls again to US\$ 8,000. Thus, if the regulator sets service price at US\$ 31.38 based on a LRIC standard, no rational firm will invest—because instead of charging US\$ 31.38 in year-1, the actual price charged to break even is US\$ 36, in year-2 US\$ 33 and so on. This example demonstrates that when regulated price is incorrect, and this form of regulation started in the US and followed in Australia, that telecommunications companies cannot break even.

Further, consider the effect of sunk costs. When sunk costs are taken into account a stochastic partial differential equation, much like the Black-Scholes formula, results. For the case of a Weiner process:

$$P^s > \frac{\beta}{\beta-1}(\delta+\lambda)I = m(\delta+\lambda)I \quad \text{for} \quad m > 1 \tag{1.4}$$

where $m > 1$ is called the mark-up term. m depends on uncertainty and other economic factors. Hausman (1997, 1999b, c) demonstrates that m is bounded by 3.2 and 3.4 for telecommunications, so assume it takes the value 3.3. This means that once sunk costs, uncertainty and the rates price decline are taken into account, the usual NPV rule needs to be adjusted by a factor of 3.

Cellular Moves to 3G

3G technology, basically allows broadband services to be delivered over the cellular spectrum. 3G has high-speed data transfer rates, and will be able to provide video and audio among its applications. Full-3G data transfer speeds are potentially available up to 2 Mbps. This speed is 200 times that of 2G and, similar to both DSL and cable modems. Current Japanese speeds are available up to 384 Kbs. An alternative technology CDMA1-X ('3G Lite') operates in Australia, Japan, Korea, New Zealand and the US, and has data transfer rates approaching 144 Kbps. A laptop computer, with a card and an antenna, works in a manner similar to WiFi, with an additional benefit of mobility. CDMA-1X is not really broadband speed, but it is substantially faster than that available to date. For 3G to succeed requires that revenue from data equal that of voice. In recent US advertisements for Sprint 3G services, an actress is showing a picture when she is calling some-body. Such data uses of 3G must be considered important by consumers, as 3G voice quality is practically the same as for 2G. Competition is already strong among 2G providers, and it is increasing as it is essentially a commodity service. Such competition in the US, and to a lesser extent in other countries, has led to a steep decline in the market value of service providers. So, if 3G works, it will be similar in it's capability to fixed-line communication services, so that many subscribers will only subscribe to mobile telephone services.

Requirements for 3G to be Successful

For 3G markets to succeed requires compelling applications be developed and provided to the market. So far only limited progress has been made. Among the bigger users in the US are mobile laptop computers with 144 Kbps data transfer speeds. Second, 3G needs sufficient spectrum to operate effectively. In the US the local governments tend to use up a lot of the spectrum. Lastly, 3G needs sufficient speed to allow for convenient broadband applications. So far 3G is only working at 144 Kbps to 384 Kbps, and the question remains whether it will operate at speeds of up to 2 Mbps.

Current 3G Market Experience

In Japan, NTT DoCoMo began service in October 2001. DoCoMo 3G is not successful to date. Its principal application in Japan is to use 3G as a picture telephone. This application seems far from compelling to a typical subscriber. DoCoMo concedes lack of content is a major problem. Hutchison, a Hong Kong company, has begun service in Europe recently with an investment of US$ 16 billion. Unfortunately, their handsets have, as yet, unresolved problems due to complicated software issues.

Spectrum Auctions

Government policy that has been successful in several countries including Australia, Europe and the US, are spectrum auctions. Government auctions did not occur because economists had convinced government of their efficiency. The US Congress needed to increase spending, and once it realized money could be gained from auctions they became economic policy. However, economists remain pleased with the outcome. Prior to auctions, governments had 'beauty contests', which meant, that proponents employed lawyers and engineers to convince government of the desirability of their plans. Next, then came lotteries. As lotteries typically resulted in sub-optimal outcomes, government policy switched to auctions.

In 2000, Germany and the UK auctioned spectrum for US$ 46 billion and US$ 35 billion, respectively, which are approximately equivalent on a per capita basis. However, governments need to design the auctions in a manner that ensures enough bidders. Economists prefer auctions as they are economically efficient, viz., rather than having a beauty contest and giving away spectrum to associates, or giving away spectrum in a lottery, economists want bidders who value the spectrum most to use it. Accordingly, firms with the higher bids are seen to value the spectrum more. However, it has been claimed the process is flawed as when firms incorrectly value 3G spectrum they incur financial difficulty. A question arises as to whether government is attempting to artificially make spectrum scarce so as to maximize financial return. Namely, is spectrum actually a scarce resource with modern technology? That is, with modern digital technology, government does not need to play a role. Instead, market participants could share the spectrum and it would not be scarce. This debate is unresolved, and its outcome will profoundly affect future government policy and telecommunications markets.

WiFi

WiFi are small radio antennae that broadcast over a 100 meter to 150 meter range, and offers high broadband speed of 1.5 megabits. The technology is based on an IEEE standard, and has operated on university campuses for the past eighteen months. WiFi is currently being deployed at hotspots. T-Mobile offers WiFi in Starbucks cafes and airport clubs located in the US, and in German airports. A difference between the technologies is that WiFi is not mobile. Also, WiFi currently cannot provide voice calling, although this functionality will likely be added in the near future. Industry support has begun for WiFi with Intel announcing its Centrino chip. Intel and Toshiba and Dell have combined with laptop computer manufacturers to widely deploy WiFi. A crucial problem for WiFi hotspots is that a successful revenue model is not yet available. That is, it is not apparent how companies are to make a return from offering the service. T-mobile charges by the minute or month. In Hong Kong when purchasing broadband from the incumbent, Hong Kong Tel (PCCW) includes WiFi use in their hotspots. However, in the US, several companies offering WiFi service are now defunct. It remains unclear that WiFi offered in hotspots will be economically viable. Perhaps Starbucks should pay for WiFi to attract customers.

A technology related to WiFi is wireless local loop (WLL). WLL is an alternative to having copper wire running through premises. This technology has the potential to solve the last mile problem. Rather than fiber to the node, fiber to the neighborhood node and wireless to the home would be provided. Speeds of 10 Mbps allow the equivalent of cable TV through WLL. As such, this technology could provide competition to cable TV. In the US, cable TV remains among the last refuges for unregulated monopoly. Cable companies charge high monthly subscription rates that rise at three times the rate of inflation. WLL technology could also provide competition to wire-line telephone companies, where government regulation has created a problem. However, WLL technology requires enough spectra for this outcome to happen, and again government has to solve this problem.

Finally, if 3G and WiFi become complements, adopters will want to be able to use their laptops or their PDAs anywhere, and the technologies will be reinforcing. Should the technologies be substitutes on the other hand, they will compete against each other, with 3G being higher cost and perhaps lower speed, but offering mobility and voice applications. The likely outcome remains an unanswered US$ 16 billion question for telecommunications markets. Hutchison is to spend probably A$ 5 billion to deploy 3G, and if it is not successful, they will incur huge losses. The fundamental question is whether consumers are willing to pay for mobility. In China a proposed new 3G standard, TDSCDMA may offer only limited mobility. Perhaps different technologies are best for different economic situations.

End to Government Regulation?

In the US, and most other countries, there is a regulatory problem, viz., governments have introduced competition in telecommunications markets, and required incumbents to lease their network elements. This situation has resulted in an acrimonious relationship between regulators and incumbents, e.g., in Australia Telstra sued the government and the government sued Telstra. The same has occurred in the US. A disappointing outcome of this regulation is the limiting of investment and associated reduction in the spread of new technology. Less innovation leads to less consumer welfare. Further, basic telecommunications research that had occurred, e.g., at Bell Labs, has contracted drastically. Bell Labs invented cellular technology, the transistor and UNIX, and has historically been the leading US research lab. Such basic research has been a fundamental cause of technology advance in telecommunications.

The solution to this problem is to put an end to regulation. Namely, local wireline service costs of provision are mainly fixed, as such for competition to be effective requires a reduction in the incumbent's market share of only 6%. So the ideal situation is to stop regulating telecommunications and let competition take over. The calculation to determine the cut-off point for sector regulation is derived from the rule:

$$(1 - MC/P)Q_1 < (1.05 - MC/P)Q_2 .$$ \hfill (1.5)

To find the 'critical market share loss', take the ratio of the marginal cost to the price, which for a telephone wire-line is about 20%, and then calculate how much market share would the incumbent have to lose, so that the incumbent could not have any market power to raise price above the competitive level. This critical share loss is approximately 6% (Hausman et al. 1996). No one knows whether 3G, WiFi or WLL are going to be successful. No one knows, but the really interesting proposition is should any of these technologies become successful the regulating of telecommunications could cease. This outcome would be a very good idea for consumers and government policy.

Conclusion

3G cellular and WiFi substitution is happening. Broadband Internet is becoming increasingly important. In the US, the FCC has used incorrect regulation, and policy needs to be designed that does not retard investment. New investment in telecommunications can lead to large consumer welfare gain. Delays in technology introduction need to be avoided. 3G and WiFi may solve the regulatory problem by offering both voice and broadband data. To succeed, 3G needs compelling applications, spectrum and high-speed access. WiFi and WLL need sufficient spectrum and changes to government regulation. Should 3G and WiFi provide sufficient competition then it is going to change the way that economists and regulators think about telecommunications. Such an outcome would provide a profound change to government policy. The result would be an acceleration of the introduction of new innovation in telecommunications markets. To conclude, two of the most consumer welfare enhancing innovations over the last fifty years, have been cellular telephones and the Internet. Both of these innovations arose from the telecommunications market. Because telecommunications is based on information and computer chips, which continue to get faster and cheaper, there is no reason to believe that there is an end to further innovation in telecommunications research. The telecommunications sector of the economy has led improved consumer welfare and productivity gains. The best outcome would be turning this into a sector of the economy where competition takes over and eliminates the need for overbearing government regulation.

Acknowledgement

Jerry Hausman is MacDonald Professor of Economics, MIT. Earlier versions of this chapter were presented in public lectures at Communication Economics and Electronic Markets Research Centre Distinguished Fellow Lecture Series at Curtin University and as the Shann Memorial Lecture at University of Western Australia.

References

Crandall RW, Alleman JH (eds) (2002) Broadband: Should we regulate high speed Internet access? AEI-Brookings Joint Center for Regulatory Studies, Washington

Hausman JA (1996) Valuation of new goods under perfect and imperfect competition. In: Bresnahan TF, Gordon RJ (eds) The economics of new goods. University of Chicago Press, Chicago, pp 209–37

Hausman JA (1997) Valuing the effect of regulation on new services in telecommunications. Brookings Papers on Economic Activity: Microeconomics 1997: 1–38

Hausman JA (1999a) Cellular telephone, new products and the CPI. Journal of Business and Economics Statistics 17(2): 188–94

Hausman JA (1999b) The effect of sunk costs in telecommunication regulation. In: Alleman J, Noam E (eds) The new investment theory of real options and its implications for telecommunications economics. Kluwer, New York, pp 191–204

Hausman JA (1999c) Regulation by TSLRIC: Economic effects on investment and innovation. Multimedia Und Recht 3: 22–6

Hausman JA (2002) Internet-related services: The results of asymmetric regulation. In: Crandall R, Alleman J (eds) Broadband: Should we regulate high speed Internet access? AEI-Brookings Joint Center for Regulatory Studies, Washington, pp 129–56

Hausman JA, Leonard G, Vellturo C (1996) Market definition under price discrimination. Antitrust Law Journal 64: 367–86

Hicks JR (1940) The valuation of the social income. Economica 7: 105–24

2 Geographic and Socially Embedded B2C and B2B E-Commerce

Charles Steinfield

Introduction

The term electronic commerce (e-commerce) provides visions of anonymous online transactions with companies like Amazon or Buy.com, where personal contact need not exist and locations of purchasers and sellers are irrelevant. Much business to consumer (B2C) e-commerce occurs this way, and the approach was prevalent among firms in the 'dot.com' era, e.g., Verticalnet and biz2biz.com. Such visions reflect the dominant perception of the role of e-commerce in economic exchange prior to the widespread failure of dot.com firms in 2000 and 2001. That is, e-commerce enables firms to access new markets, replace outmoded or inefficient supply chains and distribution channels, and achieve substantial growth in customer reach (Wigand and Benjamin 1995; Cairncross 1997; Choi et al. 1997). These perceived opportunities stem from maintained assumptions held at that time. Steinfield (2003b) describes the primitive beliefs that shaped corporate e-commerce practice and strategy:

a) Geography is irrelevant. Electronic exchange costs are the same regardless of distance (Cairncross 1997). Indeed, unless a Web site explicitly identifies the geographical location of the firm, the physical address remains unknown to most users. An implication for economic exchange is that purchasers and sellers are not required to be geographically close. The Internet had reduced transaction costs for transacting with geographically distant partners to the extent that distant sellers are competitive with local sellers (Bakos 1997, 1998).

b) Internet transactions are substitutes for transactions formerly conducted in person or through other forms of direct communication. Enhanced graphics and multimedia enable highly complex products to be sold through this medium. Customization and personalized features substitute for the activity of sales personnel who formerly interacted directly with customers (Rayport and Sviokla 1995). Modern logistics and transport systems ensure rapid fulfillment of online orders, hence the essential elements needed to effectively automate transactions are present, and enable on-line transactions that are perfect substitutes for physical transactions (Choi et al. 1997).

c) E-commerce orientated firms experience network effects. Hence, firms that accumulate a substantial customer base will achieve a sustainable competitive advantage (Choi et al. 1997; Shapiro and Varian 1999; Kaplan and Sawhney 2000; Afuah and Tucci 2001). The presence of a network effect translates into increased customer marginal utility the greater is Web site patronage. Such effects are thought crucial for sites that provide intermediary services, e.g., brokerage. To improve market liquidity such sites must attract a critical mass, or minimum efficient scale, of customers. The more customers attracted to a market, the greater the probability a customer has of finding a desired match.[1]

The motivation for this study is that such maintained assumptions concerning e-commerce do not universally hold, and firms that rigidly pursue a strategy based on them are not likely to gain desired benefits. In particular, this perspective has led to an underutilization of e-commerce in both local and regional markets, and a corresponding lack of emphasis on the means through which e-commerce works in concert with established transactions and client practices. Based on a review of research findings, an alternative 'situated' e-commerce approach is proposed. Following Steinfield (2003a), a situated view implies that: physical location matters for both B2C and B2B e-commerce; for firms with a physical retail presence e-commerce is best viewed as complementary to physical market activity; and that—especially for B2B transactions—e-commerce is best employed to strengthen existing business structures as a network-effects only focus is potentially damaging to supply chain relationships. In the following section the results of a research program on click and mortar firm dynamics establishes the relevance of a situated perspective for B2C e-commerce markets. The situated view is then employed to explain several developments in B2Bmarkets. A consideration of how a situated view may enhance the relevance of e-commerce to local and regional business clusters follows. A final section summarizes and concludes.

Toward a Situated View of B2C E-Commerce

The situated view contrasts resource-based theory explanations of firm information technology (IT) use to gain B2C market competitive advantage with those of transaction cost theory. Resource-based theory explores firm leverage of their assets to achieve maximum IT investment benefit (Teece 1986; Zhu and Kraemer 2002). Firm assets situated within a firm's immediate geographical and business context include established outlets, and supplier and customer relations. Transaction cost theory emphasizes the nature of costs incurred by firms in transacting (Williamson 1975, 1985). Relevant costs include information gathering and search costs, negotiation and settlement costs, and monitoring costs to ensure agreements are adhered to. The transaction cost view heavily influenced the development of initial B2C e-commerce business models.

Transaction Cost and Complementary Views of B2C E-Commerce

Early e-commerce analyses primarily employed transaction cost arguments to

---

[1] Another form of network effect occurs for sites that offer recommendations, as recommendation quality improves with more customer activity (Dieberger et al. 2000).

identify sources of Internet firm competitive advantage relative to that for traditional firms (Bakos 1997; Choi et al. 1997). The Internet was predicted to substantially lower transaction costs, especially search and monitoring costs (Bakos 1997). With costs reduced, purchasers could afford to search in distant geographic markets for lower prices, better service, higher quality and products that better match needs (Malone et al. 1987; Wildman and Guerin-Calvert 1991; Wigand and Benjamin 1995; Bakos 1997; Cairncross 1997; Choi et al. 1997; Wigand 1997). In particular, Internet only (virtual) enterprise was thought, relative to physical channels, to have more ready market access and ease in complementary product bundling; require holding smaller inventory; have more flexibility sourcing inputs and in bypassing intermediaries; provide continual information access; have lower menu costs ensuring rapid market response; and operate superior transaction automation facilities with an associated data mining capability (Wigand and Benjamin 1995; Choi et al. 1997; Wigand 1997; Bailey 1998; Afuah and Tucci 2001).

Many retail firms entered e-commerce markets by establishing a separate virtual channel of operation (Steinfield et al. 2002b). Recently, potential synergy available from the operation of combined physical and virtual channels is documented (Friedman and Furey 1999; Steinfield and Klein 1999; Otto and Chung 2000; Rosen and Howard 2000; Steinfield et al. 2001; Ward 2001; Steinfield et al. 2002a; Steinfield et al. 2002b). Laudon and Traver (2001) provide indirect empirical support for these propositions in reporting that only 10% of dot.com firms incorporated in 1995 survive at 2001. Moreover, Laudon and Traver also note that click and mortar retailers rapidly replaced dot.com retailers in leading e-commerce firm lists following the dot.com collapse.

Physical and Virtual Channel Synergy

Click and mortar firms have sources of synergy not usually available to virtual firms. Such synergies are situated or complementary assets that better enable them to succeed in e-commerce markets (Teece 1986; Afuah and Tucci 2001). Sources of synergy include common infrastructure, customers, marketing and operations (Porter 1985). For example, an infrastructure is 'common' when the firm can rely on the same logistical system (warehouses, trucks and so on) to distribute goods for both virtual and physical outlets. IT infrastructure is also often shared. Recent empirical analysis suggest that when an e-commerce capability is established in conjunction with existing an IT infrastructure then firm performance is more likely to improve (Zhu and Kraemer 2002). An illustration of common operation as a source of synergy is the sharing of an order processing system between virtual and physical channels. Such common operation enables, for example, improved tracking of customer movement between sales channels and cost savings. Virtual and physical channels also possess common marketing and sales assets, viz., product catalogues, product and consumer knowledge, and promotion activity that support both channels. Finally, as virtual and physical channels often target the same customer base, click and mortar firms are better placed to provide more convenient and immediate customer service and so improve customer retention.

A Situated View of Click and Mortar Benefits

Case studies of the Netherlands (Steinfield et al. 2002b) and US (Steinfield et al. 2002a) firms, identify the benefits realized when synergies associated with common assets are successfully exploited. Benefits include: lower costs; increased product differentiation through value-added services; improved trust; and broadened geographic and product market reach. These analyses reveal the importance of leveraging an existing physical presence by treating e-commerce as a complementary rather than a substitute channel to customers.

Lower Costs

When consumers conduct product information search, complete forms and seek technical assistance on-line—costs are switched from firms to consumers, and firm labor savings result. Inventory cost savings occur when firms do not stock inventory for infrequently purchased goods at local outlets while continuing to offer catalogue choice to consumers via the Internet. Marketing and promotional efficiency gains are garnered when consumer information is disseminated cross channels. Delivery cost savings are realized from physical outlets acting as collect points for Internet transactions.

Differentiation through Value-added Services

Physical and virtual channel synergy is available through the transaction cycle. Pre-purchase services, such as on-line information, assist consumers to better determine their needs and identify product testing opportunities. Purchase services include ordering, customization and reservation facilities, and access to complementary products. Post-purchase services include on-line account management, social community support and loyalty programs. Opportunity typically occurs in aspects of the installation, repair, service reminder and training processes. Although many value-added services are available to single-channel vendors, combined deployment, e.g., on-line purchase of computer with in-store repair or training does enhance differentiation and lock-in effects (Shapiro and Varian 1999).

Improved Trust

Click and mortar firms are more commonly trusted than virtual firms, as they are physically and socially situated in the markets they serve. Aspects of this enhanced sense of security includes perceived reduction in consumer risk, being affiliated and embedded in recognized local social and business networks, and an ability to leverage brand awareness. Such perceived risk reduction results from an accessible location to return goods or lodge complaints (Tedeschi 1999). Social network affiliation facilitates the substitution of contracts and associated legal fees for social and reputation governance (Granovetter 1985). DiMaggio and Louch (1998) show, particularly for risky transactions, consumers rely on social affiliations as governance mechanisms. Such ties more commonly exist between geographically proximate agents, and suggest preference for transacting with firms that have a local physical presence. Finally, branding is a means of building consumer confidence and trust in product markets (Kotler 1999). Established firms leverage familiar names to more readily gain consumers trust for affiliated on-line service markets (Coates 1998).

Extended Geographic and Product Market Reach

Many click and mortar firms attempt to extend market reach to new geographic markets via e-commerce activity. Such benefits are enhanced when firms realize the situated nature of their virtual channel. Steinfield et al. (2002a) find firms report on-line purchases from distant markets by past customers that relocated elsewhere and continue to transact with the firm. This outcome is typically observed for culturally specific products that are difficult to obtain. Moreover, an extended market reach can arise by expanding product offerings. Virtual channels extend the product scope and depth of physical channels by allowing firms to offer products not stocked that compliment current product lines. Finally, firms can add online revenue generating information services.

Managing Situated E-Commerce

A firm with multiple customer service channels may face conflicting objectives when, e.g., a Web-based outlet intentionally or otherwise competes with or bypasses established physical channels in reaching customers (Stern and Ansary 1992; Balasubramanian 1998). A common problem is when an introduced channel cannibalizes market share from an established channel. Such perceived threats can result in restricted cross-channel cooperation and perhaps sabotage, and cause confusion among customers attempting to transact when using multiple but uncoordinated channels (Friedman and Furey 1999; Useem 1999; Ward 2001). A situated view recognizes the need to manage the integration of e-commerce into existing business models. Management oversight is required to limit any potential conflict and so ensure that synergistic gains from cooperation are realized.

Difficulty in measuring the contribution made by a virtual channel to the activity of the firm creates tensions within the firm (Tedeschi 2001a). Managers must be made aware of potential intangible benefits and not, for example, evaluate e-commerce divisions solely on the basis of their sales or profits. Moreover, agreement must be reached on whether established or new customers are to be targeted by a virtual channel. Typically successful click and mortar firms align goals across physical and virtual channels, and in doing so ensure that employees realize that the firm benefits from activity originating in both channels. Further, management and employees must value existing physical assets and not replace them with e-commerce assets or expect e-commerce to function as a stand-alone operation. Firms that more successfully introduce e-commerce concepts into their business plans implement coordination and control mechanisms to exploit synergistic opportunity. Control mechanisms include IT systems integration to ensure interoperability to aid customer movement between channels. Frequently, firms demonstrate coordination through cross-channel promotion, e.g., by the allocation of e-commerce sales credit to outlets based on customer address. Finally, successful managers recognize cost structure and capabilities associated with channels and develop measures to encourage customers to use appropriate channels. In many instances, traditional firms lack the core competency required to realize synergy from e-commerce, such as Web development or logistics skills.

Empirical Analyses of Strategy and Benefit

A recent US study finds that few click and mortar retailers have an integrated strategy (Steinfield 2003a). Content analysis of almost a thousand retail Web sites, covering 9 retail categories, revealed that while most sites provided a physical address and telephone number, and two-thirds provided a map or directions to their street location, only 52% listed opening hours or in-store event information. However, very few retailers employed their site to reap any potential synergies from the establishment of a virtual channel, e.g., less than 20% of sites provided an on-line product search engine, and fewer than 10% allowed on-line purchase returns to a retail outlet. Further, only 25% of sites provided coupons or gift certificates on-line for redemption at an outlet, and only 6% of sites let on-line shoppers collect items at local outlets. Such results are surprising, given that the opportunity costs associated with this lack of an integrated strategy are substantial. For example, Steinfield (2003a) in a survey of 81 US retailers measured aspects of IT and marketing integration, as well as changes in business processes that take better advantage of firm e-commerce activity. The study identified factors that predicted the extent to which firms gained more than competitors from their e-commerce endeavors, ignoring any differences attributable to firm size or industry sector. The results suggested the presence of potentially substantial gains from situated e-commerce activity.

Towards a Situated View of B2B E-Commerce

Prior to the emergence of B2B Internet trading, open and standard-based data networks evolved into network-based markets. However, these limited and tightly-coupled electronic relationships, or electronic hierarchies, between firms were predicted to give way to exchange governed by markets (Malone et al. 1987). Accordingly, when Internet-based e-commerce appeared, vast electronic marketplace growth was forecast. Analysts argued that such markets would reduce search costs, increase transaction efficiency and, by aggregating buyers and sellers improve the likelihood of matching. By the late-1990s, the potential for B2B trade to be conducted via the Internet attracted many entrants wanting to establish virtual markets (Kaplan and Sawhney 2000; Subramami and Walden 2000; Garicano and Kaplan 2001; Laudon and Traver 2001). At 2000, the US Department of Commerce (2000) reported more than 750 B2B e-markets operate globally. However, as with B2C e-commerce, a decidedly 'not situated' view of e-commerce shaped these emerging Internet-based B2B marketplaces (Kaplan and Sawhney 2000).

Electronic Hierarchy and Small Networks

In analyzing inter-organization systems embedded within a business community, consideration should be given to the socially-enclosed nature of inter-firm transactions. A study of the French media industry illustrates that conflict between firms arise when an information system built along transaction cost rationality lines interferes with practices founded on personal relationships (Caby et al. 1998). Namely, French TV market liberalization, and the subsequent increase in private channel participation made for more complex advertising decisions. In response, the media industry created an electronic market to allow firms to search for available time slots and reserve them. This development was intended to reduce selling costs and more generally improve transaction efficiency. The market, built on France's Minitel system, required only minor premise equipment purchases. However, the marked failed. This failure is attributed to the market's prevention of relationship-based selling strategies. Accordingly, the system could not ensure the supply of preferred time and price to customers. Moreover, customers behaved strategically by reserving time slots with an intention to prevent competitors from gaining them. Such behavior led media representatives to bypass the system and return to traditional media time selling practices.

Kraut et al. (1998) investigate personal and electronic transaction co-ordination between 250 producers and industry suppliers, and support a situated view of B2B electronic trade. In particular, the study finds that electronic networks are more likely to be used when producer-supply relationships exist, while greater activity occurs when relations are closer. Moreover, to conduct electronic transactions requires equipment purchase by participants. However, suppliers are unlikely to invest unless trigger business activity levels are expected to be realized. This approach contradicts that based on network effects, whereby firms are more disposed to invest the larger the subscriber base. In this situation, membership to a B2B system is more valued when the number of participants is small. Interestingly, e-commerce is reported as complementary to, and not a substitute for personal relationships. This finding is evident from a reported positive association between the presence of personal links and extent of electronic transaction activity. Finally, the more firms attempt to substitute electronic transactions for personal coordination, the more errors and quality problems with transactions are experienced. When electronic transactions are complemented with personal coordination, such problems are mitigated.

Rise and Collapse of Third-party B2B E-Markets

Early Internet B2B electronic markets focused on improvement of procurement efficiency (Segev et al. 1999; Kaplan and Sawhney 2000; Laudon and Traver 2001). An implicit rationale, similar to that for B2C e-commerce, is that transaction efficiency arises from B2B electronic markets reducing participant search and monitoring costs (Bakos 1997, 1998; Segev et al. 1999; Steinfield et al. 2000; Garicano and Kaplan 2001). At peak dot.com euphoria, the B2B e-hub was prominent among digital economy business models (Timmer 1998). However, despite widespread optimistic projections by academics, consultants and government most third-party B2B markets failed (Katsaros et al. 2000; US Department of Commerce 2000; Laudon and Traver 2001; Tedeschi 2001b).

Despite these failures, B2B e-commerce continues to grow. However, for B2B markets the distinction between Internet-based and private industry network markets is important (Laudon and Traver 2001). B2B Internet markets are classified by how and what businesses purchase (Kaplan and Sawhney 2000). The 'how' dimension distinguishes spot purchasing for immediate need from systematic purchasing for planned long term need. Spot purchasing is often made by ephemeral market-based transactions, whereas systematic purchasing requires substantial negotiation with trusted partners for relatively large volumes. The 'what' dimension distinguish vertical (direct or manufacturing) inputs that relate to firm core products and horizontal (indirect or maintenance, operating and repair) inputs, such as office supplies, acquired by firms.

Laudon and Traver (2001) classify Internet-based B2B market as either:

a) Electronic distributors, e.g., Grainger.com that offer e-catalogues of many suppliers to support spot purchasing for horizontal inputs, and add value by reducing search costs;
b) E-procurement services, e.g., Ariba.com that offer maintenance, operating and repair supplies for systematic purchasing. These procurement services include licensed procurement software to support value-added service. E-procurement firms offer the catalogues of many suppliers in return for commission. Service providers reduce their search costs by aggregating traders, and so augment positive network effects;
c) Exchanges, e.g., the former E-Steel, focused on connecting spot manufacturing markets within an industry. Exchanges charge commissions. Purchasing services offered include support in price negotiations, auctions, and other bidding in addition to fixed-price selling. Buyers benefit by wider choice and lower price, and sellers gain access to buyers. Exchanges exhibit network effects; and
d) Industry consortia, e.g., Covisint, an electronic procurement system developed by automobile manufacturers. Consortia are jointly owned by dominant firms relying on electronic networks to support long term supply relations. Entry is by invitation only, and the importance of founder firms ensures suppliers participate.

Conversely, private industrial networks are closed user groups, linking a few strategic partners with private infrastructure (Laudon and Traver 2001). Strategic partners are usually organized through a focal organization, e.g., by a prominent manufacturing firm, which together with its suppliers and downstream channels, seeks efficiency gains by serving their common market. Inter-organization networks are currently organized by WalMart, Siemens, and Procter and Gamble (Laudon and Traver, 2001). Such private industrial networks typically encompass particular value chains to enable just-in-time inventory, efficient consumer response, and collaborative design and production. Increasingly, industry is focusing on value-networks in which strategic partner value chains are incorporated in pursuit of efficiency gain and end-customer value. Clearly, these networks span distance, and allow firms to transact business activity more efficiently with remote suppliers. That 93% of B2B electronic trade occurs through private industry networks supports a situated view of B2B e-commerce (Laudon and Traver 2001). Moreover, industry consortia, arguably the most situated Internet-based market model, exhibit the highest growth.

High failure rates by third-party B2B e-hubs, coupled with the dominance of private networks and industry consortia growth, reflect an important underlying dynamic. Namely, firms with established supplier relations have trust engendered by reliable performance and commitment that is more valuable than short term price advantage potentially gained from 'neutral' markets. Indeed, emerging in B2B electronic trade is the importance of collaborative e-commerce, whereby networks perform tasks beyond transaction support. Network tasks include joint product design and the tighter integration of inventory databases. Such developments result from firm electronic hierarchies that rely on networks to facilitate outsourcing to tightly integrated firms (Malone et al. 1987). Empirical evidence suggests that such inter-organization forms are more common and long lasting than is market exchange (Steinfield et al. 1995; Kraut et al. 1998). Finally, firm location plays an important role in determining public Internet-based B2B exchange patterns. Choi (2003) recently studied Korean public B2B e-markets in which participants actively identified other agents and completed transactions, and markets that functioned as agents by completing transactions for buyers. The spatial patterns for six months of transactions revealed that firms were more likely to make purchases from 'local' regional suppliers. However, agent-mediated purchases are more commonly directed to 'outside' region suppliers—mainly located in the Seoul metropolitan area.

Geographic Business Communities and Situated B2B E-Commerce

Firm location and the formation and maintenance of business trading communities, or clusters, is examined by Porter (1990, 1998, 2000). Porter (1998) defines a cluster as a critical mass of commonly located firms in a particular field of industry. Firms within a cluster have either specialized input, component, machinery or service supply, or are firms in downstream industries, producers of complementary products, specialized infrastructure providers, or institutions that provide specialized training and technical support. Economic benefits ascribed to business clusters closely correspond to those obtained from participation in B2B electronic markets, viz., improved access to specialized inputs, lower transaction costs and access to complementary goods. Clusters are also thought to enhance innovation among members. Rather than rely on electronic networks and automation to achieve transaction and information gains, clusters capitalize on geographic prox-imity. Concentration of skilled workers, for example, increases access to labor in-puts. Less formally, knowledge sharing through chance encounters enhances in-novation capacity (Rogers and Larsen 1984; Maskell 2001; Saxenian and Hsu 2001). Finally, Porter (1998) argues that common language, culture and social in-stitutions reduce transaction costs, and that local institutions are likely more re-sponsive to specialized cluster needs.

E-Commerce and Local Business Clusters

Local and regional clusters offer a context to conduct B2B e-commerce, through both intra-cluster co-ordination and for inter-cluster linking. Recent analysis of Internet geography suggests that the spatial concentration associated with content and infrastructure producers extends to firms that use the Internet. Namely, elec-tronic transactions flow by physical proximity, should advanced service centers be geographically concentrated, i.e. built on inter-personal networks of decision-making processes, and organized around a local network of suppliers and custom-ers (Castells 2001). Indeed, Kolko's (2000) study of Internet traffic demonstrates that most IP traffic flows within, rather across locations. B2B transactions are em-bedded in an enabling social and cultural context, yet in striving for transaction ef-ficiency gains, most efforts to create electronic networks bypass this context. More context specific analyses of IT use for commerce and co-ordination in local business clusters are required to reveal fundamental challenges to replace devel-oped social exchange processes with electronic transactions.

Evidence from inter-organization systems analysis suggests a poor fit between typical B2B market design and local business cluster needs, and suggests why pri-vate networks continue to dominate in B2B commerce. For instance, Johnston and Lawrence (1988) study a geographically defined business cluster in the Northern Italian textile industry. In particular, the research questioned how textile mills had disaggregated into smaller specialized firms that focused on only part of the value chain. Johnston and Lawrence show that networks of firms worked in concert to meet market demand for final goods, and that an inter-organization information system facilitated this co-ordination. However, Kumar et al. (1998) revisited the merchants and reported the information system essentially abandoned, as it of-fered no substantial transaction cost reductions. Apparently, regional social capi-tal, in the form of trust and personal relationships, are effective substitutes for such inter-organization systems.

Social Capital, B2B Electronic Coordination and Local Clusters

The extent to which firms realize local cluster advantage depends on effective ex-ploitation of their social capital. Social capital, or resources arising from personal relationships enhance competitive advantage, e.g., business referral to suppliers may originate through acquaintances (Huysman 2002). Recent theory emphasizes social capital has structural, relational and cognitive dimensions (Nahapiet and Goshal 1998; Tsai and Goshal 1998; Adler and Kwon 2002). A structural dimension encompasses an individual's pattern of social ties, and is a conduit to information and resource gain. Relational dimensions stress trust and obligation arising from personal contact, especially for high risk transactions, reducing opportunism and transaction costs (DiMaggio and Louch 1998). The cognitive dimension concerns information exchange through shared codes that enable common goals and understanding, and is a public good within a social system based on member interaction.[2] As described, social capital theory provides a lens to better understand and extend geographically proximate business cluster advantages. For instance, access to skilled labor is enhanced when complemented by referral from social contacts. Spontaneous interaction, which facilitates innovation, is both socially embedded and geographically closer in character, affords opportunity. Finally, common language, culture and social institutions provide a basis for shared understanding that comprises the cognitive dimension.

A sense of obligation, goodwill and reciprocity that emerges from strong relationships has important economic benefits. Social embedding models posit economic transactions are either at arms-length and characterized by short term and constantly shifting ties among loose collections of agents, or embedded and characterized by stable long term relationships (Powell 1990; Uzzi 1997, 1999). In the transaction cost view, economic exchanges dependent on social networks are inefficient, as social obligation prevents pursuit of higher quality or lower cost transactions. By contrast, social embedding arguments find advantage is gained by relying only on a few trusted relations for critical economic exchanges, e.g., reducing trading partner search. Further, cost reductions are obtained from less monitoring as trust arising from social obligation and maintaining reputation within a social structure diminish opportunistic behavior, time saving through personal referrals and an increased emphasis on joint problem solving (Granovetter 1985; Powell 1990; Uzzi 1997). To the extent that such strong ties have long gestation periods and sustained by interaction, then proximity should correlate positively with incidence. As such, social capital theory offers potential insights to understand the dynamics of local business clusters, including the relative underutilization of B2B e-marketplaces as internal and external co-ordination mechanisms.

The dramatic rise and subsequent collapse of Internet-based third party B2B markets brings into question maintained e-market assumptions held by market analysts and participants. Most B2B electronic trade occurs through private industry networks rather than via open B2B exchanges (Laudon and Traver 2001). The emergence of industry consortia and the growth of collaborative e-commerce,

---

[2]   Huysman (2002) considers the Nahapiet and Goshal (1998) dimensions are similar to the Adler and Kwon (2002) framework, whereby social capital provides opportunity, motivation and ability. Opportunity arises from network participation (structure), motivation from qualities embedded in relationships (relational) and ability from common understanding (cognitive).

suggest that a more situated view of B2B electronic trade is required. Indeed, the current almost exclusive emphasis on network effects has most probably made electronic markets less valuable to potential participants, viz., the argument implies that current trading relationships may be shared with more competitors. However, empirical research suggests that electronic transactions are more likely to occur between established trading partners in long term relations than via ephemeral spot trades (Kraut et al. 1998). A situated view is also suggested by the preference for local supply, even on public B2B exchanges (Choi 2003). An opportunity for geographically situated B2B e-commerce emerges from the focus on the importance of regional business clusters (Porter 2000). The role of social capital within these clusters also suggests that for B2B e-commerce to succeed, it also must be socially situated or embedded. Finally, available IT use in local business cluster research suggests that traditional forms of coordination often supercede electronic co-ordination, and may make it difficult for such e-commerce to gain a foothold (Kumar et al. 1998).

Conclusion

This review of B2C and B2B e-commerce illustrates the danger of solely relying on the distance insensitivity, transaction automation capability and network effect characteristics to guiding business model developments. An opposing situated perspective emphasizes the coupling of e-commerce to physical market presence, richer and off-line modes of interaction, and existing customers and supply chains. The study encourages approaching e-commerce in a manner that is sensitive to the potential complementary benefits it offers to off-line business activity.

References

Adler P, Kwon S (2002) Social capital: Prospects for a new concept. Academy of Management Review 27(1): 17–40

Afuah A, Tucci C (2001) Internet business models and strategies: Text and cases. McGraw-Hill Irwin, New York

Bailey J (1998) Internet price discrimination: Self-regulation, public policy, and global electronic commerce. Telecommunications Policy Research Conference, Washington, September

Bakos JY (1997) Reducing buyer search costs: Implications for electronic marketplaces. Management Science 43(12): 1676–92

Bakos JY (1998) The emerging role of electronic marketplaces on the Internet. Communications of the ACM 41(8): 35–42

Balasubramanian S (1998) Mail versus mall: A strategic analysis of competition between direct marketers and conventional retailers. Marketing Science 17(3): 181–95

Caby L, Jaeger C, Steinfield C (1998) Explaining the use of inter-firm data networks for electronic transactions: The case of the pharmaceutical and advertising industries in

France. In: MacDonald S, Madden G (eds) Telecommunications and socio-economic development. Elsevier, Amsterdam, pp 191–204

Cairncross F (1997) The death of distance. Harvard Business School Press, Boston

Castells M (2001) The Internet galaxy. Oxford University Press, Oxford

Choi JS (2003) Spatial analysis of transactions that use e-catalogs in public business-to-business electronic marketplaces by business model. International Geographical Union Special Symposium, Dynamics of Economic Spaces in E-Commerce, Incheon, March

Choi SY, Stahl DO, Whinston AB (1997) The economics of electronic commerce. Macmillan, New York

Coates V (1998) Buying and selling on the Internet: Retail electronic commerce. The Institute for Technology Assessment, Washington

Dieberger A, Dourish P, Hook K, Resnick P, Wexelblat A (2000) Social navigation: Techniques for building more usable systems. Interactions 7(6): 36–45

DiMaggio P, Louch H (1998) Socially embedded consumer transactions: For what kinds of purchases do people most often use networks? American Sociological Review 63(5): 619–37

Friedman LG, Furey TR (1999) The channel advantage: Going to market with multiple sales channels to reach more customers, sell more products, make more profit. Butterworth Heinemann, Boston

Garicano L, Kaplan SN (2001) Beyond the hype: Making B2B e-commerce profitable. Capital Ideas 2(4)

Granovetter M (1985) Economic action and social structure: The problem of embeddedness. American Journal of Sociology 91(3): 481–510

Huysman M (2002) Design requirements for knowledge sharing tools: A need for social capital analysis. Workshop on Social Capital and IT, Amsterdam, May

Johnston R, Lawrence P (1988) Beyond vertical integration: The rise of the value-added partnership. Harvard Business Review 66: 94–101

Kaplan SN, Sawhney M (2000) E-hubs: The new B2B marketplaces. Harvard Business Review 78: 97–103

Katsaros H, Shore M, Leathern R, Clark T (2000) U.S. business-to-business Internet trade projections. Jupiter Research, New York, September

Kolko J (2000) The death of cities? The death of distance? Evidence from the geography of commercial Internet usage. In: Vogelsang I, Compaine B (eds) Internet upheaval: Raising questions, seeking answers in communications policy. MIT Press, Cambridge, pp 73–98

Kotler P (1999) Marketing management, 10th edn. Prentice Hall, Upper Saddle River

Kraut RE, Steinfield C, Chan AP, Butler B, Hoag A (1998) Coordination and virtualization: The role of electronic networks and personal relationships. Organization and Science 10(6): 722–40

Kumar K, van Dissel HG, Bielli P (1998) The merchant of Prato-revisited: Toward a third rationality of information systems. MIS Quarterly 22(2): 199–226

Laudon K, Traver C (2001) E-commerce: Business, technology, society. Addison-Wesley, Boston

Malone TW, Yates J, Benjamin RL (1987) Electronic markets and electronic hierarchies: Effects of information technology on market structure and corporate strategies. Communications of the ACM 30(6): 484–97

Maskell P (2001) Towards a knowledge-based theory of the geographical cluster. Industrial and Corporate Change 10(4): 921–44

Nahapiet J, Goshal S (1998) Social capital, intellectual capital, and the organizational advantage. Academy of Management Review 22(2): 242–66

Otto J, Chung Q (2000) A framework for cyber-enhanced retailing: Integrating e-commerce retailing with brick and mortar retailing. Electronic Markets 10(4): 185–91

Porter ME (1985) Competitive advantage: Creating and sustaining superior performance. Free Press, New York

Porter ME (1990) The competitive advantage of nations. Free Press, New York

Porter ME (1998) The Adam Smith address: Location, clusters, and the new microeconomics of competition. Business Economics 33(1): 7–13

Porter ME (2000) Location, competition, and economic development: Local clusters in a global economy. Economic Development Quarterly 14(1): 15–34

Powell W (1990) Neither market nor hierarchy: Networked forms of organization. In: Staw B, Cummings L (eds), Research in organizational behavior, vol 12. JAI Press, Greenwich, pp 295–336

Rayport JF, Sviokla JJ (1995) Exploiting the virtual value chain. Harvard Business Review 73(6): 75–87

Rogers E, Larsen J (1984) Silicon Valley fever: Growth of high-technology culture. Basic Books, New York

Rosen KT, Howard AL (2000) E-retail: Gold rush or fool's gold? California Management Review 42(3): 72–100

Saxenian A, Hsu JY (2001) The Silicon Valley-Hsinchu connection: Technical communities and industrial upgrading. Industrial and Corporate Change 10(4): 893–920

Segev A, Gebauer J, Färber F (1999) Internet-based electronic markets. Electronic Markets 9(3): 138–46

Shapiro C, Varian HR (1999) Information rules: A strategic guide to the network economy. Harvard Business School Press, Boston

Steinfield C (2003a) Capitalizing on physical and virtual synergies: The rise of click and mortar models. In: Priessl B, Bouwman H, Steinfield C (eds) E-life after the dot.com bust. Springer, Berlin

Steinfield C (2003b) Rethinking the role of e-commerce in B2B and B2C transactions: Complementing location, personal interactions and pre-existing relations. In: Bohlin E, Levin S, Sung N, Yoon C (eds) Global economy and digital society. Elsevier, Amsterdam

Steinfield C, Adelaar T, Lai Y-J (2002a) Integrating brick and mortar locations with e-commerce: Understanding synergy opportunities. Hawaii International Conference on Systems Sciences, IEEE Computer Society Hawaii, January

Steinfield, C, Bouwman H, Adelaar T (2002b) The dynamics of click and mortar e-commerce: Opportunities and management strategies. International Journal of Electronic Commerce 7(1): 93–119

Steinfield C, Chan AP, Kraut RE (2000) Computer-mediated markets: An introduction and preliminary test of market-structure impacts. Journal of Computer Mediated Communication 5(3)

Steinfield C, DeWit D, Adelaar T, Bruins A, Fielt E, Hoefsloot M, Smit A, Bouwman H (2001) Pillars of virtual commerce: Leveraging physical and virtual presence in the new economy. Info 3(3): 203–13

Steinfield C, Klein S (1999) Local versus global issues in electronic commerce. Electronic Markets 9(1/2): 45–50

Steinfield C, Kraut RE, Plummer A (1995) The impact of electronic commerce on buyer-seller relations. Journal of Computer Mediated Communication 1(3)

Stern LW, Ansary AI (1992) Marketing Channels. Prentice Hall, Englewood Cliffs

Subramami M, Walden E (2000) Economic returns to firms from business-to-business electronic commerce initiatives: An empirical investigation. 21st International Conference on Information Systems, Brisbane

Tedeschi R (1999) Dealing with those pesky returns. New York Times, August 23

Tedeschi R (2001a) Bricks-and-mortar merchants struggling to assess Web sidelines. New York Times, September 3

Tedeschi R (2001b) E-commerce report: Companies in no hurry to buy over the Internet. New York Times, March 5

Teece DJ (1986) Profiting from technological innovation: Implications for integration, collaboration, licensing and public policy. Research Policy 15: 285–306

Timmer P (1998) Business models for electronic markets. Electronic Markets 8(2): 3–8

Tsai W, Goshal S (1998) Social capital and value creation: The role of intra-firm networks. Academy of Management Journal 41(4): 464–76

US Department of Commerce (2000) The emerging digital economy. US Department of Commerce, Washington

Useem J (1999) Internet defense strategy: Cannibalize yourself. Fortune 140(5): 121

Uzzi B (1997) Social structure and competition in interfirm networks: The paradox of embeddedness. Administrative Science Quarterly 42 (March): 35–67

Uzzi B (1999) Embeddedness in the making of financial capital: How social relations and networking benefit firms seeking financing. American Sociological Review 64 (August): 481–505

Ward MR (2001) Will online shopping compete more with traditional retailing or catalog shopping? Netnomics 3(2): 103–17

Wigand R (1997) Electronic commerce: Definition, theory, and context. The Information Society 13: 1–16

Wigand R, Benjamin R (1995) Electronic commerce: Effects on electronic markets. Journal of Computer Mediated Communication 1(3)

Wildman S, Guerin-Calvert M (1991) Electronic services networks: Functions, structures, and public policy. In: Guerin-Calvert M, Wildman S (eds) Electronic services networks: A business and public policy challenge. Praeger, New York

Williamson O (1975) Markets and hierarchies: Analysis and antitrust implications. Free Press, New York

Williamson O (1985) The economic institutions of capitalism. Free Press, New York

Zhu K, Kraemer K (2002) Electronic commerce metrics: Assessing the value of e-commerce to firm performance with data from the manufacturing sector. Information Systems Research 13(3): 275–95

3 SME International E-Commerce Activity

James H. Tiessen

Introduction

This paper examines small and medium enterprise (SME) internationalization and the Internet. The economic importance of SMEs is widely recognized through their 40% to 60% contribution to employment and value-added in most OECD countries (OECD 2002a)[1]. Further, SMEs account for 20% to 25% of global exports (OECD 2002a). The 1990's expansion of economic globalization, in concert with the not completely unrelated growth of the World Wide Web (WWW has set the stage for an even greater SME participation in global markets. Globalization, the WWW and SME internationalization became ubiquitous during the 1990s. Trade currently comprises about 20% of world GDP. Depending on the country, 40% to 60% of the population of developed nations is Internet users (International Telecommunication Union 2003). In 2002, more than 60% of Canadian, 55% of US and 45% of European Union (EU) SMEs had adopted Internet business solutions (McClean et al. 2003). Recent research on the international activities of SMEs shows that start-up companies, especially those in high-tech sectors, are internationalizing at increasingly faster rates (Oviatt and McDougall 1997; Schrader et al. 2000; Knight 2000).

Analysts note that the Internet allows SMEs to acquire base foreign market intelligence, establish international networks and achieve improved economies of scale through inexpensive advertising and the provision of their offerings to niche markets (Quelch and Klein 1996; Hamill 1997; Prasad et al. 2001). Further, forecasts of global e-commerce revenue remain robust. eMarketer recently predicted world B2B Internet sales to reach US$ 1.4 billion by end-2003 and US$ 2.7 billion by 2004 (eMarketer 2003). The confluence of these factors led Tiessen et al. (2001) to explore in 1999-2000 how and why Canadian SMEs used the WWW to undertake international business. This case study research provided a model that relates environmental factors and firm capability to how SMEs use the WWW, and in particular a better understanding of SMEs, the Internet and international business activity.

Since undertaking that study fundamental change has occurred. Importantly, hype concerning the Internet has dissipated. Additionally, a high-tech equity bubble burst in March 2000 and WWW technology adoption slowed. Most SMEs in Canada, Europe and the US that employ Internet business solutions initially did so

---

[1] SMEs are typically defined by number of employees, though this does vary. In Canada and the US firms employing less than 500 persons are considered SMEs. In Australia and the EU the upper bounds are 200 and 250 persons, respectively.

in the period 1998 through 2000. Further, the OECD reports that rates of SME sales and purchases on the WWW have not substantially grown, and are declining for smaller and growing for larger firms (OECD 2002b: 11). Another, change is a slowing in the rate of globalization. World trade grew an average 6.7% p.a. during the 1990s, but expanded by only 2.5% in 2002—a recovery from the 2001 decline of 1% p.a. (WTO 2003). That said, in Canada there is evidence that firms are using the WWW to export—17% of e-commerce sales are exports, and this share is more than 50% for accommodation and food services (UNCTAD 2002: 68). Accordingly, this study examines the stability of the Tiessen et al. (2001) model. An aspect of the study is gaining an understanding how and why Internet use by firms has changed. To do so case firm activity, and their related variables are compared for the 1999-2000 and 2003 periods.

SME Export Behavior and the Internet

Studies of international SME e-business behavior, and more generally e-business adoption, are typically undertaken based on the maintained assumptions that: SMEs should export; e-business reduces barriers to exports; and e-business adoption progresses in stages. Jointly, these assumptions imply that SMEs either should, or will, use the WWW to sell abroad. Large sample research, however, suggest that SMEs are yet to effectively exploit the WWW, especially for international business.

Maintained Assumptions

The view that SMEs should export arises from the literature on small business internationalization, e.g., Bilkey (1978), Leonidou and Katsikeas (1996) and Coviello and McAuley (1999), and is reflected by governments that promote exports. These studies argue that exporting expands revenue opportunities, enables exploitation of economies of scale, and importantly, is associated with firm competitiveness.[2] Since exporting is viewed so positively, a research focus is the identification of barriers to internationalization, and how barriers can be addressed.

The Internet reduces the economic consequences of geographic distance and, real and perceived export barriers (Prasad et al. 2001; Madden and Coble-Neal 2002). Perceptions affect firms' judgment concerning their export orientation. For instance, a study of UK and Australian SME WWW users shows that UK firms considered the Internet reduced the need for costly foreign representation, while Australian firms did not (Hornby et al. 2002). Modeling SME Internet use in stages is a common treatment, similar to the use of stages to understand SME ex-

---

[2] The direction of the link between competitiveness and exporting is typically not questioned in the literature. Recent work suggests that 'good' firms export, while evidence of benefits of exporting is more ambiguous (Bernard and Jensen 1999).

port behavior. It is recognized that firms differ in their implementation of Web technology. For instance, Morrison and King (2002) find alternative levels of engagement from Wilderness to Techno-Whizzo firms. Daniel et al. (2002) identify the firm clusters or stages of: Developers, Communicators, Web Presence and Transactors. Tiessen et al. (2001) employ functional sophistication categories that suggest increasing levels of commitment.

SME E-Business

Direct e-commerce transactions are currently only a small proportion of commercial activity, e.g., New Zealand (0.3%), Canada (0.5%), Australia (0.7%), the US (1.3%) and Sweden (2%) (McClean et al. 2003). Further, for instance, EU and Canadian SMEs transact only 22% (12%) and 15% (5%) of their purchases (sales) on-line, respectively. The OECD, which assembles e-commerce data for 16 developed countries, find that 12.5% of firms make Internet sales, while 25% make WWW purchases (OECD 2002c). Such low SME WWW transaction volumes are not indicative of wider WWW use by firms. Approximately 75% of EU SMEs use the Internet for the conduct of business activity, compared to 70% for Canadian firms, and more than 90% for Danish and Japanese firms (OECD 2002c). However, SMEs do not typically embrace the WWW's potential. For example, Sparkes and Thomas (2001) in seeking to establish that Welsh agri-food SMEs benefit from WWW use, report that WWW technology use is not typically viewed a critical success factor. Similarly, Morrison and King (2002) examine Victorian Tourism Online, a State government-sponsored Web recruitment project for small tourism operators, and find that six months after program introduction only 13% of targeted operators had participated. To sum, the sentiment stated in a recent UNCTAD (2002: 147) report on e-commerce and development resonates:

> The majority of SMES still limit their activities to maintaining a web page, with various levels of links and advertising. On the Internet they also gather information about markets and competitors, as well as searching for partners, with further negotiations taking place either through e-mails or offline, while successful deals are generally completed in a traditional manner, that is, with traditional paperwork or through the use of cash.

Method

The Tiessen et al. (2001) sample is comprised of 12 Canadian firms. The case studies of the firms addressed the questions: (a) How are SMEs using the Web internationally, and (b) Why are SMEs using the Web internationally? Firms are selected that differ in terms of their variables linked to the research focus (Eisenhardt 1989; Yin 1989; Miles and Huberman 1994). The current sample contains only nine of these cases. Two firms refused to participate, citing time constraints, and the remaining firm, a family business, ceased operations in the interim. Case studies are employed as a research tool, because the phenomena of the Internet, internationalizing SMEs and greater globalization are contemporary issues. Yin (1989) writes, case studies are advantageous when a, 'how' or 'why' question is being asked about a contemporary set of events, over which the investigator has little control.

Data are collected by semi-structured key informant interviews (McCracken 1990). In both periods, interviewers focused on firm business environment and Web activity. Interviews are supplemented by firm Web site assessments. In 1999-2000, interviews are conducted in person at firm locations (8) and by telephone (4). Four interviewers are used. Interviews ran one to three hours. In 2003, Tiessen conducted the interviews by telephone. 2003 interviews are shorter, twenty minutes to an hour. For both waves, notes are written within 24 hours, and sent to the informants who were invited to make corrections and additions. Variables and categories are derived from a broad range of research streams of SME internationalization (Aaby and Slater 1989; McDougall et al. 1994; Leonidou and Katsikeas 1996; Reuber and Fischer 1997), e-business (Berthon et al. 1999; Evans and Wurster 1999; Kassaye 1999; Sahlman 1999; Mahadevan 2000), the resource view of the firm (Barney 1991), technology adoption (Dishaw and Strong 1999), institutional theory (DiMaggio and Powell 1983; Oliver 1991), economics of industrial organization (Porter 1985) and innovation (Schumpeter 1961). The categories evolved with data collection and analysis (Strauss and Corbin 1990).

Participant Firms

Firms are chosen from the hospitality/tourism and high-tech industry. 32% (64%) of firms in the Canadian hospitality/tourism (manufacturing) sector had Web sites in 1999, and only 8% (15%) used the Web to sell services (Statistics Canada 2000). However, approximately 1.3% of hospitality/tourism operating revenues are Web generated, compared with 0.2% for manufacturing. Hospitality/tourism firms have both consumer and business clients. Manufacturers also operate in the B2B markets. It is anticipated that managers and owners of hospitality/tourism firms are relatively less technically skilled than high-tech firm managers.

To control for foreign market focus, SMEs interested in Japanese markets are targeted. The cases are selected from the Canada-Japan Trade Council (1998) directory of firms doing business with Japan. In addition, business associations are asked to identify firms that had sold to Japan or are interested in doing so. In all cases the firms employed less than 500 persons and had a Web site. Table 3.1 summarizes the main characteristics of the sampled companies studied. The hospitality/tourism firms are: Avonlea, a tour operator in Prince Edward Island; Le Bateau-Mouche, a tourist cruise boat operator; Le Père Saint-Vincent, a restaurant and inn; Manoir Harvard Montreal, a bed and breakfast establishment; Ski Banff/Lake Louise, a ski-hill marketing joint venture; and Times Square Travel, a travel agent mostly serving persons of Japanese descent and run by a Japanese-Canadian married to a Web consultant. The high-tech firms are: CIDtech Research, a custom pharmaceutical maker; Gennum, a supplier of advanced hearing aids and semi-conductors; Semiconductor Insights, which offers competitor analysis for semi-conductor makers; SiGEM, a maker of location tracking devices used mostly in the taxi industry; StressGen Biotechnologies, a pharmaceutical development firm; and Videospheres, a 2000 start-up selling video delivery platforms for the Internet.

Firms are based across Canada: from the West (Alberta and British Columbia), from the most populous province Ontario and Quebec, and in the East, Prince Edward Island. Three firms started in the 1970s, three in the 1980s, five in the 1990s and one in 2000. Their size ranges from 4 to 450 employees. High-tech firms typically had more employees and served B2B markets. The hospitality/tourism firms are oriented toward consumer markets. The share of sales to Japan ranged from 1% to 75%. Japanese customers tended to be more important for the hospitality/tourism firms, though for two high-tech companies, Japanese revenues exceeded 20% of sales.

Table 3.1. Firm Characteristics

	Business	Place[a]	Start	Market[b]
Avonlea	Tour Operator	PEI	1989	B2B, B2C
Le Bateau-Mouche	Boat cruise	QC	1992	B2C
Le Père Saint-Vincent	Restaurant / Inn	QC	1987	B2C
Manoir Harvard Montreal	Bed and breakfast	QC	1997	B2C
Ski Banff/Lake Louise	Ski resort	AL	1978	B2B, B2C
Times Square Travel	Travel agent	BC	1970	B2B, B2C
CIDtech Research	Pharmaceutical manufacture	ON	1984	B2B
Gennum Corp	Semi-conductor manufacture	ON	1973	B2B
Semiconductor Insights	IP support and analysis	ON	1997	B2B
SiGEM	Location tracking systems	ON	1997	B2B
StressGen Biotechnologies	Pharmaceutical developer	BC	1990	B2B
Videospheres	Internet video systems	ON	2000	B2B

Note. a. AL is Alberta, BC is British Columbia, ON is Ontario, PEI is Prince Edward Island and QC is Quebec. b. B2B is business-to-business; B2C is business-to-consumer.

Model

Tiessen et al. (2001) develop a model (Fig. 3.1) relating levels of international Web use (how firms use the Web) to firm characteristics and environmental factors (why firms use the Web). In a sense the model resembles frameworks used to describe electronic data interchange adoption (Iacovou et al. 1995). Firm variables—capability and the nature of business relationships—are found associated with Web use. Salient environmental factors identified are changes in the market and the norms prevalent in a firms industry.

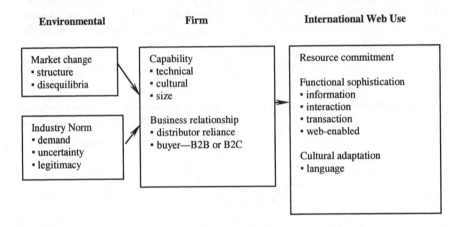

Fig. 3.1. Model of SME International E-Commerce

International Web Use: Commitment, Sophistication and Adaptation

Web-use variables reflect the importance of management behavior to SME export behavior, e.g., see Leonidou et al. (1998). Firm resource commitment to international Web initiatives, relative to available resources, is important because plans, of themselves, do not lead to export success; they need to be acted upon (Leonidou et al. 1998). Consistent with the early research literature on Web marketing, an implicit stage model is used to identify how firms used their Web sites. Functional sophistication is the complexity of customer interaction facilitated by the Web sites. Levels of functional sophistication considered are: information (brochureware); interaction with customers and stakeholders; transactions; and Web-enabled business (Ho 1997; Seybold 1998; Mahadevan 2000). A final type of use, Web-enabled, denotes companies such as Amazon.com for which the Web is an integral aspect of their business model.

A narrow approach cultural adaptation—the use of a foreign language, Japanese, on the Web site—is to indicate the degree to which firms adjust to foreign markets. Of the originally 12 sampled firms, five of six hospitality/tourism firms offer more than a single language (three had Japanese), while only one high-tech firm, Gennum, did so (it had a Japanese site).

Firm Factors: Capability and Business Relationships

Firm characteristics are found associated with how an SME uses the Web. Two competencies, technical familiarity with the Internet and cultural know-how, tend to enlighten and enable. That is, employees who understand the Web and other cultures know what can be done, and are better able to implement their ideas. SMEs studied varied in their familiarity and experience with the Internet and its potential. This capability has been recognized by researchers as linked to new technology use, including the Internet (Davis et al. 1989; Dishaw and Strong 1999; Mehrtens et al. 2001). In 1999-2000 most in the hospitality/tourism employees saw Internet technology as relatively new. Knowing how the Web can be used is associated with how the Web is used, whether or not these activities are contracted out. That is, familiarity is positively associated with international Web use. Another capability is cultural, and related to familiarity with and language of the target market. SME internationalization research established ties between this capability and export sales (McDougall et al. 1994; Reuber and Fischer 1997). In this study firms interested in exporting to Japan are selected, and so familiarity with Japan is used as a proxy for this capability. In short, firms Japanese-familiar staff are more likely to orient their Web activity to that market.

Another firm capability is the availability of firm resources. Resource availability is a real and perceived barrier to successful SME export activity. Empirical links between firm size and export propensity are not well established (Bonaccorsi 1992; Calof 1994). Tiessen et al. (2001), however, finds evidence that a firms' relative size—ratio of firm size to customers served—is linked positively to its international Web use. The current study revises this view. Firm relative size includes stakeholders such as investors, as well as customers, in identifying the Web site's targeted audience. This view is especially important for capital-orientated SMEs such as biotech-firm StressGen, contained in the sample. A further firm factor identified is the nature of business relationships, viz., firm reliance on distributors. Firms, such as those selling travel packages dependent on wholesalers are reluctant to sell direct around their distributors. Such activity threatens vital supply chain relationships. Whether firms operated in B2B or B2C markets is also important. B2B high-tech firms typically make fewer inquiries, so e-business is more easily manageable, compared to their B2C counterparts. However, the value and complexity of many B2B transactions tends to increase the importance of close, trust-based relationships that are typically developed in camera.

Environmental Factors: Market Change and Industry Norms

Environmental factors that influence international Web use relate to economic and institutional factors. It is widely accepted that the Internet has created market disequilibria by lowering the costs of distributing and collecting market information, increasing the availability of substitute suppliers and services and increasing rivalry in markets (Evans and Wurster 1999; Sahlman 1999; Mahadevan 2000). The cumulative effect can establish Schumpeterian 'creative destruction' conditions, viz., an uncertain business environment. Schumpeter (1961) identifies the innovation types: a new product that can enable new ways of production, make new markets accessible, increase the number of new suppliers and potentially lead to the breaking-up of concentrated industries. The Internet's disruptive effects are clearly felt by hospitality and travel firms. Another economic factor, a liberal market environment occurs through firms directing their activity to acquire resources. In this study, e.g., firms with a substantial or potentially substantial Japanese customer base are more likely to incur costs of adding a Japanese language facility to their Web sites.

A set of environmental influences on the SMEs are industry norms or institutional factors. These norms are non-economic responses that lead firms to undertake organizational and market initiatives that resemble those of firms in their industry (DiMaggio and Powell 1983). The effects can be pervasive for SMEs as they are less able to shape their environment (Oliver 1991). Under conditions of uncertainty, such as those introduced by the Web, organizations tend to imitate the actions of competitors, especially market leaders. Alternatively firms may be coerced by firms they depend on to act in certain ways. For example in the late-1990's automobile parts suppliers had to transfer their selling process on-line because of demands by GM and Ford. A further institutional mechanism affecting Web use is the need to establish legitimacy, viz., it is not plausible that a high-tech business would not have a Web site. Further, when a firm's business is internet-related, it is imperative to demonstrate technical proficiency on the corporate Web site. Analysis of the relationships led to the development of propositions describing relations between variables (Tiessen et al. 2001). Four propositions describe influences of environmental factors, and another three describe how firm characteristics are associated with aspects of international Web use, viz., commitment, functional sophistication and cultural adaptation. The propositions are listed in Table 3.6.

Participant Firm Change

To consider the robustness of the Tiessen et al. (2001) model data are gathered on changes to environment, firm factors and international Web use. These data are summarized in Table 3.2 through Table 3.4. The tables show change and activity since 2000, core firm characteristics, current international e-commerce use, and environment and firm characteristics. Changes in the dependent and independent variables since 2000 are noted. These data allow the stability of the seven research propositions to be assessed (reported in Table 3.6). The participant firms demonstrate the dynamism typically associated with SME populations, and is reflected by changes—limited, moderate or substantial—in their business or market.

Table 3.2 shows that a third of firms experience substantial change or creative destruction, another third exhibit moderate change, and for the remainder things remain essentially unchanged. Hospitality/tourism sector firms saw either substantial or limited change. High-tech firms are mostly only subjected to moderate

Table 3.2. Participant Firm Change, 1999-2000 and 2003

Firm Name	Key Change	Employment		Japan Sales (%)	
		1999-2000	2003	1999-2000	2003
Avonlea	Acquired Prince Edward Tours in 2001	2-12[a]	35	75	70
Le Bateau-Mouche	No change	4-50[a]	4-50	5	2-3
Le Père Saint-Vincent	Closed in 2001	12	n.a.	1-5	n.a.
Manoir Harvard Montreal	Did not participate	4	n.a.	25	n.a.
Ski Banff/Lake Louise	No change	10-80[a]	20-90	5	2
Times Square Travel	Acquired by Richmond Times Square Travel 2002	9	12	70	20
CIDtech Research	New facility late-1999 custom chemical synthesis focus	6	9	10-15	<5
Gennum Corp	No change	450	600	20	50
Mobile Knowledge (formally SiGEM)	New name 2002. Insolvent and sold early-2003	89	n.a.	<1	n.a.
Semiconductor Insights	Polish facility	125	120	20	50
StressGen Biotech	San Diego office 2001	65	90	7-11	0
Videospheres	Start-up from product to market orientation	7	12	n.a.	0

Note. *a.* seasonal employment. *n.a.* is not applicable.

change. Three firms subject to substantial disruption are in the hospitality/tourism sector, viz., Le Pere Saint Vincent (restaurant) closed down, Times Square Travel is purchased by Avonlea (related entity) in partnership with a regional travel agency, purchased its main provincial competitor for Japanese tourism, Prince Edward Tours, and so tripled size. Location device maker SiGEM receives recognition as an important Canadian start-up company (September 2001) and experiences a tenfold sales increase to 2002. However, the firm entered receivership in early-2003, despite a name change to Mobile Knowledge in 2002.

Other high-tech companies, subject to moderate change, include CIDtech who transformed from a virtual company to having a bricks-and-mortar facility. Firm employment grew from six to nine persons. Further, Gennum established leadership in their niche markets and Japanese sales revenue expanded from 20% to 50%. This increase led to a 33% growth in employment. Biotech developer StressGen opened their San Diego office in 2001 and increased employment to 90 persons. Profitability is still pending. A senior StressGen executive noted that the necessary investment capital for biotech development has tightened since 2000, rendering it a difficult environment. Semiconductor Insights opened a Polish facility in 2002 to reduce production costs. Employment remained the same, though the relative importance of their sales to Japan grew from 20% to 50%. Videospheres did not operate a Web site at summer 2000. It survives and has 12 employees. Not atypical for a start-up company, it changed its Web video delivery technology business model dramatically in the interim. The main reason cited for this change was that, "People don't want to pay for that". A new emphasis stresses direct financial benefits from delivering Web-based video and other sales collateral (information) to customers, especially in B2B markets.

Current E-Business Use

A comparison of 1999-2000 and 2003 international Web use, summarized in Table 3.3, indicates much dynamism in international Web use deployment. Seven of the SMEs report substantial change by increasing at least an aspect international Web use, e.g., when the Japanese language content of Web pages is introduced or augmented. Tour operator Avonlea, which had limited Japanese content, intends to supply more Japanese content to advertise offerings and support sales to wholesalers. Avonlea also implemented a system enabling employees to schedule tours via the Web. Further, Semiconductor Insights also substantially increased the Japanese content of its site to enabling Japanese engineers to more readily understand the nature of services (including complex intellectual property related matters). Semiconductor Insights did translate technical reports, but realized that Japanese engineers accepted or preferred English technical information. More generally, Semiconductor Insights is focusing on increasing the effectiveness of the site as a marketing tool. This approach resulted from direct e-mail campaigns being less effective due to anti-spam filters. Semiconductor Insights is also considering implementing on-line transactions for the open market (as opposed to custom) market reports.

Table 3.3. International E-Business Use, 2003

	Commitment	Sophistication	Adaptation[a]	Change
Avonlea	Moderate	Information	Full Japanese	Positive. Site updated. Expanding Japanese content
Le Bateau-Mouche	Low	Information	None	Limited
SkiBanff/ Lake Louise	High	Information Interaction Transaction	Part Japanese Part French Part German Part Spanish	Positive. Site updated. Greater on-line effort
Times Square Travel	High	Information Interaction Transaction	None	Positive (on-line transactions plan and negative (drop Japanese)
CIDtech Research	High	Information Interaction Transaction	None	Limited
Gennum Corp	High	Information Interaction	None	Positive (dedicated Web staff) and negative (drop Japanese)
Semiconductor Insights	High	Information Interaction	Part Japanese	Positive. More and expanded Japanese content
StressGen Biotechnologies	Moderate	Information Interaction Transaction	None	Positive. Site updated. More investor information.
Videospheres	High	Information Interaction Transaction Web enabled	None	Positive. Video delivery

Note. *a*. Languages on site other than English.

At 1999-2000 SkiBanff used the Web to enable international customers to purchase ski-passes and packages on-line. The company also offered basic information in four non-English languages. At late-2002 SkiBanff further developed an on-line transaction capability. The installed system allows customers to assemble personalized ski packages—including a pass for hills represented by SkiBanff—as well as reserve lodging at associated hotels and motels. Their foreign language offerings however, remain limited. Videospheres' international Web activity increased with the implementation Web-enabled business in 2000-2001. Currently, Videospheres is developing a North American presence as a sales tool, prior to full international deployment. Finally, StressGen has a developed Web strategy that enables on-line transactions for the chemical reagent sales, and makes the site more useful for current and potential stakeholders. The site does not target overseas investors or customers, and focuses on seeking legitimacy from US capital sources.

SMEs, Times Square and chipmaker Gennum, adjusted their international Web activity by reducing foreign language (Japanese) site content and devoting resources to other Web functions. That is, Times Square when it acquired the identically-named travel agency firm surveyed three years ago, chose not to buy the associated Web site as it did not generate many sales. Conversely, Times Square is beginning to promote on-line package tours, and considering linguistically focused, e.g., Chinese and Japanese pages, as 80% of TST's promotion spending is aimed at the Web site. Gennum, though dropping its Japanese language pages and increased its Web focus. The company hired technical staff to implement ideas proposed by marketing staff and initiate product line branding through the Web. The firm's IT Director indicates the Web site is especially useful in providing for easily assessable product specifications and support information concerning Gennum products. Implementing an on-line transactions facility is unlikely as the cost of these systems is too great given the small global customer base (less than 200).

Bateau Mouche and CIDtech report little change to their international e-business commitment. CIDtech had a relatively advanced e-business facility at 2000. The site, enabled on-line quotation of custom synthesis services, and transactions are finalized by e-mail. Their site essentially remained unchanged, with only periodic tweaking. The site accounts for 30% of the firm's new business, most of it overseas, though it only offers English. Bateau Mouche, like CIDtech continues to offer a simple Web site, with little change since 1999-2000. The site does not have features such as animation, and presents basic cruise and contact information. Like CIDtech, Bateau Mouche has an effective and inexpensive way to use the Web.

Environmental and Firm Factor Change

Most sampled SMEs experience changes in environmental or firm factors that are linked by the Tiessen et al. (2001) model. Table 3.4 and Table 3.5 report evidence that shifts in markets and industry norms are more important. Interview responses, however, do not reveal apparent changes in the character of the business relationships. This finding is surprising, given predictions that the Internet will reduce reliance on distributors and intermediaries. Four firms indicate increases in uncertainty in their markets. Avonlea and Times Square e.g., participated in industry restructuring through the purchase of competitors to gain size economies and market power. StressGen highlight uncertainty in biotech investment capital supply.

Table 3.4. Environmental Factors, 1999-2000 and 2003

Firm	Market Change	Industry Norm
	Uncertainty, U / Market opportunity, O	Customer demand, D / Legitimacy, L
Avonlea	Moderate U, Much change Much O	Moderate D, Much change Moderate L
Le Bateau-Mouche	Low U, Little change Low O	Moderate D Moderate L
SkiBanff/Lake Louise	Much U, Moderate change Moderate O	Much D Much L
Times Square Travel	Much U, Little change Much O	Moderate D, Much change Moderate L, Much change
CIDtech Research	Low U, Moderate change Moderate O	Moderate D Much L
Gennum Corp	Moderate U, Little change Much O	Much D Much L
Semiconductor Insights	Moderate U, Little change Much O	Moderate, Much change Much L
StressGen Biotechnologies	Moderate U, Much change Low O	Moderate D Much L
Videospheres	Much U, Little change O n.a., Much change	Much D Much L

CIDtech, which operates in related markets, notes an exit of competitors and a pressure to reduce costs. Videospheres as it moved from start-up status reported an increased real domestic and international market opportunity.

The growing ubiquity of the Web in consumer and business (especially high-tech) markets, means that firm Web sites are common business features. At 2000, SkiBanff realized its high-end customers demanded travel information on the Web. Avonlea and Times Square both indicate that clients want detailed trip information prior to purchase. A Web site is also important for legitimacy, especially for long-distance travel transactions. Semiconductor Insights similarly mentioned an increased demand for Web-based material—especially their open market reports. Only four firms indicated substantial change in the Tiessen et al. (2001) firm factors. Times Square agency at 2000 had considerable in-house Internet and Japanese cultural capability; the firm that purchased it does not. However, the resultant larger firm is able to better outsource these capabilities or alternatively hire persons with the required skills. Avonlea has an increased in-house cultural capability, with 12 Japanese employees.

Firm size, relative to stakeholder numbers, changed little in the interim. Exceptions are Avonlea and StressGen. Avonlea's employment rose three-fold since 2000, while the firm continues to sell to wholesalers rather than buyers directly. StressGen increased employment 50%, while continuing to use the Web to promote its products and the company as an investment. Further, it is notable that distributor reliance has not changed for these sampled SMEs. Large destination travel firm SkiBanff is conducting more direct work while continuing to distribute for partner ski hills and hotels. Smaller, Avonlea and La Bateau Mouche are reluctant to sell more directly as they want to preserve their ties with wholesalers. Not surprisingly, these companies had not changed final buyers, business or consumers.

Table 3.5. Firm Factors, 1999-2000 and 2003

	Firm Capability			Relationship	
	Technical	Cultural	Size ^{a}	Supply	Buy ^{b}
Avonlea	Moderate	High Little change	Moderate Much change	High	Cons
Le Bateau-Mouche	Moderate	Low	Moderate	High	Cons
SkiBanff/LakeLouise	Moderate Much change	Low	High	High	Cons
Times Square	Moderate Much change	Moderate Much change	Low	Moder-ate	Cons
CIDtech Research	High	Low	Moderate	Low	Bus
Gennum Corp	High	High	High	Low	Bus
Semiconductor Insights	Moderate Much change	Moderate	High	Low	Bus
StressGen Biotech	Moderate	Low	Moderate Much change	Low	Bus
Videospheres	High	Low	Low	Low	Bus

Note. a. Size is employee to stakeholder numbers. b. Cons is consumer. Bus is business.

Model Stability

The stability of the Tiessen et al. (2001) model is assessed by revisiting the case firms and examining their current Web use and variables linked to Web use. The intention is to establish whether model propositions are consistent with case activity. The analysis indicates that the model is robust. The firms comprise those that demonstrate change and those that report no change in Web use. As no change is reported in terms of reliance on distributors and final buyers, only seven of the propositions are assessed. The current survey finds support for five, weak support for one and none for the other proposition. Firms that increase their commitment to international e-commerce report changes in related environmental and firm factors. Increased uncertainty (Proposition 1) is important for Avonlea and Stress-Gen, while a greater market opportunity (Proposition 2) matters for Videospheres.

Table 3.6. E-Commerce Propositions, 1999-2000 and 2003

Proposition	Supported	Unclear
P1: Perceived industry disequilibria or un-certainty are positively related to e-commerce resource commitment	Avonlea StressGen	CIDtech SkiBanff
P2: Perceived foreign market opportunity is positively related to e-commerce re-source commitment and cultural adaptation	Videospheres	
P3: Customer demand for e-commerce and capability is positively related to e-commerce commitment and the Web site functional sophistication	Avonlea Semiconductor Insights Times Square	
P4: Industry technological legitimacy need is positively related to e-commerce re-source commitment	Times Square	
P 5: Internet technical capability is posi-tively related international resource com-mitment and Web site functional sophisti-cation	SkiBanff Semiconductor Insights	Times Square
P 6: Cultural capability is positively related to international resource commitment and directed cultural adaptation	Avonlea Times Square	Gennum Times Square
P 7: Relative capacity is positively related to international resource commitment and Web site functional sophistication	Avonlea StressGen	

Industry norms, especially increased customer expectations (Proposition 3) are acknowledged by Avonlea, Semiconductor Insights and Times Square as linked to their Web use. That said, Gennum increased its Web commitment despite no apparent customer demand. Using the Web to demonstrate technological legitimacy (Proposition 4) is not apparent—only the Web-enabled firm Videospheres had a relatively simple Web site. The posited association between in-house technical capability, Web commitment and site sophistication (Proposition 5) had little support. Though Times Square does not have in-house technical or cultural expertise that the firm it acquired has, the technical ambition of the site has increased. Conversely, developing in-house competency at SkiBanff and Semiconductor Insight is reflected by commitment to more sophisticated Web sites. Gennum sustains a commitment as it increases in-house capability.

The sampled firms provide tentative support for the link between cultural capability and language adaptation (Proposition 6). Avonlea increased its employment of Japanese staff and Web site content. Times Square, while losing technical capability also lost their Japanese market association entirely. Conversely, Gennum removed their Japanese content despite the market's importance. They find Japanese engineers are comfortable using English for the technical information the site provides. The changed relative size of firms and their stakeholders (Proposition 7) is indeed linked to an increased resource commitment and sophistication of the Web site. It is apparent that the substantial growth of Avonlea and StressGen is related to these aspects of their Web use. Both a high-tech (CIDtech) and hospitality/tourism (Bateau Mouche) firm, more or less, maintained a simple Web strategy. Both firms indicate their Web strategy is effective. These firms regularly consider making their sites more sophisticated—by adding transaction capability or information—but are yet to do so. Further, excluding an increase in CIDtech market disequilibrium or uncertainty, factors linked in the model to international Web use remain reasonably stable over the interim.

Discussion

To sum, revisiting the firms that gave rise to the Tiessen et al. (2001) model of e-commerce use by internationalizing SMEs provided findings related to the model and to assumptions underlying research on SME internationalization via the Web. The results suggest that the model remains sound. Unstable markets, customer demand and relative firm size are linked to the propensity of firms to increase their commitment to international Web use. Firms experiencing turbulent environments, appear more likely to embrace their technology to sell abroad, especially when turbulence is Internet induced. Market opportunities in cyberspace appear linked to resource commitment. Size is important, with size related to the relative size of the firm and the number of financial stakeholders–customers and, for public firms, investors. The higher is the firm size to stakeholder ratio, the greater is commitment to on-line business. The influence of technical and cultural capability is less clear. In-house competency enlightens managers to possibilities and facilitates implementation. However, this experience also teaches managers what is unnecessary, e.g., Times Square and Gennum both removed Japanese pages on their sites. Similarly, Times Square's loss of Web support has not inhibited its commitment to the Web. Namely, the owner knows what can be done but chooses to outsource. It is apparent that as Web familiarity grows and costs of outsourcing decline, in-house capability is less important. Similarly, experience-based knowledge of the need to adapt Web strategy to serve foreign markets is more important than first-hand familiarity with that culture. When adaptation is shown profitable appropriate staff are hired or outsourced.

More generally the findings suggest that two assumptions that underlie research should be revisited. That exporting is good is not unequivocal. Gennum and Semiconductor Insights expanded sales to Japan and are among the most competitive of the sampled firms. However, substantial Japanese revenue is not sufficient for Times Square Travel to survive. Conversely, CIDtech survived and grew with substantially lower Japanese exports. Videospheres did not sell abroad until a sound North American customer base is established. Evidence suggests that firm offerings, not exporting per se, leads to viability. Sampled firms suggest that the Web can assist firms seek international markets. In particular, technology—e-mail and the Web—is essential for the sampled firms. No firms contemplated closing their Web sites. A Web presence helps to establish legitimacy. For foreign sales, language adaptation appears to matter more for the travel firms targeting Japanese clients (Avonlea) than for high-tech firms. Finally, the validity of the assumption that firms progress through stages of Web use is questioned. This study suggests that firms use the Internet in a manner appropriate to their business. Personal links and distribution channels matter. Several years of Web experience show managers the degree to their markets are disrupted and how to respond. That said, since 2003, the creative destruction witnessed in the small slice of the travel industry is substantial. However, there is no technological inevitability that firms will do more on-line.

Conclusions

Internet offers promise as a tool enabling SMEs to conduct business abroad. The model developed by Tiessen et al. (2001) provides insight as to how and why SMEs use the Web to export globally. That said it remains unclear whether the Web leads to further SME internationalization. As usual, it appears contingency matters. Competitive firms able to serve international niche markets will continue to do so, but in somewhat different ways due to the Web. The business proposition of Videospheres is based on this notion, rather than claiming the Web eliminates personal contact, their technology is reducing the volume of personal calling necessary to secure new sales. It cannot be assumed that inevitability most firms will become sophisticated Web users. This study is limited by a small sample size, which limits generalizability. Further research will expand the sample size to gain an understanding of how SMEs, including those acting internationally, embrace the Web. Longitudinal analysis will allow further insight into Internet and SME co-evolution.

References

Aaby N, Slater SF (1989) Management influences on export performance: A review of the empirical literature. International Marketing Review 6(4): 7–26

Barney J (1991) Firm resources and sustained competitive advantage. Journal of Management 17(1): 99–120

Bernard AB, Jensen JB (1999) Exceptional exporter performance: Cause, effect or both? Journal of International Economics 47(1): 1–25

Berthon P, Pitt L, Katsikeas CS, Berthon JP (1999) Virtual services go international: International services in the marketplace. Journal of International Marketing 7(3): 84–105

Bilkey WJ (1978) An attempted integration of the literature on the export behavior of firms. Journal of International Business Studies 9(1): 33–46

Bonaccorsi A (1992) On the relationship between firm size and export intensity. Journal of International Business Studies 23: 605–35

Calof JL (1994) The relationship between firm size and export behavior revisited. Journal of International Business Studies 25: 367–87

Canada-Japan Trade Council (1998) Directory of companies in Canada doing business with Japan. Canada-Japan Trade Council, Ottawa

Coviello NE, McAuley A (1999) Internationalization and the smaller firm: A review of contemporary empirical research. Management International Review 39(3): 223–56

Daniel E, Wilson H, Myers A (2002) Adoption of e-commerce by SMEs in the UK, towards a stage model. International Small Business Journal 20(3): 253–68

Davis FD, Bagozzi RP, Warshaw PR (1989) User acceptance of computer technology: A comparison of two theoretical models. Management Science 35(8): 982–1003

DiMaggio PJ, Powell WW (1983) The iron cage revisited: Institutional isomorphism and collective rationality in organizational fields. American Sociological Review 18: 122–35

Dishaw MT, Strong DM (1999) Extending the technology acceptance model with task-technology fit constructs. Information and Management 36: 9–21

Eisenhardt KM (1989) Building theories from case study research. Academy of Management Review 14: 532–50

eMarketer (2003) Worldwide B2B revenues to pass one trillion. Available at: http://www.nua.com/surveys/index.cgi?f=VS&art_id=905358753&rel=true.

Evans P, Wurster TS (1999) Getting real about virtual commerce. Harvard Business Review 77(6): 85–94

Hamill J (1997) The Internet and international marketing. International Marketing Review 14(5): 300–23

Ho J (1997) Evaluating the World Wide Web: A global study of commercial sites. Journal of Computer-Mediated Communication [online] 3(1)
    Available at: http://jcmc.huji.ac.il/vol3/issue1/ho.html

Hornby G, Goulding P, Poon S (2002) Perceptions of export barriers and cultural issues: The SME e-commerce experience. Journal of Electronic Commerce 3(4):213–26

Iacovou CL, Benbasat I, Dexter AS (1995) Electronic data interchange and small organizations: Adoption and impact of technology. MIS Quarterly 19(4): 465–85

International Telecommunication Union, Telecommunications Free Statistics Homepage, Available at: http://www.itu.int/ITU-D/ict/statistics/

Kassaye WW (1999) Sorting out the practical concerns in World Wide Web advertising. International Journal of Advertising 18: 339–61

Knight G (2000) Entrepreneurship and marketing strategy: The SME under globalization. Journal of International Marketing 8(2): 12–32

Leonidou LC, Katsikeas CS (1996) The export development process: An integrative review of empirical models. Journal of International Business Studies 27(3): 517–51

Leonidou LC, Katsikeas CS, Piercy NF (1998) Identifying managerial influences on exporting: Past research and future directions. Journal of International Marketing 6(2): 74–102

Madden G, Coble-Neal G (2002) Internet economics and policy: An Australian perspective. Economic Record 78(242): 343–57

Mahadevan B (2000) Business models for Internet-based e-commerce: An anatomy. California Management Review 42(4): 55–69

Matthews G (2003) Local firm gets latitude: Longitude fund acquires mobile knowledge. Ottawa Sun March 18: 47

McClean RJ, Johnston DA, Wade M (2003) Net impact Canada: The international experience. Report prepared for the Canadian e-Business Initiative

McCracken G (1990) The long interview. Sage, Newbury Park

McDougall PP, Shane S, Oviatt BM (1994) Explaining the formation of international new ventures: The limits of theories from international business research. Journal of Business Venturing 7: 469–87

Mehrtens J, Cragg PB, Mills AM (2001) A model of Internet adoption by SMEs. Information and Management 39: 165–76

Miles MB, Huberman AM (1994) Qualitative data analysis. Sage, Thousand Oaks

Morrison AJ, King BEM (2002) Small tourism businesses and e-commerce: Victorian tourism online. Tourism and Hospitality Research 4(2): 104–15

OECD (2002a) Small and medium enterprise outlook. OECD, Paris

OECD (2002b) Electronic commerce and SMEs. ICT and electronic commerce for SMEs: A progress report DSI/IND/PME (2002) 7, OECD, Paris

OECD (2002c) Measuring the information economy. OECD, Paris

Oliver C (1991) Strategic responses to institutional processes. Academy of Management Review 16(1): 145–79

Oviatt BM, McDougall PP (1997) Challenges for internationalization process theory: The case of international new ventures. Management International Review 37(2): 85–99

Porter ME (1985) Competitive strategy: Techniques for analyzing industries and competitors. Free Press, New York

Prasad VK, Ramamurthy K, Naidu GM (2001) The influence of Internet-marketing integration on marketing competencies and export performance. Journal of International Marketing 9(4): 82–110

Quelch JA, Klein LR (1996) The Internet and international marketing. Sloan Management Review 37(3): 60–75

Reuber AR, Fischer E (1997) The influence of the management team's international experience on the internationalization behavior of SMEs. Journal of International Business Studies 28(4): 807–25

Sahlman WA (1999) The new economy is stronger than you think. Harvard Business Review 77: 99–106

Schrader RC, Oviatt BM, McDougall PP (2000) How new ventures exploit trade-offs among international risk factors: Lessons for the accelerated internationalization of the 21st century. Academy of Management Journal 43(6): 1227–47

Schumpeter JA (1961) Theory of economic development. Oxford University Press, New York

Seybold P (1998) Customer.com: How to create a profitable business strategy for the Internet and beyond. New York Times Press, New York

Sparkes A, Thomas TB (2001) The use of the Internet as a critical success factor for the marketing of Welsh agri-food SMEs in the twenty-first century. British Food Journal 103(5): 331–37

Statistics Canada (2000) E-commerce and business use of the Internet. Available at: www.statcan.ca/daily/English/000810/d000810a/htm

Strauss AL, Corbin J (1990) Basics of qualitative research. Sage, Newbury Park

Tiessen JH, Wright RW, Turner I (2001) A model of e-commerce use by internationalizing SMEs. Journal of International Management 7: 211–33

UNCTAD (2002) E-commerce and development report 2002. United Nations, New York

WTO (2003) World trade figures 2002: Trade recovered in 2002, but uncertainty continues. WTO, Geneva

Yin RW (1989) Case study research. Sage, Newbury Park

4 SME Interaction in Supply Chains

Phil Malone

Introduction

E-business activity involves the electronic transfer of critical business information. In particular, e-business activity encompasses e-procurement, supply chain management and the transformation of firms' internal functions so as to allow the seamless transfer of information along supply chains. The role of the National Office for the Information Economy (NOIE), in the development of Australian e-business markets, involves the brokering of business relationships and the playing of a catalytic role in innovative projects, so as to: promote the effective use of e-business that results in productivity growth for the Australian economy; facilitate firm e-business adoption—especially by small business; remove impediments to small business participation in e-business by reducing associated costs, complexity and risk; and ensuring government e-business activity is consistent with broad industry contexts. In October 2001 NOIE released the report, *B2B e-Commerce: Capturing Value Online*, which provided independent assessment of contemporary developments in Australian e-business markets. The core emerging issues identified by that report for Australian e-business growth revolved around necessary cooperation on standards, greater interoperability between e-business frameworks and the effective integration of new and established technology.[1]

Small Business E-Market Entry Barriers

Because the size of the Australian economy is relatively small when compared to average OECD Member Country GDP, the necessary e-business participation to achieve minimum efficient scale or critical mass in on-line markets is higher. That is, greater firm e-business engagement is required for the operation of viable on-line trading environments. Improved interoperability provides a better opportunity through which to achieve required critical mass. Namely, current interoperability standards do not allow the seamless (or 'plug and play') transfer of business information. Figure 4.1 illustrates the challenges facing business intending entry into on-line markets. That scenario indicates suppliers with several customers are compelled to make multiple technology investments, and this is not feasible for many small firms. Accordingly, on-line market entry is inhibited by proprietary-based technology systems and confusion over standards. Additionally, there is a

---

[1] Interoperability means the ability to transfer and use critical business information across organisations and technology systems.

lack of strategic engagement among stakeholders to decide on shared business processes that underpin effective e-business solutions.

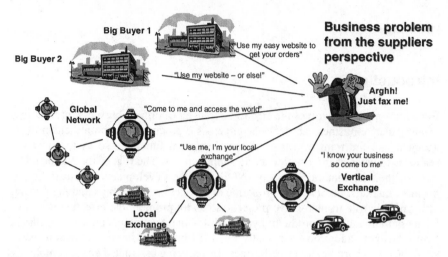

Fig. 4.1. Barriers to On-line Market Entry

Accordingly, firms must devote resources to adapt their back office operations, such as invoicing and purchasing, to better meet e-market transaction requirements. Competitive forces tend to resolve problems relating to any lack of interoperability through market rationalization. However, firm attrition of itself may fail to deliver an environment that fully engages small firms in e-markets. While the number of firms operating in the market may fall, the cost and complexity of solutions remain unchanged. The experience of electronic data interchange, a pre-Internet technology developed to automate the exchange of business information, provides an example that resulted in such an outcome.

Figure 4.2 illustrates the evolutionary cycle of Australian e-business markets that commenced in 1995. Despite a surge in e-business market entry during the mid-1990s that resulted in a dot.com boom, entry has remained sluggish due to difficulty encountered by small firms. That is, improved firm interoperability that allows low cost engagement in e-business markets by small firms is required to accelerate economy-wide adoption. Other challenges to e-business implementation by small firms include those concerning the cost-effectiveness and risks associated with e-business market operations. Namely, small firms, whether they choose to purchase or build their e-business solutions, must confront issues created by complex technology and confusion about standards. Examples include the customization of product numbers and description systems to better suit electronic catalogue requirements; uncertainty that standards used to generate business documents, such as purchase orders and invoices will gain general acceptance in electronic markets; and the integration of financial management information systems with e-business applications is often difficult, and require additional technology investment.

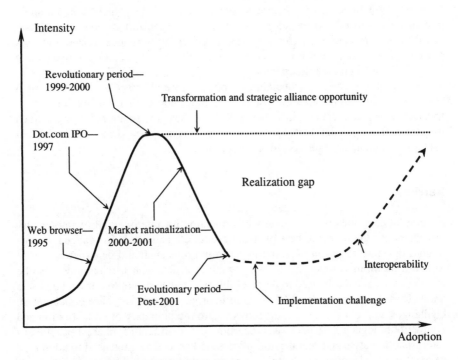

Fig. 4.2. E-Business Market Evolutionary Cycle

Finally, Figure 4.2 suggests that once these issues are addressed, improved interoperability will propel e-business entry. Australian Bureau of Statistics data identify a lack of engagement by very small firms and underutilization of the Internet to conduct e-business transactions as characteristic of Australian markets. While 72% of firms have on-line presence, 90% of firms employing more than 10 persons are on-line. Therefore, very smallest Australian firms are less likely to have an on-line presence. Further, only 35% of on-line firms use the Internet to purchase goods and services, while 26% of firms use the Internet to make on-line payments.

Drivers and Barriers to Interoperability

Drivers

Several factors, from both public policy and private sector viewpoints, are pursuing an agenda for greater interoperability. From a public policy perspective, greater e-business diffusion, enabled through improved interoperability, leads to both macroeconomic and microeconomic efficiency gain, innovation and transformation effects across the economy. For instance, supply chains are genuinely transformed with e-business adoption beyond the organization, both in upstream and downstream markets. Also improved interoperability reduces uncertainty surrounding investment decisions. Namely, firm information technology investment is more likely when the interoperability of the asset is assured. From a private sector perspective, greater interoperability implies more efficient supply chain management. Seamless information transfer among supply chains result in greater transformation and value added beyond that a firm in isolation achieves. Such transformation includes better response to business cycles, from more efficient data collection, and greater technical integration.

Barriers

Several hurdles must be overcome when pursuing for interoperability among supply chains for the potential benefits from e-business activity to be realized. In particular, business-to-business (B2B) integration projects should be structured so as to avoid prohibitive investment costs; creation of technical barriers must also be avoided, viz., standards need to be carefully selected so as not to burden end users and software programmers; B2B integration projects commonly face collaboration challenges, i.e., obtaining clear agreement as to the boundary of collaboration and competition; legacy systems issues regarding the treatment of inherent proprietary interests often arise and require resolution; and knowledge gaps need to addressed as there is a tendency to associate e-business with dot.com hype.

E-Business Principles

This analysis of the current Australian e-business market environment has provided an understanding of the drivers and challenges to improved interoperability among supply chains. In doing so, several principles for application to e-business models have emerged. A prevalent theme uniting these principles is the need to encourage collaboration.

Problem Identification

An important business challenge is to identify problems causing difficulty in core business operation for most firms in the supply chain that require immediate resolution. To be successful solutions must, in the long term, ensure supply chains attract and retain a critical mass of transactions. Such critical mass is only achieved when e-business solutions add substantial value to stakeholder groups.

Process Separation

Competitive, for example, product quality and price, and cooperative or common business processes need to be effectively identified and separated. Collaborative e-business models are intended to deliver improved sector outcomes rather than provide competitive advantage in an exclusive and selective manner. When e-business efficiencies are realized for an entire sector, market participants are able to reallocate scarce resources and actively pursue more innovative activity.

Business Solutions

A popular misconception among firms is that an important e-business objective is the adoption of particular technology solutions. However, common business problems should be identified prior to searching for potential solutions (that may need further technical expertise to make them operational). Failed e-business initiatives, particularly those developed during the dot.com period typically follow a technology-led agenda. Without first gaining a proper understanding of underlying business processes such attempts usually fail. Namely, it is important to align technology to core business goals and not allow technology to drive rather than support or improve business processes.

Stakeholder Support

Stakeholders support is required for e-business market activity to expand. Industry associations provide valuable channels through which to attract and maintain industry buy-in, assist in the piloting of solutions and their eventual adoption. Further, industry leaders and champions are needed to provide e-business project legitimacy among supply chain members. Such legitimacy is required for sustained commitment.

Governance and Funding

A central governance group, comprising important supply chain stakeholders, is necessary for both effective guidance and oversight. Additionally, written agreements delineating responsibility for pecuniary and non-pecuniary contributions of the parties are required so that a clear understanding between project participants, as a mechanism for ongoing delivery and conflict resolution, is provided. An equitable funding model should be entered into so that supply chain participants can better determine their value propositions and assess their intended contributions.

Supply Chain Interoperability Initiatives

NOIE has recently established several initiatives that focus on supply chain interoperability issues. For e-business markets to be successful, especially in the exchange of critical business information among supply chain members, requires that open and inclusive standards are adopted. Further, such adoption must be made by small and large firms alike, as small companies comprise the majority of Australian firms. Small firms have legacy systems which do not necessarily interoperate well with supply chain systems. Therefore, an interface to existing applications must be created so that seamless interoperability occurs, irrespective of firm size or information technology capability. The NOIE (2001) report, *B2B E-Commerce: Capturing Value Online*, identified that improved interoperability was required for e-business activity among supply chains to expand. This project developed from the National Interoperability Workshop of April 2002. The workshop formulated projects to build virtual infrastructure for supporting widespread interoperability. The forum was attended by representative stakeholder groups including banks, end users, governments, industry associations, standards organizations, third-party providers, and vendors and systems integrators. Projects to arise from the workshop are the B2B Registry and the Integration Toolkit.

B2B Registry and Integration Toolkit

The NOIE B2B Registry promotes e-business open standards use to reduce compliance costs. This service stores data and schema necessary for automating collaborative business documentation, e.g., invoices, purchase orders, remittance advices and consignment notes. To do so, a company profile information register allows, e.g., firms to locate other firms with compatible invoice processes. The e-business extensible mark-up language and architecture is employed to support standards that suit simple processes. A standard stored by the Registry is for an Integration Toolkit aimed at providing small business with a low cost connector to e-business activity with larger firms. That is, the Registry allows Financial Management Information Software (FMIS) to be used by small business, e.g., MYOB and Quicken, to interoperate with the FMIS software of larger corporations and government, e.g., SAP and Oracle. As the Registry allows trading partners to transact directly, and so avoid the use of hubs and e-markets, e-business processing costs are reduced. The project is currently in pilot phase mode, i.e., registry architecture is being constructed as is the technology blueprint for the integration toolkit. NOIE funded the development stage, and is partnered with Standards Australia to provide governance and industry awareness. The building of a fully functional registry is to be funded by industry so as to ensure long term sustainability.

Information Technology Online Program

NOIE is also responsible for the Information Technology Online (ITOL) small competitive grants program (http://www.noie.gov.au/itol), which provides grants to consortia to facilitate innovative e-business adoption. Projects supported by the ITOL program address supply chain issues that support open standards and include small business participation.

Appliance Industry

This project is supported by a consortium of appliance industry participants, and intends to devise and implement Message Implementation Guidelines across the industry supply chain to transfer cost reductions via standardized messaging.

Mining Industry

The mining industry project is a consortium comprised of IBM, Mincom, MSA Australia, Newmont Mines and Quadrem, and is led by XML Yes. The project concerns backend integration and intends to develop B2B infrastructure to allow the exchange of data, business documents and messages between small firms and large corporations.

Textile Supply Chain Interoperability

This project integrates production supply chain systems used by Australian Weaving Mills suppliers through to important retailers. The facilitation of the seamless movement of data through the textiles production system allows a leveling of production cycles. By providing data transfer through a bureau system operated by Tasmanian Business Online (TBO), small firms can enter e-business markets at lower costs than presently available with electronic data interchange data transfer systems. TBO is a translation hub that provides interoperability between different systems.

NOIE and the Path Ahead

Most firms currently conduct e-business functions in-house. Soon increased e-business activity will require firms to collaborate more with other firms in their supply chain. Accordingly, firms will come to view their supply chains as important sources of innovation and competitive advantage. However, to realize this potential requires that firm's implement standards, particularly for cooperative business data exchange. To date, standards development investment has mostly related to industry specific standards. Industry-wide adoption to ensure critical mass is achieved is now required. However, first small firms must have confidence in the Internet as a transaction medium. The NOIE guide, *Trusting the Internet*, considers Internet related business security issues. This initiative forms part of a national approach to creating a trusted and secure electronic operating environment for private and public sectors. More specifically, the integration of freight management systems of transport and logistics firms with e-business applications of their clients will assist the development of e-business among supply chains. With this in mind, NOIE is working with the Australian Logistics Council to forge standards adoption. Finally, Web services that allow firms to link applications with those of their partners, suppliers and customers via the Internet, will also facilitate e-business adoption. The services are an emerging Internet-based technology that contributes to productivity improvement, innovation and transformation by enabling more effective collaboration. Web services help reduce the complexity, and so cost, of collaboration by providing a standards-based framework to describe, discover, transfer and use information.

Conclusion

E-business market activity and its adoption by Australian firms continue to grow. A challenge is to ensure that e-business markets develop in a fully integrated fashion among industry supply chains in a manner that encourages participation by small firms. It is only through active participation by Australian small firms will e-business be pervasive throughout the economy.

Part II: Technology and Productivity

5 Deciding on Network Architecture for 3G Wireless Services

Hak Ju Kim

Introduction

The demand for high speed data services, such as multimedia and interactive information services, is increasing. Since such data services are a potential source of network service provider profit, third generation (3G) technology is being developed to support them. Further, this emerging wireless network technology will ultimately result in a much more complex network and service environment. With increasing technological complexity and uncertainty wireless network operators require more flexible network architectures to achieve and maintain competitive advantage. However, since the complete replacement of existing wireless network architecture is not practicable, technology choice in the migration of existing networks requires that carrier management examine options available as a strategic decision. To allow strategic analysis of the possible evolution of wireless networks and their architectures—innovation (ITh, Henderson and Clark 1990) and real options (RO, Dixit and Pindyck 1994) theory—are introduced. Henderson and Clark (1990) examine why some firms are able to leverage innovations. RO allows the linking of strategic planning and financial strategy by determining a value for an uncertain opportunity. Clearly, the potential exists to employ ITh to analyze wireless network technology evolution and architecture, and RO to support strategic decision-making. The principal motivation for this study is to develop a conceptual framework through which wireless network operators can support their strategic decisions concerning next generation network architecture. Accordingly, typical network architecture migration path alternatives, global systems for mobile communications (GSM) and code division multiple access (CDMA) based network scenarios, are examined using ITh. Strategic options for network migration as a means to manage wireless network architecture evolution based on the RO are also discussed.

Wireless Network Architectures

During the past decade, wireless networks have evolved, rapidly moving from first-generation (1G) analog—voice-only communications—to second generation (2G) digital—voice and data communications—to 3G wireless networks, a convergence of wireless and the Internet.

1G Wireless Networks

1G wireless networks are based on analog technology or electronic transmission techniques accomplished by adding signals of varying frequency or amplitude to carrier waves of a given frequency of alternating electromagnetic current. This current is usually represented as a series of sine waves because the modulation of a carrier's wave is analogous to voice fluctuations. Figure 5.1 shows generic transport architecture for a 1G cellular radio network, which includes mobile terminals, base stations (BS) and mobile switching centers (MSC). A MSC maintains mobile related information and controls mobile hand-off. The centre also performs necessary network management functions, such as call handling and processing, billing and fraud detection. An MSC is interconnected with the public switched telephone network (PSTN) via trunks and a tandem switch.

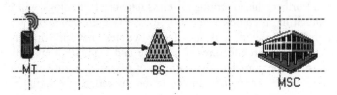

Fig. 5.1. 1G Wireless Network Architecture

The main 1G wireless network technology standards are advanced mobile phone system (AMPS), total access communication system (TACS) and Nordic mobile telephony (NMT) (Dahlman et al. 1998).

AMPS

In the United States (US), the AMPS system, released in 1983, is a 1G wireless technology standard using the 800 MHz to 900 MHz frequency band and the 30 kHz bandwidth with 666 channels (Garg and Rappaport 2001). AMPS is also the first standardized cellular service and currently the most widely used standard for cellular communications, e.g., the technology is employed in China, South America and the US. However, since the AMPS uses frequency modulation (FM) for radio transmission, it exhibits a low calling capacity, limited spectrum, poor data communication and minimal privacy protection.

TACS

TACS is a mobile telephone standard originally used in United Kingdom for the 900 MHz frequency band (Rappaport 2002). TACS is essentially the European version of AMPS. The standard operates on the 900 MHz frequency band, allowing up to 1320 channels using 25 kHz channel spacing (Garg and Rappaport 2001). The system is designed to consider the very high subscriber density in large urban centers, and also sparsely populated rural areas. TACS is obsolete in Europe, and is replaced by the more scalable and digital GSM system.

NMT

NMT, developed by Ericsson, is a cellular standard that uses 12.5 kHz channel spacing, and is deployed in 30 countries (Garg and Rappaport 2001). NMT is the common Nordic standard for analog mobile telephony as established in the early-1980s by telecommunications administrations in Denmark, Finland, Norway and Sweden. The NMT system is also installed in several European countries—including parts of Russia—and in Asia and the Middle East.

2G Wireless Networks

Second generation (2G) digital wireless networks replaced 1G analog networks, allowing data transmissions to form part of wireless traffic. 2G systems had their genesis with the advent of digital technology.[1] Using digital technology, voice signals are digitized and sent as bits of data over radio waves. As shown in Fig. 5.2, 2G introduced network architectures that are different in several respects from 1G architecture. First, the 2G system reduced computational burdens of an MSC by introducing a base station controller (BSC), which is an advanced call processing mechanism. The BSC, also called a radio port control unit, allows data interface between a base station and the MSC. Second, 2G systems use digital voice coding and digital modulation. Finally, 2G technology provides dedicated voice and signaling between MSCs, and between an MSC and the PSTN. In contrast to 1G systems designed for voice transmission, 2G technology is specifically developed to transmit data.

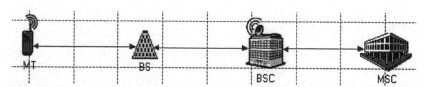

Fig. 5.2. 2G Wireless Network Architecture

Several wireless technologies, including time division multiple access (TDMA), GSM, cdmaOne and PDC, are classified as 2G. However, prior to the universal introduction of 3G wireless systems, a hybrid 2.5G technology—that allows 2G users to sample 3G-like services—is introduced. 2.5G systems refer to

---

[1] Digital technology transmits or stores data with a string of zeros and ones. A digit is referred to as a bit and a string of bits that a computer can address individually as a group is a byte. Digital technology made its most fundamental technological change in telecommunications industry (Garg and Rappaport 2001).

enhanced data rates for GSM evolution (EDGE), general packet radio service (GPRS) and high speed circuit switched data (HSCSD) systems (Dahlman et al. 1998). 2.5G technology is commonly viewed as an upgraded 2G networks, and mostly involves 2G infrastructure undergoing simple software or hardware development.

TDMA

TDMA is a digital transmission technology that allows several users access to a radio frequency channel without the restriction of having to allocate unique time slots to users within a channel (Garg and Rappaport 2001). The current TDMA standard for cellular divides a channel into six time slots, with a signal using two slots, i.e., providing a 3-to-1 improvement in capacity compared to AMPS. Operationally, callers are assigned particular time slots for transmission. TDMA has three users sharing a 30 MHz carrier frequency (Rappaport 2002). Further, TDMA has a capacity to carry data rates of 64 kbps to 120 Mbps. This capacity allows operators to offer personal communications services including facsimile, voiceband data and short message service, as well as, bandwidth intensive multimedia and video conferencing applications.

GSM

GSM is a globally accepted standard for digital cellular communication, and is a common European mobile telephone standard operating at 900 MHz in 1982 (Rappaport 2002). Current GSM networks transmit data at 9.6 kbps with a circuit-switched data transmission. The system allows as many as eight users to share a 200 kHz radio channel by allocating a unique time slot to users (Garg and Rappaport 2001).

GPRS

Currently many GSM carriers are employing GPRS service as a 2.5G technology. GPRS permits packet-switched instead of circuit-switched data transmission at high speed using GSM technology (Rappaport 2002). Depending on the coding schemes used, much faster data transmission rates can be realized per time slot with GPRS, than 9.6 kbps with GSM. The principal infrastructure elements of GPRS are the gateway GPRS support node (GGSN), and the serving GPRS support node (SGSN) (Garg and Rappaport 2001). GGSN provides interconnection to other networks such as the Internet or private networks, while SGSN tracks the location of mobile devices and routes their packet traffic.

EDGE

EDGE is a phase technology after GPRS. Based on the GSM standard, EDGE permits faster data transfer rates, and is another transition technology from GSM technology toward 3G. EDGE is a radio based high-speed mobile data standard that attains a 384 kbit/s data transmission speed when the available eight time slots are in use (Garg and Rappaport 2001). The technology was initially developed in response to the demands of mobile network operators that had failed to gain access to spectrum for 3G network transmission. The intent in EDGE technology design is to achieve higher data rates than that available from 200 kHz GSM radio carriers. The technology changes the type of modulation employed, while still working with existing circuit switches (Rappaport 2002). The result is that EDGE delivers data transmission rates nearing 500Kbps with GPRS infrastructure.

HSCSD

HSCSD is an enhancement of data services or circuit switched data provided through existing GSM networks (Garg and Rappaport 2001). The technology allows access to non-voice services at a speed three times faster than that currently available. This transition speed enhancement means subscribers are able to send and receive data from their portable computers at speeds to 28.8 kbps, and is being upgraded in many networks to 43.2 kbps (Rappaport 2002).

cdmaOne

cdmaOne employs CDMA spread spectrum technology to allow the simultaneous use of multiple frequencies (Garg and Rappaport 2001). CDMA technology codes the digital packets for transmission with a unique key. Receivers respond only to this key, and identify and demodulate associated signals. CDMA proponents claim a bandwidth efficiency approaching 13 times that for TDMA and from 20 to 40 times analog transmission rates. cdmaOne describes not a single technology but a complete wireless system based on the TIA/EIA IS-95 CDMA standard, including the IS-95A and IS-95B systems. The IS-95A system utilizes at least one 1.25 MHz carrier, and operates within the 800 MHz and 1900 MHz frequency bands. cdmaOne provides circuit-switched data connection at 14.4 kbps. Due to achievable data speeds for the IS-95B system, IS-95B is categorized as a 2.5G technology. IS-95B is an interim standard change that drives higher data rates for packet- and circuit-switched CDMA data transmission. Data transmission rates approaching 115 kbit/s are realized by using, as many as eight, 14.4 kbit/s or 9.6 kbit/s data channels (Rappaport 2002).

3G Wireless Networks

Currently, wireless network architectures are evolving toward 3G wireless technology. Ultimately, successful transition will allow carriers to provide high-rate voice and data service. 3G systems also enable multi-megabit Internet access, with both an 'always on' feature and data rates near 2.048 Mbps suitable for multimedia services. These wireless systems are commonly categorized into the universal mobile telecommunication system (UMTS) and the cdma2000 groups. The 3G partnership project (3GPP) is collaboration between organizational partners that study W-CDMA/TD-SCDMA/EDGE standards, while the 3G Partnership Project 2 (3GPP2) is collaboration between organizational partners that cdma2000 standards.

UMTS

UMTS, developed in 1996 through the sponsorship of the European Telecommunications Standards Institute, is a wideband CDMA (WCDMA) technology. WCDMA is a key technology, especially as an air interface standard (Dalal 2001). The technology uses a direct spread with a chip rate of 3.84 Mcps and a nominal bandwidth of 5 MHz. Technically UMTS is an upgraded GSM/GPRS technology with spectral efficiency enhanced six fold. As a 3G standard, UMTS offers a packet-based wireless service with transmission of nearly 2.048 Mbps, and an expectation of reaching an 8 Mbps platform. A 5 MHz UMTS channel can support 100 to 350 voice calls simultaneously through a more efficient modulation technique.

In Fig. 5.3 UMTS network architecture is separated into a radio access network (RAN) and core network. The RAN contains user equipment—that includes terminal equipment and mobile terminals—and the UMTS terrestrial radio access network—that includes the Node-B and radio network controller (RNC). The core network, focusing on packet domain, is similar to the GPRS core network, and includes the network nodes: the serving GPRS support node (SGSN) and the gateway GPRS support node (GGSN). The SGSN monitors user locations, and performs security functions and access control. The GGSN contains routing information for packet switched (PS) attached users and provides inter-working with external PS networks such as the packet data network.

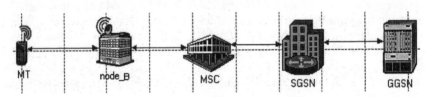

Fig. 5.3. 3G Wireless Network Architecture

cdma2000

cdma2000 is another wireless standard designed to support 3G services, as defined by the International Telecommunication Union and the IMT-2000 (Carsello et al. 1997). The technology supports mobile data communication at speeds from 144 kbps to 2 Mbps as WCDMA technology (Garg and Rappaport 2001). cdma2000 uses the same base chip rate of 1.2288 Mcps as cdmaOne (Dalal, 2001). Individual carriers are modulated with separate orthogonal codes, and have an optional overlay mode. The coding distinguishes cdmaOne and cdma2000 users.

The cdma2000 standard is a high data rate upgrade of the interim standard-95 (IS-95), a 2G CDMA standard that is strictly devoted to traditional CDMA infrastructure. A 2G mobile carrier adapted to a 3G cdma2000 network does not need new base stations or channel bandwidth reorganization. The bandwidth of a radio channel remains 1.25 MHz. The difference is that, at most, three channels are jointly utilized to provide data transmission speeds exceeding 2.048 Mbps per user. Currently cdma2000 standards are cdma2000-1xRTT, cdma2000-1xEV, data and voice (DV), data only (DO) and cdma2000-3xRTT. cdma2000-1xRTT (Radio Transmission Technology) technically supports twice as many users as 2G CDMA with data rates up to 153.6 Kbps. cdma2000-1xEV is a high data rate packet standard originally developed by Qualcomm, Inc. An advantage of the standard is that it provides both, DO and DV radio channels to deliver transmission rates approaching 2.48 Mbps. Cdma2000-1xEV-DO is obtained to suit applications such as video streaming and large file downloads.

Innovation and Wireless Networks

Technology typically develops in an evolutionary manner, i.e., from a change in conceptual framework along particular paths called technological paradigms or trajectories (Dosi 1982). Such technological evolution is not easily incorporated into traditional explanation of product developments. Such explanations rely on the consideration of periods of disruptive innovation followed by less drastic incremental change (Nelson and Winter 1977). To better understand the realized evolution of wireless network technology and their architectures the Henderson and Clark (1990) ITh framework is introduced. This theory of innovation, allows investigation of core architectural features of competing wireless network technology. Isolating these features allows insight into the evolutionary path of wireless network technology to be gained.

Henderson and Clark develop a model to demonstrate that architectural innovation makes obsolete established firm's architectural knowledge. The model provides an explanation as to why only some firms are able to leverage their innovations. The argument distinguishes between components or modules (the physically distinct portion of a product that embodies the core design concept) and linkage or interface (that describes how a firm's product interacts with other products and any synergies that may arise). The model's conceptual framework consists of classifications for technological innovation as either incremental, modular, architectural or radical. Incremental innovation refines and extends an established design in an individual component without changing core design concepts and links between them. Such incremental design shows a relatively minor change to the existing product or its design, and often reinforces the dominance of an established firm. Modular innovation only changes core design concepts, and not linkages between components. Architectural innovation involves reconfiguration of an established system to link existing components in a new and deliberate manner, and creates new interaction and linkages with other components within established products while not changing the core design concepts. Finally, radical innovation is a new dominant design and set of core design concepts embodied in components that are linked through new architecture.

Analog to Digital

The transition from 1G analog to 2G digital technology involves a change of core network architecture design, and alters the link between BS and MSC components. As discussed, analog technology is based on a series of sine waves that are similar to those of voice fluctuations. With digital technology, analog voice signals are digitized and transformed into 64 kbps pulse code modulated bit streams by vocoders. Vocoders reside at an originating BS. In using digital cellular compression techniques at a mobile station (MS), it is no longer economically feasible to convert voice to 64 kbps speech at a BS and use a DS0 to carry the call, therefore vocoders are moved to MSCs. Digitization by vocoders alters the core modular innovation concept. Further, installation of a digital-to-analog converter (DAC) provides a key interface between analog and digital systems. A DAC is a necessary component for high speed Internet access, i.e., xDSL. Accordingly, the introduction of DACs is considered a change in linkage in terms of architectural innovation. Finally, a distinct feature of the analog to digital technology transition is the modulation method. For an analog system, simple FM, amplitude modulation (AM), double sideband AM (DSB), single sideband AM (SSB) and vestige sideband AM (VSB) techniques are employed. However, modulation techniques for digital signal transmission are more complex and numerous, and include binary phase shift keying (BPSK), differential phase shift keying (DPSK), quadrature phase shift keying (QPSK), binary frequency shift keying (BFSK), minimum shift keying (MSK), Gaussian minimum shift keying (GMSK), m-ary phase shift keying (MPSK), m-ary quadrature amplitude modulation (QAM), m-ary frequency shift keying (MFSK) (Rappaport 2002). Accordingly, the transition from analog to digital is viewed as a radical innovation.

GSM to GPRS

GPRS is essentially based on 'regular' GSM (with the same modulation) and designed to complement existing circuit-switched cellular telephone services such as SMS and cell broadcast. GPRS should improve the peak time capacity of a GSM network as it simultaneously transports traffic previously sent using circuit switched data through a GPRS overlay, and reduce SMS center and signal channel loading. GPRS packet-based service should prove less costly than circuit-switched services to consumers as communication channels are on a shared use basis, viz., on a 'packets needed' basis, rather than dedicated to a user. Then, to incorporate GPRS data functionality into existing GSM systems, wireless network operators must upgrade the equipment without changing architectures, i.e. incrementally innovate. BTS must also upgrade their software, as does MSC to be able to respond to new forms of data request, viz., modularly innovate.

GPRS to EDGE

EDGE provides an evolutionary migration path from GPRS to UMTS by more expeditiously implementing necessary modulation changes. As such, EDGE does not radically alter the core network as it uses GPRS/GSM. Rather, the technology concentrates on improving capacity and efficiency over the air interface by introducing a more advanced coding scheme whereby time slots transport more data. Additionally, EDGE adapts coding to current conditions, which means that transmission speeds are higher when radio reception is clear. The implementation of EDGE by network operators is designed to be simple, with only the addition of an extra EDGE transceiver unit to cells, a modular innovation, required. For most vendors, it is envisaged that BSC and BS software upgrades will be carried out remotely, viz., on an incremental (innovation) basis. New EDGE capable transceivers can also handle standard GSM traffic and automatically switch to EDGE mode as required. Further, terminals must be upgraded to be EDGE network functional, as existing GSM terminals do not support new modulation techniques, and provide another example of incremental innovation.

Moving to 3G

Clearly, a move to 3G is a radical innovation requiring change in core design concepts and most components, i.e., 3G networks require new radio and core network elements. For example, a 3G radio access network comprises an RNC and Node B. A RNC replaces the BSC, and the RNC includes support for legacy system connection and provides efficient packet connection with the core network packet devices (SSGN or equivalent). The RNC performs radio network control functions that include call establishment and release, handover, radio resource management, power control, diversity combination and soft handover. Another required piece of network infrastructure for 3G is a media gateway that resides at the boundary between networks to process end user data, such as voice coding and decoding, convert protocols and mapping service quality.

Alternative 3G Paths

A salient feature of wireless network evolution is that it encompasses an innovation portfolio, i.e., several mixed or hybrid innovations are involved at a migration stage. For instance, the evolution from GSM to GPRS is an architectural innovation as well as an incremental innovation. Clearly, the distinction between architectural, incremental, modular and radical innovation is matter of degree. Further, developments, from cdma2000 1X to cdma2000 3X, can viewed as a radical rather incremental innovation. Another example is provided by wireless network operators and carriers that prefer steady change to minimize risk. As a result, most of migration paths approximate a stepping-stone or evolutionary rather than revolutionary approach. An exception is the successful Qualcomm (2002) CDMA technology choice.

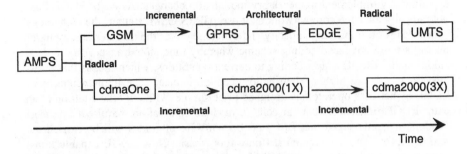

Fig. 5.4. A Typology of Wireless Technology Innovation

GSM-based Network Migration

UMTS does not support the reuse of GSM BS equipment. CDMA signals require linear amplification and additional filtering. Operators must install new hardware cabinets adjacent to existing systems. Additionally, operation in GSM and TRA modes within a 5 MHz band is not possible. Figure 5.5 shows for GPRS service to be provided in a GSM network required the addition of several components, e.g., SGSN and GGSN. Further, transition from a GSM/GPRS to UMTS access network section requires fundamental change or additional networks.

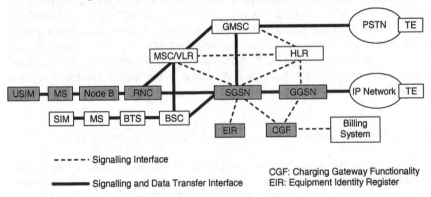

Fig. 5.5. A GSM-based Network Architecture Migration Path

Table 5.1 summarizes components that must be upgraded or replaced in GSM networks. In the case of provisioning GPRS service, only software upgrade and little hardware (HW) or software (SW) replacement are required. For UMTS, however, most network access facilities are replaced as the technology in GSM/GPRS (TDMA-based) is fundamentally different from CDMA-based UMTS technology. Accordingly, substantial investment is required for the transition from a GSM to 3G network architecture.

Table 5.1. Component Requirements to Upgrade GSM Networks

Category	GSM to GSM / GPRS		GSM / GPRS to UMTS	
	HW	SW	HW	SW
MS / SIM	Upgrade	Upgrade	New	New
BTS	Upgrade	No Change	New	New
BSC	Upgrade	PCU Interface	New	New
MSC / VLR	Upgrade	No Change	No Change	Upgrade
HLR	Upgrade	No Change	No Change	No Change
SGSN	New	New	No Change	Upgrade
GGSN	New	New	No Change	No Change

CDMA-based Network Migration

Since cdma2000 evolved from IS-95-based systems, only minor upgrade and so minimal capital investment is required. Accordingly, the transition from cdmaOne to cdma2000 1X is relatively easy to implement and transparent to consumers. Service providers can gradually migrate from cdmaOne to cdma2000 at the 1X (1.2288 Mcps) rate. As consumers migrate, network operators can shift to cdma2000 1X radios by inserting a cdma2000 3X radio to increase cell capacity. Operators can also choose from three cdma2000 1X radios or convert to a single cdma2000 3X radio. cdma2000 uses the same 9.6 kbps vocoder as cdmaOne.

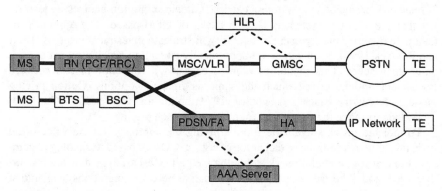

Fig. 5.6. A CDMA-based Network Architecture Migration Path

Table 5.2 shows the transition from cdmaOne to cdma2000 requires both channel card and software upgrades to cdmaOne BSs (older BSs may require hardware upgrades) and new handsets. cdma2000 1X, implemented within existing spectrum, delivers approximately twice the voice capacity of cdmaOne, and provides average data transmission rates of 144 kbps. The cdma2000 3X standard signifies three times 1.25 MHz or approximately 3.75 MHz. The cdma2000 3X multi-carrier approach, or wideband cdmaOne, is important for the evolution of IS-95-based standards. In short, cdma2000 3X, with data transmission rates approaching 2Mbps, offer more capacity than cdma2000 1X. So, unlike UMTS, cdma2000 does not require substantial investment to achieve 3G service capability.

Table 5.2. Component Requirements to Upgrade CDMA Networks

Category	cdmaOne to cdma2000 1x		Cdma2000 1x to cdma200 3x	
	HW	SW	HW	SW
MS	New	New	No Change	Upgrade
BTS	No Change	Upgrade	No Change	Upgrade
BSC	No Change	Upgrade	No Change	Upgrade
MSC / VLR	No Change	Upgrade	No Change	Upgrade
HLR	No Change	No Change	No Change	No Change
HA / FA	New	New	No Change	No Change
AAA Server	New	New	No Change	No Change
PDSN	New	New	No Change	No Change

Real Options and Risk Neutrality

The RO approach offers an opportunity for 2G and 3G network operators to review and assess their technology transition choices when faced with an upgrade to 3G network architectures. Wireless carrier strategic migration choices depend on whether the existing system is CDMA-based or GSM-based. The approach attempts to quantify management's flexibility in strategic investment projects (Dixit and Pindyck 1994). Such values are not typically taken into account by net present value approaches. This flexibility value is basically a collection of real network investment options priced much the same as financial options. Option pricing models such as the Black and Scholes (1973) option and Cox et al. (1979) binomial model provide the theoretical foundation for such pricing.

To indicate insights available from applying the approach, let the value of an investment in revolutionary technology, e.g., a CDMA-based technology, compared to that for an evolutionary technology, e.g., GSM-based, be denoted H. Further, let P and B be the net values for alternative network migrations available at time t. For expository purposes, P relates to a revolutionary technology and is associated with the greater risk and investment (an aggressive strategy), while the remaining choice value B relates to a stepping-stone technology with the smaller risk and investment (a conservative strategy). Assuming that required investment for network performance improvement is proportional to revenue, the choice of strategic options reduces to quantifying the trade-off between the performance improvement and the risk neutral premium value. Risk neutrality means comparing, say, a portfolio containing a stepping-stone architecture investment with a premium to another portfolio containing a revolutionary architecture investment with a potentially higher value. Effectively, the choice between scenarios reduces to a comparison of wireless network technology migration portfolios.

Within this framework the investment scenarios are defined as: revolutionary portfolio, $W_{REV} = v_P P$, i.e., a CDMA-based architecture; and evolutionary portfolio, $W_{EVO} = v_B B$, i.e., a GSM-based architecture, where v_P and v_B are the amounts invested in a scenario. To compare the portfolios, introduce the quantity $H(P, B)$ which is defined as:

$$v_H H + W_{EVO} = W_H + W_{EVOt} = W_{REV}.$$ (5.1)

Totally differentiating Eq. 5.1 provides:

$$v_H dH(P, B) = v_p dP - v_B dB.$$ (5.2)

Combining Eq. 5.1 and Eq. 5.2, and rewriting gives:

$$W_H \frac{dH}{H} = W_{REV} \frac{dP}{P} - W_{EVO} \frac{dB}{B}.$$ (5.3)

With $H(P, B)$ interpreted as the value of the option of investing in revolutionary rather than evolutionary technology, $W_H = W_{REV} - W_{EVO}$ (obtained from rearranging Eq. 5.1) is the premium to be paid to achieve improved network performance, assuming risk neutrality. $H(P, B)$ is the maximum premium required to reduce uncertainty associated with this technology migration. That is, as long as the actual value of a premium paid for improved network performance is less than $H(P, B)$, it is advantageous to invest in revolutionary technology.

Conclusion

This paper discusses the evolution of wireless network technology employing ITh architectural innovation concepts to construct a framework for the assessment of wireless network technology using a RO approach to assist in network architecture investment decisions toward 3G. Preliminary analysis shows that wireless network technology evolution has occurred through combining architectural, incremental, modular and radical innovations. Based on this framework, GSM-based network architecture has short-term advantages as data services can be readily deployed, but radical change is required for 3G upgrade, e.g., WCDMA technology is not so flexible. By contrast, CDMA-based network architecture has the greater long-term advantage because of its capacity and ready migration to 3G. From a strategic perspective, network service providers need to consider potential challenges that may hinder such migration, e.g., uncertainty associated with markets and emerging technology. By identifying such challenges, network service providers are better prepared for transition pitfalls, and will more likely select appropriate transition technology. Within this wireless network technology and architecture framework, a test of wireless network migration scenarios using the RO is proposed. Comparison is complex because available technology offers different data rates, capacity and user interface. The next step in this research program is to measure the performance of wireless network alternatives and conduct an assessment of the behavior of network elements within wireless networks. This study will assist wireless network service providers in their strategic decision-making when upgrading or migrating to next generation network technology and architectures by resolving ambiguity as to the nature of network evolution.

References

Black F, Scholes MS (1973) The pricing of options and corporate liabilities. Journal of Political Economy 81: 637–54

Carsello RD, Meidan R, Allpress S, O'Brien F, Tarallo JA, Ziesse N, Arunachalam A, Costa JM, Berruto E, Kirby RC, Maclatchy A, Watanabe F, Xia H (1997) IMT-2000 standards: Radio aspects. IEEE Personal Communications. August: 30–40

Cox JC, Ross SA, Rubinstein M (1979) Option pricing: A simplified approach. Journal of Financial Economics 7: 229–63

Dixit A, Pindyck R (1994) Investment under uncertainty. Princeton University Press, Princeton

Dahlman E, Gudmundson B, Nilsson M, Sköld J (1998) UMTS/IMT-2000 Based on Wideband CDMA. IEEE Communications Magazine September, pp 70–80

Dalal N (2001) A comparative study of UMTS (WCDMA) and cdma2000 networks. IEEE MetroCon, Fort Worth, September 19

Dosi G (1982) Technological paradigms and technological trajectories. Research Policy 11: 147–62

Garg V, Rappaport TS (eds) (2001) Wireless network evolution: 2G to 3G. Pearson Education, New York

Henderson R, Clark K (1990) Architectural innovation: The reconfiguration of existing product technologies and the failure of established firms. Administrative Science Quarterly 35: 9–30

Nelson RR, Winter SG (1977) In search of useful theory of innovation. Research Policy 6: 36–76

Prasad R, Ojanpera T (1998) An overview of CDMA: Evolution toward wideband CDMA. IEEE Communication Surveys, Fourth Quarter. 1(1): 2–29

Rappaport TS (2002) Wireless communications: Principles and practice, 2nd edn. Prentice-Hall, Englewood Cliffs

Qualcomm (2002) Economics of mobile wireless data. Qualcomm Incorporated, San Diego

6 Measurement of TFP Growth for US Telecommunications

Jeffrey I. Bernstein and Charles J. Zarkadas

Introduction

Gains in productivity reflect the extent to which output from production processes grow at a faster rate than the inputs to these processes. Thus productivity growth is an indicator that measures productive efficiency over time. Various sources directly contribute to productivity gains, notably among them is technological progress. The significance of technological advance to national welfare implies that estimates of the rate of productivity growth shape views of the long-term productiveness of a firm, industry or national economy. The main focus of this study is to measure productivity growth for the US telecommunications industry for the period 1985 to 2001. Estimates of productivity growth rates also appear prominently in the regulation of the telecommunications industry. In the last decade and a half price-cap regulation has been adopted in the telecommunications industry by the US Federal Communications Commission (FCC) and in over thirty states.[1] Price-cap regulation typically specifies an average rate at which the prices that a firm charges for its regulated services must decline, after adjusting for inflation. This rate is called the X factor, or offset. An important element in determining the rate of change in inflation-adjusted regulated prices is the productivity growth of the industry.

This study develops two sets of total factor productivity indexes. The first method is based on the original FCC total factor productivity study used to calculate X factors for the interstate services of the incumbent local exchange carriers (ILECs).[2] That FCC study is updated here to the year 2001. The methodology used to calculate the total factor productivity index (TFPI) is based on the ILECs regulated books of account, which includes three output categories: local, intrastate, and interstate services. The second TFPI computed contains two modifications. First, intrastate services are disaggregated into switched access and toll services. The FCC method measures intrastate output using dial equipment minutes (DEMS) relevant to intrastate services. DEMS are a measure of overall traffic volume taken from telephone company's central office switches. However, there are two physical intrastate outputs provided by the ILECs: intrastate toll minutes and intrastate switched access minutes. Intrastate toll minutes are provided directly to the ILECs end user, while intrastate access minutes are sold to long-

---

[1] See Sappington (2002).

[2] See FCC (1997), and Gollop (2000) for an updated study to 1998.

distance carriers that provide long-distance services to their end users.[3] The revised measure of intrastate output accounts for toll and access price changes, which are regulated by state regulatory agencies and the changing pattern of respective volumes over time. The second modification to the FCC productivity study involves the introduction of a broadband service into the TFPI. Included in ILEC output is the high bandwidth portion of existing ILEC access lines used to provide digital subscriber line (DSL) services. ILECs supply high bandwidth transmission services to carriers, who then offer broadband DSL or high-speed internet services, to their end users. Higher frequency DSL services share the line or 'ride' alongside regular lower frequency telephone services on existing lines. There is a paucity of data available for these services. The FCC introduced line sharing rules in 1999, at which time the Commission starting tracking such services.[4]

Indexes of productivity involve the aggregation of multiple outputs and multiple inputs associated with a production process. In the following section the concept of total factor productivity and its characterization as a ratio of an output quantity index to an input quantity index is discussed. This characterization leads to the development of some of the main alternative indexes used in productivity calculations. These are the Paasche, Laspeyres and Fisher indexes. A review of the guidelines used to choose among alternative index number characterization is then provided. The Fisher TFPI is then shown to satisfy the most comprehensive set of tests governing index number selection. The FCC study, which is based on a Fisher TFPI, is then updated for the period 1985 to 2001. A penultimate section modifies this work to account for disaggregated intrastate services and DSL services. A final section concludes the study.

The Concept of Total Factor Productivity

Consider a production process that uses a single input to produce a single output. In period t, the quantity used of the single input is denoted by v_t, and the quantity produced of the single output is y_t. Factor productivity (*FP*) in period t measures the quantity of output per unit of the input quantity and is provided by:[5]

$$FP^t = (y^t / v^t). \tag{6.1}$$

A problem with the concept of FP is that it is not dimension-free. Suppose there is no change in the production process, while there is a change in the measurement of the input quantity, e.g., suppose labor, formerly measured as numbers of em-

---

[3] Long-distance providers need to interconnect with an ILEC to either originate or terminate the long-distance call on ILEC lines to individual telephone locations.

[4] See FCC (1999).

[5] Factor productivity is also referred to as the average product of a factor.

ployees, is presently measured by numbers of hours. This change alters FP, although the production process remains unaltered. The same issue occurs when output quantities change dimension. A solution to this problem is to define a factor productivity index *(FPI)* between period t and period s, ($s < t$), as the ratio of factor productivities between the two periods:

$$FPI(s,t) = \left(\frac{y^t}{v^t}\right) / \left(\frac{y^s}{v^s}\right) = FP^t / FP^s. \tag{6.2}$$

As Eq. 6.2 shows, the FPI is dimension-free. It embodies the growth in FP between period t and period s. If more output per unit of input is produced in period t compared to period s, then the FPI exceeds 1. In this case there are productivity gains. Conversely, if the FPI is less than 1 there are productivity losses. Thus the FPI introduces time-wise, or intertemporal, comparisons of FP. Eq. 6.2 reveals that the FPI equals an index of output quantity divided by an input quantity index. This is seen from rearranging Eq. 6.2 as:

$$FPI(s,t) = \left(\frac{y^t}{y^s}\right) / \left(\frac{v^t}{v^s}\right). \tag{6.3}$$

If output growth is greater than (respectively less than) input growth from period s to period t, then the FPI is greater than (respectively less than) 1, and there are productivity gains (respectively losses).

To generalize the concept of the FPI to the case of multiple outputs and multiple inputs some additional notation is introduced. For a general n-input, m-output production process, the period t output and input quantity vectors are denoted respectively by $y^t = [y_1^t,...,y_m^t]$ and $v^t = [v_1^t,...,v_n^t]$, while $p^t = [p_1^t, p_2^t,..., p_m^t]$ and $w^t = [w_1^t,...,w_n^t]$ represent respectively the period t output and input price vectors. Define an output quantity index between period t and period s as $Q^y(p^s, p^t, y^s, y^t)$, which is a function of the four m variables relating to the output prices and quantities pertaining to the two periods under consideration. Similarly, an input quantity index between the two periods is $Q^v(w^s, w^t, v^s, v^t)$, which is a function of four n variables relating to the input prices and quantities pertaining to the periods under consideration. In the multiple output and multiple input case, the generalization of the factor productivity index is called the TFPI. Following Eq. 6.3, TFPI is defined as a ratio of output and input quantity indexes:[6]

$$TFPI(s,t) = Q^y(p^s, p^t, y^s, y^t) / Q^v(w^s, w^t, v^s, v^t). \tag{6.4}$$

---

[6] In the m output / n input case there are $m \times n$ factor productivities, and therefore for any two periods, there are $m \times n$ factor productivity indexes. Some researchers refer to the total factor productivity index just as TFPI.

If the output quantity index is greater than the input quantity index between period s and period t, then aggregate output growth exceeds input growth from period s to period t. In this case TFPI exceeds 1, and there are productivity gains. Conversely, if the output quantity index is less than the input quantity index between period s and period t, the TFPI is less than 1, and there are productivity losses. The next section demonstrates that different index number choices for $Q^y(p^s, p^t, y^s, y^t)$ and $Q^v(w^s, w^t, v^s, v^t)$ result in different formulas for the TFPI, although in practice these alternative formulations may not lead to substantial differences in measured productivity.

Indexes of Total Factor Productivity

To begin the discussion of the indexes of total factor productivity, define nominal total cost and revenue in period $\tau = t$, s as C^τ and R^τ respectively, and are provided by:

$$C^\tau = \sum_{i=1}^{n} w_i^\tau x_i^\tau = V(w^\tau, v^\tau) \tag{6.5}$$

and

$$R^\tau = \sum_{g=1}^{m} p_g^\tau y_g^\tau = Y(p^\tau, y^\tau) \qquad \tau = s, t. \tag{6.6}$$

It is also possible to construct four hypothetical aggregates, each of which can be thought of as a cost or revenue under hypothetical price conditions. When period τ quantities are evaluated using period $v \neq \tau$ prices:

$$V(w^v, x^\tau) = \sum_{i=1}^{n} w_i^v x_i^\tau \tag{6.7}$$

and

$$Y(p^v, y^\tau) = \sum_{g=1}^{m} p_g^v y_g^\tau \qquad v, \tau = t, s \qquad v \neq \tau. \tag{6.8}$$

The aggregates defined in Eq.6.5 through Eq. 6.8 are linear price-weighted sums of quantities. These aggregates are sufficient to define the Paasche, Laspeyres and Fisher quantity, and total factor productivity indexes.

Quantity index numbers are defined as ratios of quantity aggregates for two periods, and thereby they compare quantities in two situations. More specifically, these indexes measure the growth in quantity between two periods under consideration. Using the quantity aggregates given in Eq. 6.6 and Eq.6.8, the definitions for the Paasche, Laspeyres and Fisher output quantity indexes are given, respectively, by:

$$Q_P^y(p^s, p^t, y^s, y^t) = \sum_{g=1}^{m} p_g^t y_g^t / \sum_{h=1}^{m} p_h^t y_h^s$$
$$= Y(p^t, y^t) / Y(p^t, y^s),$$
(6.9)

$$Q_L^y(p^s, p^t, y^s, y^t) = \sum_{g=1}^{m} p_g^s y_g^t / \sum_{h=1}^{m} p_h^s y_h^s$$
$$= Y(p^s, y^t) / Y(p^s, y^s)$$
(6.10)

and

$$Q_F^y(p^s, p^t, y^s, y^t) = (Q_P^y(p^s, p^t, y^s, y^t) Q_L^y(p^s, p^t, y^s, y^t))^{1/2}.$$
(6.11)

Similarly, using Eq. 6.5 and Eq. 6.7, the Paasche, Laspeyres and Fisher input quantity indexes are given by:

$$Q_P^v(w^s, w^t, v^s, v^t) = \sum_{i=1}^{n} w_i^t v_i^t / \sum_{j=1}^{n} w_j^t v_j^s$$
$$= V(w^t, v^t) / V(w^t, v^s),$$
(6.12)

$$Q_L^v(w^s, w^t, v^s, v^t) = \sum_{i=1}^{n} w_i^s v_i^t / \sum_{j=1}^{n} w_j^s v_j^s$$
$$= V(w^s, v^t) / V(w^s, v^s)$$
(6.13)

and

$$Q_F^v(w^s, w^t, v^s, v^t) = (Q_P^v(w^s, w^t, v^s, v^t) Q_L^v(w^s, w^t, v^s, v^t))^{1/2}.$$
(6.14)

From the output and input quantity indexes, and using Eq. 6.4, the Paasche, Laspeyres and Fisher TFPIs are provided by:[7]

$$TFPI_P(s,t) = Q_P^y(p^s, p^t, y^s, y^t) / Q_P^v(w^s, w^t, v^s, v^t)$$
$$= \frac{Y(p^t, y^t)}{Y(p^t, y^s)} / \frac{V(w^t, v^t)}{V(w^t, v^s)},$$
(6.15)

---

[7] See the work by Balk (1995) and Diewert (1992).

$$TFPI_L(s,t) = Q_L^y(p^s, p^t, y^s, y^t) / Q_L^v(w^s, w^t, v^s, v^t)$$

$$= \frac{Y(p^s, y^t)}{Y(p^s, y^s)} / \frac{V(w^s, v^t)}{V(w^s, v^s)} \qquad (6.16)$$

and

$$TFPI_F(s,t) = Q_F^y(p^s, p^t, y^s, y^t) / Q_F^v(w^s, w^t, v^s, v^t)$$

$$= (TFPI_P(s,t)TFPI_L(s,t))^{1/2}. \qquad (6.17)$$

Choice of TFPI

There are three approaches to the determination of the 'best' index number: the axiomatic or test approach, the equivalence or exact approach and the econometric approach. The axiomatic approach to index number selection delineates mathematical properties that an index number should satisfy. These properties are called tests or axioms. Index number formulae are analyzed to determine whether they satisfy these tests.[8] Second, the equivalence approach, in the context of producer behavior, involves the delineation of a functional form for a production process, e.g., by a production or cost function. Index number formulae are analyzed to determine their equivalence to the production process characterization.[9] Third, the econometric approach involves the estimation of a functional form for the production process. With the estimated functional form the appropriate index number can be constructed.[10]

There are several advantages, as well as, limitations to each approach. The axiomatic approach does not involve any assumptions regarding the production process or producer behavior. However, with the axiomatic approach, the axioms or tests are based on consistency properties for index numbers not on the delineation of a production process. Thus the only outcome is the determination of an index number. No other features of the production process can be gleaned from the analysis. The equivalence approach does require a description of the production process and assumptions on the producer optimizing behavior, e.g., cost minimization. However, these additional assumptions permit the investigation of the contribution to productivity gains and losses, since the TFPI can be decomposed into its components such as technical change and returns to scale. Lastly, the econometric approach involves the specification of equations describing the production process

---

[8] Recent contributors to the axiomatic approach include Eichhorn and Voeller (1976), Funke and Voeller (1978), Diewert (1976, 1992) and Balk (1995).

[9] The equivalence or exact approach is developed by Diewert (1976).

[10] The econometric approach is used extensively in several contexts. See e.g., the survey by Jorgenson (1986), and also Denny et al. (1981) and Bernstein and Mohnen (1998).

and producer optimizing behavior through factor demand equations. Although the econometric approach typically requires the most extensive set of assumptions it also provides for a wide range of results. It is possible to determine the rates of productivity gains and losses, their decomposition and additionally the various elasticities of factor demand reflecting the responses of producers to changes in such variables as prices, scale and technology. However, a practical constraint of the econometric approach is the limitation on the feasible number of outputs and inputs in the analysis. Since the objective here is to measure productivity growth for the US telecommunications industry, the axiomatic approach is applied to select the appropriate TFPI. Specifically, this approach is applied to the quantity indexes that comprise the TFPI, and therefore indirectly to the choice of productivity index. Once the best functional forms for the output and input quantity indexes are determined, then by definition of the TFPI provided by Eq. 6.4, the functional form for the TFPI is also resolved.

The first test that an output quantity index is assumed to satisfy is the Product Test. This test is given by:

$$P_I^y(p^s, p^t, y^s, y^t) Q_i^y(p^s, p^t, y^s, y^t) = R^t / R^s \qquad (6.18)$$

where $P_I^y(p^s, p^t, y^s, y^t)$ is an output price index and the subscript i or the quantity and price indexes convey that they are of the same type. This test states that the product of the output price index and the companion output quantity index should equal the nominal revenue ratio between period t and period s. Notice that the Product Test implies that given a functional form for the output quantity (respectively price) index, the output price (respectively quantity) index must have the same functional form.

Next, the Identity Test is given by:

$$Q^y(p^s, p^t, y, y) = 1. \qquad (6.19)$$

This means that if $y^s = y^t = y = (y_1, ..., y_m)$ so that all quantities are equal in the two periods, then the quantity index should be one regardless of the prices for period s and t.

Lastly, the Time Reversal Test is:

$$Q^y(p^t, p^s, y^t, y^s) = 1 / Q^y(p^s, p^t, y^s, y^t). \qquad (6.20)$$

This test is satisfied when the prices and quantities for period s and period t are interchanged; the resulting quantity index is the reciprocal of the original quantity index. Three functional forms for the output quantity index are introduced: the Paasche index defined by Eq. 6.9 and Eq. 6.12, the Laspeyres index defined by Eq. 6.10 and Eq. 6.13, and the Fisher index defined by Eq. 6.11 and Eq. 6.14. The Fisher index satisfies these tests, while the Paasche and Laspeyres indexes fail the product and time reversal tests.[11] Thus, from the viewpoint of the axiomatic approach, the Fisher quantity indexes, and thereby the Fisher TFPI are good choices.

TFPI: Updating the FCC Study

In the preceding discussion the TFPI is based on comparison of two points in time, and so a single bilateral index number is developed involving price and quantity information for period s and t. In constructing the TFPI over several time periods, and in order to make period-to-period comparisons over these periods, requires multiple bilateral index numbers. Constructing multiple bilateral indexes involves the selection of the base period with which other periods are compared. The base period is either fixed or chained. As the name suggests, for a fixed-base index the comparison period does not change with each component in the index. In a chained index the comparison period is moved forward by one period for each component in the chain. During periods when prices are relatively stable fixed-base and chained indexes yield similar results. However, a chained index is preferable during periods of relatively extensive price changes because the quantity index is calculated using the price-weights of adjacent years. Thus a chained quantity index avoids the problem of arbitrarily updating the weights in a fixed-base index in order to accommodate significant price changes. As a consequence, the preferred index is a chained Fisher TFPI.[12] Two measures of TFPI for US telecommunications carriers are developed. The first follows the approach of the FCC and classifies output into three major categories: local, intrastate and interstate.[13] The FCC's study measures productivity for 1985 to 1995, and Gollop (2000) provides an update to 1998. Here the FCC productivity study is updated to 2001. This study develops productivity indexes for a common group of firms, namely the ILECs that today are grouped into four companies—SBC, Verizon, Quest and BellSouth. The second TFPI developed below, disaggregates intrastate services into switched access and toll, and introduces output associated with the high bandwidth portion of existing ILEC access lines used to provide DSL services.

Calculation of Output Quantity Index

Following the FCC's productivity study, interstate output is measured as the aggregation of access lines, interstate switched access minutes and special access lines. Access lines, measured by the sum of the number of business, public and residential access lines, and special access lines provide connectivity to the network, while the volume of interstate activity is measured by interstate switched

---

[11] When a more extensive list of tests is developed, the Fisher price index continues to satisfy more tests than other index number formulations. See, Funke and Voeller (1978).

[12] See Landefeld and Parker (1997) for a discussion of chained index numbers.

[13] See FCC (1997).

access minutes. The initial step is to construct an interstate quantity index from the three measures of interstate output.

Table 6.1. Interstate Revenues (US$ millions), 1985–2001

Year	End User	Switched Access	Special Access	Total
1985	1,499	10,906	1,961	14,366
1986	2,400	10,484	2,575	15,460
1987	3,091	9,612	2,658	15,360
1988	3,604	9,663	2,540	15,806
1989	4,399	9,093	2,254	15,745
1990	4,679	8,596	2,209	15,484
1991	4,828	8,514	2,119	15,461
1992	4,963	8,651	2,154	15,768
1993	5,244	8,999	2,098	16,341
1994	5,590	9,294	2,217	17,101
1995	5,770	9,333	2,530	17,633
1996	5,931	9,410	3,071	18,411
1997	6,268	8,764	3,851	18,883
1998	7,808	7,275	4,815	19,898
1999	8,428	6,895	6,150	21,472
2000	9,087	5,506	8,211	22,804
2001	9,631	3,897	10,632	24,159

Source: Frank Gollop, USTA Report (1985–98); ARMIS Report 43-02, Table I1 (1999–2001)

Table 6.2. Local, Intrastate and Interstate Revenues (US$ millions), 1985–2001

Year	Local Service	Intrastate Toll and Access	Interstate	Total
1985	26,961	13,047	14,366	54,374
1986	28,626	13,539	15,460	57,625
1987	29,151	14,167	15,360	58,678
1988	29,227	14,995	15,806	60,028
1989	29,973	14,868	15,745	60,587
1990	30,699	15,015	15,484	61,198
1991	32,059	14,522	15,461	62,043
1992	33,360	14,225	15,768	63,353
1993	34,599	14,497	16,341	65,437
1994	35,759	14,356	17,101	67,215
1995	37,685	13,123	17,633	68,441
1996	40,523	12,987	18,411	71,922
1997	42,461	12,309	18,883	73,652
1998	44,993	11,978	19,898	76,870
1999	47,759	11,136	21,472	80,367
2000	48,831	9,950	22,804	81,585
2001	47,429	9,334	24,159	80,922

Note. Revenues exclude Miscellaneous Revenues, which include revenue from Directory, Rent, Corporate Operations, Special Billing Arrangements Customer Operations, Plant Operations, Other Incidental Regulated Revenues and Other revenue settlements. Source: Frank Gollop, USTA Report (1985–98); ARMIS Report 43-02, Table I1 (1999–2001)

Table 6.3. Fisher Interstate Output Quantity Index Calculation, 1985–2001

Year	Revenue Share			Quantity			Output Index				
	End User (%)	Switched Access (%)	Special Access (%)	Access Lines (mill.)	Switched Access Minutes (bill.)	Special Access Lines (mill.)	Las.	Pa.	Fis.	Fis. Ch.	Growth (%)
1985	10.4	75.9	13.7	93	157	1.2	1.00	1.00	1.00	1.00	
1986	15.5	67.8	16.7	95	157	1.7	1.05	1.05	1.05	1.05	5.14
1987	20.1	62.6	17.3	98	173	1.8	1.08	1.08	1.08	1.14	7.78
1988	22.8	61.1	16.1	98	188	2.7	1.14	1.12	1.13	1.29	12.19
1989	27.9	57.8	14.3	101	210	2.4	1.07	1.06	1.06	1.37	6.05
1990	30.2	55.5	14.3	104	232	3.5	1.13	1.11	1.12	1.53	11.49
1991	31.2	55.1	13.7	107	259	5.2	1.14	1.13	1.13	1.74	12.52
1992	31.5	54.9	13.6	109	274	6.0	1.06	1.06	1.06	1.84	5.69
1993	32.1	55.1	12.8	112	290	10.2	1.13	1.10	1.12	2.06	11.15
1994	32.7	54.3	13.0	115	308	13.8	1.09	1.08	1.09	2.23	8.22
1995	32.7	52.9	14.4	120	332	16.1	1.08	1.08	1.08	2.40	7.38
1996	32.2	51.1	16.7	125	359	19.6	1.09	1.09	1.09	2.62	8.54
1997	33.2	46.4	20.4	132	385	22.9	1.08	1.08	1.08	2.83	7.80
1998	39.2	36.6	24.2	137	405	29.8	1.10	1.10	1.10	3.11	9.40
1999	39.3	32.1	28.6	141	421	42.6	1.13	1.12	1.13	3.50	11.96
2000	39.9	24.1	36.0	141	435	59.5	1.12	1.12	1.13	3.94	11.75
2001	39.9	16.1	44.0	134	406	72.1	1.04	1.04	1.04	4.10	4.02

Note. Las. = Laspeyres, Pa. = Paasche, Fis. = Fisher, Fis. Ch. = FisherChained. Column 7 = Total Switched Access Lines subtracted from Total Access Lines (Switched & Special). Fisher Index = the square root of the product of the Laspeyres and Paasche indices. Source: Access Lines—Frank Gollop, USTA Report (1985–96); ARMIS Report 43-08, Table 3 (1997–2001); Special Access Lines—Frank Gollop, USTA Report (1985–98); ARMIS Report 43-08, Table 3 (1999–2001); Switched Access Minutes—Frank Gollop, USTA Report (1985–90); ARMIS Report 43-08, Table 4 (1991–2001)

A chained Fisher quantity index is constructed for interstate output, with each service share of interstate revenue used to weight the respective interstate service. End user common line revenue relative to total interstate revenues constitutes the weight for the number of access lines, the switched access revenue share is the weight for the number of switched access minutes and special access revenue share is the weight for the number of special access lines. The revenue data required for the output quantity index are presented in Table 6.1 and Table 6.2. The interstate revenue shares, interstate quantities, and the resulting chained Fisher index for interstate output are presented in Table 6.3. In Table 6.3 the column presenting the chained index for period t is derived by taking the Fisher index in period t and multiplying this index by the chained value in period $t-1$. Taking the natural logarithm of the ratio of the chained index in period t over the chained index in period $t-1$ develops the column showing the growth rate.

The overall chained Fisher output quantity index is computed using the quantities for local service, intrastate service and the interstate service output index with their respective revenue shares. The chained Fisher output quantity index and the data required for its construction are shown in Table 6.4. The physical units associated with local service are the number of local calls, while for intrastate services the physical units are intrastate DEMS. Table 6.4 shows that the revenue composition changed for the ILECs over time as local service revenue share grew from 50% in 1985 to about 59% in 2001, while the intrastate revenue share declined from 24% to about 12% for the same period. The interstate revenue share is reasonably constant over the whole period. From the last column in Table 6.4, the average annual growth in total output is 4.1%, with the ILEC growth rate falling since 1996.

Table 6.4. Fisher Output Quantity Index Calculation, 1985–2001

Year	Revenue Share			Quantity			Output Index				
	Local Service (%)	Intra-state Toll & Access (%)	Inter-state (%)	Local Calls (bill.)	Intrastate DEMs (bill.)	Inter-state Quantity Index	Las.	Pa.	Fis.	Fis. Ch.	Growth (%)
1985	49.6	24.0	26.4	311	164	1.00	1.00	1.00	1.00	1.00	
1986	49.7	23.5	26.8	316	173	1.05	1.04	1.03	1.04	1.04	3.45
1987	49.7	24.1	26.2	321	184	1.14	1.04	1.04	1.04	1.08	4.22
1988	48.7	25.0	26.3	319	192	1.29	1.04	1.04	1.04	1.12	3.98
1989	49.5	24.5	26.0	330	207	1.37	1.05	1.05	1.05	1.18	5.23
1990	50.2	24.5	25.3	342	218	1.53	1.06	1.06	1.06	1.26	5.98
1991	51.7	23.4	24.9	353	220	1.74	1.05	1.05	1.05	1.32	4.93
1992	52.7	22.4	24.9	365	224	1.84	1.04	1.04	1.04	1.37	3.67
1993	52.9	22.1	25.0	377	228	2.05	1.05	1.05	1.05	1.44	4.74
1994	53.2	21.4	25.4	393	235	2.23	1.05	1.05	1.05	1.51	4.96
1995	55.0	19.2	25.8	409	247	2.40	1.05	1.05	1.05	1.59	5.13
1996	56.3	18.1	25.6	422	264	2.62	1.05	1.05	1.05	1.67	5.14
1997	57.7	16.7	25.6	433	287	2.83	1.05	1.05	1.05	1.76	4.96
1998	58.5	15.6	25.9	445	295	3.11	1.04	1.04	1.04	1.84	4.35
1999	59.4	13.9	26.7	448	317	3.50	1.05	1.05	1.05	1.92	4.70
2000	59.8	12.2	28.0	429	333	3.94	1.01	1.01	1.01	1.95	1.25
2001	58.6	11.5	29.9	409	350	4.10	0.99	0.99	0.99	1.93	-1.11
										Avg	4.10

Note. Las. = Laspeyres, Pa. = Paasche, Fis. = Fisher, Fis. Ch. = Fisher Chained. The Fisher Index = the square root of the product of the Laspeyres and Paasche indices. Source: Local Calls—Frank Gollop, USTA Report (1985–98); ARMIS Report 43-08, Table 4 (1999–2001); Intrastate DEMs—Frank Gollop, USTA Report (1985–96); http://www.fcc.gov/wcb/iatd/neca.html, see file netwo00.zip (stdems.xls) (1997–2000); 2001 is estimated using average annual growth rate from 1997 to 2000

Calculation of Input Quantity Index

There are three input categories in this TFPI, labor, capital and intermediate inputs or materials. In a similar manner to aggregating different types of outputs to construct the Fisher output quantity index, the three input quantities are aggregated to construct the Fisher input quantity index.

Table 6.5. Labor and Materials Input, 1985–2001

Year	Employ-ees ('000)	Labor growth (%)	Mater-ials Price Index	Operating Expense (US$b)	Dep. & Amort. Expense (US$b)	Employee Compen-sation (US$b)	Materials Expense (US$b)	Materials Quantity ('000)	Materials Quantity Index	Materials Quantity Growth (%)
1985	504.1		1.00	41	10	17	14	14	1.00	
1986	482.7	-4.3	1.02	42	12	17	14	14	0.99	-1.00
1987	477.7	-1.0	1.05	44	13	17	14	13	0.95	-3.71
1988	466.8	-2.3	1.09	47	14	17	16	15	1.06	10.85
1989	461.1	-1.2	1.13	49	14	17	18	16	1.13	6.28
1990	443.1	-4.0	1.17	50	14	18	18	15	1.10	-2.74
1991	414.5	-6.7	1.22	51	14	17	20	17	1.19	7.91
1992	411.2	-0.8	1.25	51	14	17	20	16	1.14	-4.90
1993	395.6	-3.8	1.28	53	14	18	21	16	1.16	1.84
1994	367.2	-7.5	1.30	56	15	17	24	18	1.30	12.10
1995	345.8	-6.0	1.33	57	16	16	25	19	1.34	2.93
1996	338.0	-2.3	1.36	58	16	18	23	17	1.22	-9.86
1997	338.2	0.0	1.38	60	17	17	26	18	1.32	8.35
1998	338.4	0.1	1.40	61	17	18	25	18	1.30	-1.69
1999	344.9	1.9	1.42	63	18	19	26	18	1.29	-0.85
2000	345.2	0.1	1.45	65	19	19	26	18	1.30	0.69
2001	329.8	-4.5	1.48	66	21	20	25	17	1.19	-8.65

Note. Column 8 = Column 5 – Column 6 – Column 7. Column 9 = Column 8 divided by Column 4. Source: Total Employees, Operating Expense, Depreciation & Amortization Expense, Compensation—Frank Gollop, USTA Report (1985–98); ARMIS Report 43-02 Table I1 (1999–2001); Materials Price Index—GDPPI: Bureau of Economic Analysis, Table 7.1 (1985-2001)

Labor and Materials Input

Labor quantity is based on annual accounting data for the number of employees. These data are reported in Table 6.5. The materials quantity is computed as materials expense divided by a materials price index. The materials price index is taken to be the gross domestic product price index (GDPPI).[14] Materials expense is a re-

---

[14] The materials price index used by the FCC is based on those categories of expenditures from the input-output tables compiled by the Bureau of Economic Analysis (BEA) of the US Department of Commerce that focus on materials purchases by communications industries. Over the common period, 1985 to 1998, the material price index used here and that used by the FCC are similar, and result in a comparable TFPI. Given the similarity of

sidual. It is the difference between total operating expense and the sum of labor compensation and depreciation and amortization expense.

Materials expenses for 1985 through 1987 are adjusted for two accounting changes that became effective in 1988. First, beginning in 1988 all expenses from non-regulated services that had common costs with regulated services is reported in operating expenses. Second, certain plant investment that is capitalized began to be expensed in the year incurred. Accordingly, expenses for 1985 through 1987 are adjusted upward to make them comparable to the expenses recorded from 1988 onward. The adjustment factor is computed by dividing the sum of annual operating expense plus the additional materials expense resulting from both the regulated/non-regulated change and the capital/expense shift by reported operating expenses for the period 1985 to 1987. These percentages are used to adjust 1985 through 1987 operating expenses of the LECs.

Capital Input

The data requirements for the capital input include an initial (or benchmark) capital stock, measures of annual investment (in inflation-adjusted prices) and a rate of depreciation. Beginning with the benchmark capital stock, the estimate of capital stock for subsequent years is derived by depreciating the previous year's capital stock and then adding the constant dollar gross capital additions in that year. The benchmark capital stock is based on the end-1985 net (of depreciation) book value in constant dollars. The benchmark capital stock is that developed by the FCC. Capital additions are routinely reported to the FCC but, because of the aforementioned capital/expense shift, capital additions for 1985, 1986 and 1987 are developed as the product of the unadjusted capital additions and the adjustment factor of 0.888.

A composite asset price index is constructed from BEA asset prices obtained from the National Income and Product Accounts (NIPA) complied by the BEA to deflate capital additions. Three asset categories are used: communications equipment, telecommunication structures and a composite asset price for producer durables. Capital additions data are then grouped into categories corresponding to NIPA asset categories, and category expenditure share are calculated.[15] From these

---

results, the fact that the input-output tables are produced with a significant time lag, and the BEA price index relates to all communications industries, not solely telecommunications, the more current GDPPI is used. In addition, a material price index from the BEA for the telephone and telegraph industry is also used. There is a split in this index, as pre-1988 the index is based on the 1972 Standard Industrial Classification, and post-1988 the index is based on the 1987 SIC. Using this index, the TFPI shows lower gains compared to the use of the GDPPI as the materials price index. These results are available from the authors.

[15] NIPA tables are respectively, Communications Equipment (NIPA Table 7.8: Chained Type Price indexes for Private Purchases of Producers' Durable Equipment by Type, Line 7), Telecommunication Structures (NIPA Table 7.7: Chained Type Price Indexes for Private Purchases of Structure by Type, Line 12), and a composite asset price for

data, a chained Fisher price index is computed and used to deflate adjusted capital additions used in the construction of the capital stock.

A time-invariant depreciation rate for the single asset class is computed as an arithmetic average of annual depreciation rates. The depreciation rate for period t is, $(DEPR.ACRLS_t)/((TPIS.BOY_t + TPIS.EOY_t)/2)$, where $DEPR.ACRLS_t$ denotes depreciation accruals, $TPIS.BOY_t$ and $TPIS.EOY_t$ are telephone plant in service at the beginning and end of year respectively. Data used to construct the capital stock and the depreciation rate is presented in Table 6.6. The construction of the capital input is shown in Table 6.7.

Table 6.6. Capital Adjustments and Depreciation, 1985–2001

Year	TPIS.BOY (US$b)	Unadj. Additions (US$b)	TPIS.EOY (US$b)	Retires (US$b)	Adj. Factor	Adj. Additions (US$b)	Adj. TPIS.EOY (US$b)	Dep. Accrual ('000)	Adj. Dep. Rate (%)
1985	139	15	149	5	0.89	13	147	10	7.16
1986	149	15	159	5	0.89	13	157	12	7.72
1987	159	14	168	5	0.89	13	166	13	8.19
1988	169	14	176	7	1.00	14	176	13	7.63
1989	176	13	183	6	1.00	13	183	13	7.48
1990	183	14	187	10	1.00	14	187	13	7.26
1991	187	15	192	10	1.00	15	192	13	6.97
1992	192	15	196	10	1.00	15	196	13	6.87
1993	196	15	203	8	1.00	15	203	14	7.03
1994	203	15	209	8	1.00	15	209	15	7.21
1995	209	15	217	7	1.00	15	217	15	7.20
1996	217	18	227	8	1.00	18	227	16	7.27
1997	227	18	237	9	1.00	18	237	17	7.13
1998	237	19	249	7	1.00	19	249	17	7.12
1999	249	21	262	9	1.00	21	262	18	6.95
2000	262	27	281	7	1.00	27	281	19	6.87
2001	281	27	301	6	1.00	27	301	20	6.87
								Avg	7.23

Note. Column 5 = Column 2 + Column 3 – Column 4. Column 7 = Column 3 times Column 6. Column 8 = Column 2 + Column 7 – Column 5. Source: TPIS.BOY—Frank Gollop, USTA Report (1985–94, 1996–98); ARMIS Report 43-02, B1b (1995, 1999–2001): Unadjusted Additions—Frank Gollop, USTA Report (1985–92, 1996); ARMIS Report 43-02, B1b (1993–95, 1997–2001): TPIS.EOY—Frank Gollop, USTA Report (1985–90, 1993, 1995–98); ARMIS Report 43-02, B1b (1991, 1992, 1994, 1999–2001): Depreciation Accruals—Frank Gollop, USTA Report (1985–89, 1996–98); ARMIS Report 43-02, Table B6 (1990–95, 1999–2001): Adjustment Factor—Frank Gollop, USTA Report (1985–98); 1999–2001 is continued at a value of 1.00

Producer Durables (NIPA Table 7.1, Line 39). In addition, the capital/expense shift adjustment factor previously noted has no effect on the capital expenditure shares because it is multiplicative and applies equally to all asset categories.

Table 6.7. Capital Input Quantity, 1985–2001

Year	Adj. Capital Additions (US$b)	BEA Composite Asset Price	Capital Stock Quantity (US$b)	Capital Input Quantity Index	Capital Input Quantity Growth (%)
1984			114		
1985	13	1.00	117	1.00	
1986	13	1.00	130	1.13	12.06
1987	13	1.01	143	1.26	10.65
1988	14	1.01	157	1.38	9.15
1989	13	1.05	170	1.51	9.41
1990	14	1.07	183	1.63	7.76
1991	15	1.08	197	1.76	7.69
1992	15	1.07	210	1.89	7.11
1993	15	1.08	224	2.02	6.68
1994	15	1.08	238	2.16	6.35
1995	15	1.07	252	2.29	5.92
1996	18	1.08	269	2.43	5.87
1997	18	1.08	286	2.59	6.42
1998	19	1.05	304	2.75	6.09
1999	21	1.03	324	2.92	6.09
2000	27	1.02	350	3.12	6.57
2001	27	1.01	377	3.37	7.75

Note. Capital Benchmark (1984) = $US 103,903 million. Source: BEA Composite Asset Price—1985-2001: Derived using method described in Appendix D, Estimation of TFP Under FCC Rules, C. Anthony Bush and Lori Huthoefer, CC Docket No 94-1 and 96-262, May 21, 1997

Input Quantity Index

Using shares of payments to the factors as weights, a Fisher input quantity index is constructed. Factor payment shares are based on the input cost developed as the sum of labor compensation, materials expense and the payment to capital, which is property income inclusive of depreciation. The factor payment shares and the Fisher input quantity index and growth rates are presented in Table 6.8. This table shows the capital cost share of around 44% is reasonably stable for the period from 1985 to 2001. The labor cost share fell from slightly over 31% in 1985 to about 25% in 2001. Materials input is substituted for labor in production, as the materials cost share increased from around 26% in 1985 to about 31% in 2001. This substitution is due in part to contracting out of input requirements. Over the period, Table 6.8 shows that the average annual input quantity growth rate of 1.24%. Annual growth, fluctuated over the period and post-1989, reached a maximum in 1997 but input growth is yet to return to that rate.

Table 6.8. Fisher Input Quantity Index Calculation, 1985–2001

Year	Share			Quantity			Quantity Index				
	Labor Compensation (%)	Material Payment (%)	Property Inc. w/ dep. (%)	Labor	Materials (US$m)	Capital	Las.	Pa.	Fis.	Fis. Ch.	Growth (%)
1985	31.3	25.6	43.1	504	14	1.00	1.00	1.00	1.00	1.00	
1986	29.0	24.5	46.5	483	14	1.06	1.01	1.01	1.01	1.01	0.88
1987	28.9	23.9	47.2	478	13	1.11	1.01	1.01	1.01	1.02	0.99
1988	28.4	26.9	44.7	467	15	1.15	1.04	1.04	1.04	1.06	3.73
1989	27.9	29.4	42.7	461	16	1.20	1.03	1.03	1.03	1.09	3.39
1990	28.7	29.5	41.8	443	15	1.23	0.99	0.99	0.99	1.09	-0.72
1991	27.7	32.6	39.7	414	17	1.28	1.02	1.02	1.02	1.11	1.91
1992	27.1	31.1	41.8	411	16	1.31	0.99	0.99	0.99	1.10	-0.60
1993	27.4	31.4	41.2	396	16	1.35	1.01	1.01	1.01	1.11	0.64
1994	25.5	35.3	39.2	367	18	1.38	1.03	1.03	1.03	1.14	3.09
1995	23.9	36.4	39.7	346	19	1.42	1.00	1.00	1.00	1.15	0.45
1996	25.7	32.0	42.3	338	17	1.45	0.97	0.97	0.97	1.11	-2.90
1997	23.7	34.7	41.6	338	18	1.51	1.04	1.05	1.04	1.16	4.38
1998	23.6	33.0	43.4	338	18	1.56	1.01	1.01	1.01	1.18	0.93
1999	23.5	31.8	44.7	345	18	1.62	1.02	1.02	1.02	1.20	1.83
2000	23.4	32.2	44.4	345	18	1.70	1.02	1.02	1.02	1.23	2.42
2001	24.6	30.5	44.9	330	17	1.83	1.00	0.99	0.99	1.22	-0.55
										Avg	1.24

Note. Las. = Laspeyres, Pa. = Paasche, Fis. = Fisher, Fis. Ch. = Fisher Chained. The Fisher Index = the square root of the product of the Laspeyres and Paasche indices

Calculation of the TFPI

The chained Fisher TFPI and growth rates are presented in Table 6.9. The TFPI is calculated as the ratio of the chained Fisher output quantity index from Table 6.4, and the chained Fisher input quantity index from Table 6.8. The growth rate of total factor productivity (TFPG) is defined as the difference between output quantity growth and input quantity growth. From Table 6.9 the annual average rate of TFPG for the period from 1986 to 2001 is 2.86%. This rate is more than two percentage points greater than the TFPG for the US non-farm business sector that annually averaged 0.77% over the same period.[16] Annual TFPG for the ILECs fluctuated, reaching a peak rate in 1996 of 8.05%. Since 1996 productivity growth has declined, and in fact turned negative in 2000 and 2001. This negative growth arose from the precipitous drop in output growth.

---

[16] For the TFPG of the US nonfarm business sector see, Multifactor Productivity: Bureau of Labor Statistics, Table 4.

Table 6.9. Fisher TFPI and Growth, 1985–2001

Year	Output Quantity Index	Output Quantity Growth	Input Quantity Index	Input Quantity Growth	TFPI	TFP Growth
1985	1.00		1.00		1.00	
1986	1.04	3.45	1.01	0.88	1.03	2.57
1987	1.08	4.22	1.02	0.99	1.06	3.23
1988	1.12	3.98	1.06	3.73	1.06	0.25
1989	1.18	5.23	1.09	3.39	1.08	1.84
1990	1.26	5.98	1.09	-0.72	1.16	6.70
1991	1.32	4.93	1.11	1.91	1.19	3.02
1992	1.37	3.67	1.10	-0.60	1.24	4.26
1993	1.44	4.74	1.11	0.64	1.30	4.10
1994	1.51	4.96	1.14	3.09	1.32	1.87
1995	1.59	5.13	1.15	0.45	1.38	4.68
1996	1.67	5.14	1.11	-2.90	1.50	8.05
1997	1.76	4.96	1.16	4.38	1.51	0.57
1998	1.84	4.35	1.18	0.93	1.56	3.42
1999	1.92	4.70	1.20	1.83	1.61	2.87
2000	1.95	1.25	1.23	2.42	1.59	-1.17
2001	1.93	-1.11	1.22	-0.55	1.58	-0.55

TFPI: Intrastate Disaggregation and Broadband

The TFPI calculated in this section modifies the update of the FCC's measure presented in the previous section in two respects. First, intrastate output quantity is disaggregated into switched access and toll minutes. The previous FCC method measured intrastate output using intrastate DEMS. Subsequent to the 1984 AT&T divestiture into long-distance and regional local exchange companies—long-distance carriers had provided increasing numbers of toll calls, thus ILECs are providing increasing numbers of access minutes and relatively fewer toll calls. The revised measure of intrastate output more fully accounts for toll and access price changes, and the changing pattern of volumes through time. The second modification to the FCC productivity study includes the high bandwidth portion of existing ILEC access lines into the output mix. ILECs supply high bandwidth transmission services to carriers, who then offer broadband DSL or high-speed internet services to their end users. There is a paucity of data available for these services, as the FCC began data collection with the introduction of its line sharing rules in 1999.

Intrastate Output Index

Intrastate output consists of intrastate switched access minutes (ISAMS) and intrastate toll minutes (ITMS). With respect to ISAMS, the FCC's Automated Reporting Management Information System (ARMIS) records these data for the period

1991 to 2001. ISAMS for 1985 to 1990 are estimated according to the change over time in intrastate DEMS (IDEMS), for which data are available for the entire period. The procedure followed is: $ISAMS_{90} = (DEMS_{90} / DEMS_{91}) \times ISAMS_{91}$. The ITMS measure used in this study is developed from data on ISAMS and IDEMS. IDEMS are not split exactly between toll minutes and the measure of access minutes because toll calls require time to setup. Setup time is charged to long-distance carriers as part of the measure of access minutes, but not to long distance customers. The ITMS is estimated as: $ITMS = IDEMS - (ISAMS \times 1.07)$, where 1.07 is the factor generally relied upon by the industry to account for the call setup time included in the measure of access, but not toll, minutes.

Table 6.10. Fisher Intrastate Output Quality Index Calculation, 1985–2001

| Year | Revenue Share | | Quantity | | Output Index | | | | |
	Access (%)	Toll (%)	Switched Access Minutes (mill.)	Toll Minutes (mill.)	Las.	Pa.	Fis.	Fis. Ch.	Growth (%)
1985	22.8	79.2	59	101	1.00	1.00	1.00	1.00	
1986	22.3	77.7	62	107	1.05	1.05	1.05	1.05	5.33
1987	23.9	76.1	66	113	1.06	1.06	1.06	1.12	5.85
1988	25.5	74.5	69	118	1.05	1.05	1.05	1.17	4.43
1989	27.1	72.9	74	128	1.08	1.08	1.08	1.26	7.72
1990	29.5	70.5	78	134	1.05	1.05	1.05	1.33	4.99
1991	30.7	69.3	79	136	1.01	1.01	1.01	1.34	0.82
1992	31.7	68.3	82	136	1.02	1.02	1.02	1.36	1.73
1993	32.1	67.9	89	132	1.01	1.00	1.00	1.37	0.45
1994	33.6	66.4	98	131	1.02	1.02	1.02	1.40	2.29
1995	37.6	62.4	112	127	1.03	1.02	1.03	1.44	2.56
1996	38.8	61.2	131	123	1.05	1.04	1.04	1.50	4.15
1997	42.0	58.0	148	129	1.08	1.08	1.08	1.61	7.42
1998	44.4	55.6	161	123	1.01	1.01	1.01	1.63	0.87
1999	46.3	53.7	187	117	1.04	1.03	1.04	1.69	3.64
2000	49.7	50.3	209	109	1.02	1.01	1.02	1.71	1.67
2001	53.9	46.1	194	142	1.12	1.08	1.10	1.89	9.55
								Avg	3.97

Note. Las. = Laspeyres, Pa. = Paasche, Fis. = Fisher, Fis. Ch. = Fisher Chained. The Fisher Index = the square root of the product of the Laspeyres and Paasche indices. Source: Switched Access Minutes—FCC ARMIS 43-08, Table IV, Telephone Calls, Column ei. (1991–2001): Intrastate DEMs—Frank Gollop, USTA Report (1985–96); http://www.fcc.gov/wcb/iatd/neca.html, see file netwo00.zip (stdems.xls) (1997–2000); 2001 is estimated using average annual growth rate from 1997 to 2000.

Table 6.10 presents the chained Fisher intrastate output index and growth rate. This table shows that intrastate output grew at an average annual rate of 3.97%. This growth differs from the rate when IDEMS alone measures intrastate output in the preceding section. From Table 6.4 the average annual growth in IDEMS for the period is calculated as 4.73%, or about 20% more than the average annual growth of the intrastate toll and access minute index. Inspection of the annual growth rates in the two intrastate output quantities shows the measures are clearly different for the period 1991 to 2001 (the period for which data existed for both IDEMS and ISAMS). The main reason for the difference is due to ISAMS growing at an average annual rate of 7.47%, which is substantially faster than IDEMS growth at 4.73%. This difference implies that ITMS, which, until recently, accounted for the majority of intrastate services (see the output shares in Table 6.10) grew considerably slower than IDEMS. Indeed ITMS growth averaged only 2.12% per year.

High Bandwidth Output Index

The FCC makes data available for high-speed lines and advanced service lines. High-speed lines include all lines with transmission speeds over 200 Kbps in at least one direction for categories including DSL, wireline other than DSL, coaxial cable, fiber and satellite and fixed wireless. The advanced service lines are reported for the same categories but are capable of transmission speeds of over 200 kbps in both directions. Data about both types of lines provided to residential and small business customers are reported separately. The measure of broadband lines included in this productivity analysis are based on the count of advanced service DSL lines provided to residential and small business customers. Between 1999 and 2001 these lines accounted for about 71% of all advanced lines.

Table 6.11. DSL Quantities and Revenues, 1999–2002

	High-Speed		Advanced Service		Residential and Small Business						
Year	Lines (mill.)	Line Growth	Lines (mill.)	Line Growth	HSL (mill.)	HSL Growth	ASL (mill.)	ASL Growth	ASL Index	LSR (US$)	LSR (US$m)
1999	0.4		0.2		0.3		0.1		1.00	1.03	
2000	2.0	1.68	0.7	1.29	1.6	1.70	0.4	1.21	3.36	1.03	27.1
2001	3.9	0.69	1.4	0.71	3.6	0.82	1.2	1.15	10.63	1.03	15.4
2002	6.4	0.49	2.8	0.72	5.5	0.42	2.2	0.57	18.78	1.03	27.1
Avg		0.95		0.91		0.98		0.98			

Note. HSL = High speed line, ASL = Advanced service line, LSR = Line sharing revenue. Source: FCC, Trends in Telephone Service, August 2003, Tables 2.1, 2.2, 2.3 and 2.4.

Table 6.11 shows the number of high-speed and advanced service lines used to provide DSL services. Also shown are the prices (line sharing rate) charged for the use of advanced service lines used to provide DSL service to residence and small business customers. From 1999, the first year that DSL data is available, the average annual rate of line growth for the service categories is around 100%. Data about prices (the line sharing rate) charged is not systematically recorded and available. These prices are determined by negotiation between the ILECs and competitive local exchange carriers (CLECs) who offer DSL services to their own end users. Trade press articles, however, do provide an indication of the prices CLECs pay ILECs to use the high bandwidth portion of existing access lines. These articles indicate that the average monthly rate charged by SBC and Quest between 1999 and 2001 fluctuated between about US$ 5 to US$ 8 per month. It is estimated here that the average monthly charge is about US$ 6.83. There is no indication in the trade press that Verizon or BellSouth charge CLECs for the use of the high bandwidth portion of existing lines.

Table 6.12. Fisher Alternative Output Quantity Index Calculation, 1985–2001

Year	Revenue Share				Quantity				Output Index				
	Local Service (%)	Intra- state Toll & Access (%)	Inter- state (%)	DSL (%)	Local Calls (mill.)	Intra- state Output Quant- ity Index	Inter- state Quant- ity Index	DSL Lines Index	Las.	Pa.	Fis.	Fis. Ch.	Growth (%)
1985	49.6	24.0	26.4	0.00	311	1.00	1.00	1.00	1.00	1.00	1.00	1.00	
1986	49.7	23.5	26.8	0.00	316	1.05	1.05	1.00	1.04	1.03	1.04	1.04	3.45
1987	49.7	24.1	26.2	0.00	321	1.12	1.14	1.00	1.04	1.04	1.04	1.08	4.22
1988	48.7	245.0	26.3	0.00	319	1.17	1.29	1.00	1.04	1.04	1.04	1.12	3.98
1989	49.5	24.5	26.0	0.00	330	1.26	1.37	1.00	1.05	1.05	1.05	1.18	5.23
1990	50.1	24.5	25.3	0.00	342	1.33	1.53	1.00	1.06	1.06	1.06	1.26	5.98
1991	51.7	23.4	24.9	0.00	353	1.34	1.74	1.00	1.05	1.05	1.05	1.32	4.93
1992	52.7	22.5	24.9	0.00	365	1.36	1.84	1.00	1.04	1.04	1.04	1.37	3.59
1993	52.9	22.2	25.0	0.00	377	1.37	2.05	1.00	1.05	1.05	1.05	1.43	4.52
1994	53.2	21.4	25.4	0.00	393	1.40	2.23	1.00	1.05	1.05	1.05	1.50	4.72
1995	55.1	19.2	25.8	0.00	409	1.44	2.40	1.00	1.05	1.05	1.05	1.57	4.67
1996	56.3	18.1	25.6	0.00	422	1.50	2.62	1.00	1.05	1.05	1.05	1.65	4.69
1997	57.7	16.7	25.6	0.00	433	1.61	2.83	1.00	1.05	1.05	1.05	1.73	4.76
1998	58.5	15.6	25.9	0.00	445	1.63	3.11	1.00	1.04	1.04	1.04	1.80	4.05
1999	59.4	13.9	26.7	0.00	448	1.69	3.50	1.00	1.04	1.04	1.04	1.88	4.19
2000	59.9	12.2	28.0	0.01	429	1.71	3.94	3.36	1.01	1.01	1.01	1.89	0.82
2001	58.6	11.5	29.9	0.02	409	1.89	4.10	10.63	1.00	0.99	0.99	1.88	0.55
												Avg	3.95

Note. Las. = Laspeyres, Pa. = Paasche, Fis. = Fisher, Fis. Ch. = Fisher Chained. The Fisher Index = the square root of the product of the Laspeyres and Paasche indices

To find the average monthly rate, the count of high-speed service lines provided by the four ILECs is relied upon. Table 2.5 of the FCC's *Trends in Telephone Service* provides a state-by-state distribution of high-speed lines. Assuming this distribution holds for advanced service lines, and assigning the carriers to the states in which they operate, leads to the line proportions: SBC 48.8%, Verizon 19.9%, BellSouth 19.8% and Quest 11.5%. Using these weights, the weighted average monthly line sharing rate over the period is US$ 1.03. The brief time period for which such broadband services exist, along with their relatively low price, imply that the introduction of these services into the productivity study should not affect the TFPI or productivity growth. Nevertheless, it is instructive to observe how fast these services grew in such a short period.

TFPI

Using the index of intrastate output formed from intrastate access minutes and toll minutes, along with the index for high bandwidth services leads to an output quantity index that differs from that developed above. The output quantity index is shown in Table 6.12. Average annual output growth for the period 1986 to 2001 is 3.95%. This is somewhat less than the growth rate of 4.1% obtained by updating the FCC study. Nevertheless, annual total output growth rate is similar for different characterizations of intrastate output. Total output growth is not substantially affected by disaggregating intrastate output into access and toll minutes because the share of intrastate revenue as a proportion of total revenue averaged only 20% over the period. This result implies that the TFPI does not dramatically change with the adjustment in intrastate output. As shown in Table 6.13, productivity growth is only slightly below the annual rate computed in the updated FCC study. Based on the new output quantity index, productivity growth annually averaged 2.71%.

Table 6.13. Fisher TFPI and Growth, 1985–2001

Year	Output Quantity Index	Output Quantity Growth (%)	Input Quantity Index	Input Quantity Growth (%)	TFPI	TFP Growth (%)
1985	1.00		1.00		1.00	
1986	1.04	3.45	1.01	0.88	1.03	2.57
1987	1.08	4.22	1.02	0.99	1.06	3.23
1988	1.12	3.98	1.06	3.73	1.06	0.25
1989	1.18	5.23	1.09	3.39	1.08	1.84
1990	1.26	5.98	1.09	-0.72	1.16	6.70
1991	1.32	4.93	1.11	1.91	1.19	3.02
1992	1.37	3.59	1.10	-0.60	1.24	4.19
1993	1.43	4.52	1.11	0.64	1.29	3.88
1994	1.50	4.72	1.14	3.09	1.31	1.63
1995	1.57	4.67	1.15	0.45	1.37	4.23
1996	1.65	4.69	1.11	-2.90	1.48	7.60
1997	1.73	4.76	1.16	4.38	1.48	0.38
1998	1.80	4.05	1.18	0.93	1.53	3.12
1999	1.88	4.19	1.20	1.83	1.57	2.36
2000	1.89	0.82	1.23	2.42	1.54	-1.59
2001	1.88	-0.55	1.22	-0.55	1.54	0.01

Conclusion

The main purpose of this study is to measure productivity growth for the major local exchange carriers operating in the US telecommunications industry. Two sets of TFPIs are produced. The first is based on the FCC study used to calculate the offset applicable to regulated interstate services. Using this approach it is estimated that the average annual rate of productivity growth over the period 1986 to 2001 is 2.86%. The second method disaggregates intrastate services into switched access and toll, and introduces high bandwidth services provided by the ILECs. With these changes in output quantities, productivity growth annually averaged 2.71%. Under either output configuration, telecommunications productivity growth outpaced the growth for the national economy as a whole by approximately 2 percentage points per year.

There are a number of issues for future research regarding the measurement of telecommunications productivity growth. First, a more accurate measure of the labor input would rely on hours worked as opposed to number of employees. If carriers cut their workforce, but require extended working hours, the number of employees underestimates labor input growth. Such under estimation is further exacerbated if extended hours are accompanied by overtime pay. In addition, the labor input should be constructed from a Fisher-type index of types or categories of worker hours. This procedure results in a more accurate input measure, because the weights used to construct the index are based on the appropriate unit costs for the specific labor category. For example, if higher wage groups are growing faster (respectively slower) than lower wage groups then labor input index is underestimated (respectively overestimated) by simply using an index based on total hours worked.

The second issue pertains to the capital input. Instead of relying on a single class of asset to construct the capital stock, it should be constructed from a Fisher-type index of types or categories of assets. The Fisher index typically leads to a more accurate input measure, because the weights used to construct the index are based on the appropriate unit costs (or user costs of capital) for specific assets. However, user cost measurement is difficult. For instance, many capital inputs cannot be adjusted, i.e., bought and sold, instantaneously. Therefore, a producer must form a priori expectations about the purchase and disposal prices as well as future interest rates, depreciation rates and tax rates in order to calculate the ex ante user costs of a capital input. However, it is only possible to observe ex post prices, interest rates, depreciation rates and tax rates. Different assumptions about expectations formation may lead to substantially different user costs of capital and therefore Fisher capital quantity indexes.

References

Balk B (1995) Axiomatic price theory: A survey. International Statistical Review 63: 69–93
Bernstein JI, Mohnen P (1998) International R&D spillovers between US and Japanese R&D intensive sectors. Journal of International Economics 44: 315–38
Denny M, Fuss M, Waverman L (1981) The measurement and interpretation of total factor productivity in regulated industries, with an application to Canadian telecommunications. In: Cowing T, Stevenson M (eds) Productivity measurement in regulated industries. Academic Press, New York, pp 179–218

Diewert WE (1976) Exact and superlative index numbers. Journal of Econometrics 4: 115–46

Diewert WE (1992) The measurement of productivity. Bulletin of Economic Research 44: 163–98

Eichhorn W, Voeller J (1976) Theory of the price index. Lecture Notes in Economics and Mathematical Systems 140. Springer-Verlag, Berlin

FCC (1999) Third report and order. FCC 99-355

FCC (1997) Price cap performance review for local exchange carriers. Fourth report and order. CC Docket No. 94-1. Second report and order. CC Docket No. 96-262

FCC (2003) Trends in telephone service. Table 2.1, Table 2.2, Table 2.3 and Table 2.4, August

Funke H, Voeller J (1978) A note of the characterization of Fisher's ideal index. In: Eichhorn W, Henn R, Opitz O, Shephard RW (eds) Theory and applications of economic indices. Physica-Verlag, Wurzburg

Gollop FM (2000) The FCC X-factor: 1996-1998 update. Attachment 4 USTA comments, CC Docket No. 94-1 and 96-262

Jorgenson DW (1986) Econometric methods for modeling producer behaviour. In: Griliches Z, Intriligator MD (eds) Handbook of econometrics, vol III. Elsevier Science Publishers BV, New York, pp 1842–915

Landefeld JS, Parker RP (1997) BEA's chain indexes, time series, and measures of long-term economic growth. Survey of Current Business 77: 58–68

Sappington DEM (2002) Price regulation. In: Cave ME, Majumdar SK, Vogelsang I (eds) Handbook of telecommunications economics. Elsevier, Amsterdam, pp 227–98

7 Measuring TFP for an Expanding Telecommunications Network

Russel Cooper, Gary Madden and Grant Coble-Neal

Introduction

Typically, multiple-output econometric models of Australian telecommunications production maintain an assumption that carriers are in long-run equilibrium (Bloch et al. 2001a; Bloch et al. 2001b; Madden et al. 2002). This assumption implies all factors of production are adjusted in a costless manner to determine carrier long-run factor demands instantaneously (Nadiri and Prucha 1999). It has long been recognized that this assumption is problematic if carriers incur adjustment costs in altering their capital stock. Internal adjustment costs are commonly characterized in the form of output foregone due to quasi-fixed factor changes (Nadiri and Prucha 1990). Also, short-run network expansion costs (that relate to the addition of quasi-fixed network capital) vary with subscriber density, e.g., extending service to remote areas is more costly than upgrading the metropolitan local-loop network. Productivity growth is an important indicator of the nature of production technology, and is related to changes in cost as technology change occurs. Accurate measurement is important as productivity growth is often employed to evaluate past and forecast future performance of monopoly carriers, and the effect on industry performance of the introduction of competition (Fuss 1994). Further, conventional measures of total factor productivity (TFP) assume both constant returns to scale and full static equilibrium, and they are biased measures when either of these conditions is violated (Schankerman and Nadiri 1986).

In addressing issues concerning the structure of telecommunications production, such as measuring TFP response to technology change, the role of R&D capital has been explored extensively, see, e.g., Nadiri and Prucha 1990. However, reliable R&D data are seldom available. Other frequently used telecommunications technology change variables include the percentage of telephones with access to distance direct dialing and the percentage of telephones with access to electronic switching facilities. However, in experimenting with these variables (in dynamic cost function estimation), Bernstein (1989: 272) reports that best results are obtained with a binary technology change proxy. Further, difficulty is often reported in obtaining sensible estimates for associated economic variables, e.g., marginal costs and returns to scale, as well as in satisfying regularity conditions. Such outcomes are not surprising considering the potential for radical shifts of the production (cost) surface in response to underlying technology change (that may enable delivery of new services and service combinations, as well as affecting the costs of provision of extant services).

Conventional response to such difficulties is the embedding of an optimizing paradigm within adjustment cost models. Typical expressions incorporate adjustment costs in a continuous framework that enable continual disequilibrium adjustment toward long-run equilibrium. Technology change is sometimes drastic, and that this change is more realistically reflected by a series of radically different technological regimes. An alternative approach proposes that occasional, and possibly drastic, adoption costs are better characterized as entry into a new technological regime. Utilizing a generalization of the translog cost specification, which allows for structural breaks across long time series, this chapter decomposes TFP into continuous and discontinuous effects. This represents a different dichotomization to standard short-run (disequilibrium) and long-run (equilibrium) classifications. When structural change is not allowed for TFP is underestimated by not accounting for occasional substantial qualitative change in outputs. To illustrate the approach, results are presented within the context of an equilibrium model. This chapter reports first stage results for an ongoing project that aims to embed a continuous/discontinuous dichotomization within the familiar equilibrium/disequilibrium modeling approach. Employing this paradigm, the study calculates TFP for the Australian telecommunication industry for the period 1922 through 1983. In explaining the impact on costs of disequilibrium associated with ongoing network geographic expansion for much of the 20th century, the approach is consistent with predictable (or continuous) aspects of the expansion being factored into carrier plans. That is, massive technology change during the period suggests an overlay of steady technological progress with periodic discontinuity rather than disequilibrium adjustment. Should these discontinuities not be taken into account rapid equilibrium adjustments may be mistaken for disequilibrium adjustments under technology which is inadvertently characterized as more stable than plausible. The approach focuses on estimating parameters associated with cost minimization, and hence may be stable through substantial periods even while the underlying production function exhibits periodic shifts due to radical technology change. This outcome is achieved by concentrating attention on the manner in which technology (and output) variables enter the cost function. This concern is addressed by allowing technology and output arguments to exhibit occasional major change in their impact, while enforcing some degree of continuity on those economic parameters, representing the outcome of an optimization decision. In particular, non-linear output (and technology) terms enter a modified translog cost function to provide a better approximation of the evolution of technology that arises from multiple-regime shifts, and at times, radical technology change.

Another innovation employed in this study is the analysis of data and voice joint production. Questions potentially addressed by this approach include the impact of network expansion on carrier performance and the impact that joint supply of data (Telegraph and TELEX) and voice (local and toll) on industry cost structures. Perceptions of recent performance are likely to be guided by reference to prior benchmarks. However, performance statistics do not typically account for large fluctuations in investment that occur during successive waves of network extension and upgrade. Network expansion is likely to affect overall operational efficiency through adjustment costs. Currently, no attempt has been made to measure variation in the operational cost from these influences. In this respect, tele-communications carriers are both providers of telecommunications services and network suppliers. Despite these activities being distinct operations, typically financial reports are amalgamated. Thus, developing a fair measure of TFP performance for the service division has proven difficult. This paper employs a cost function model to analyze time-series (1922-1983) in terms of continuous expansion overlaid with occasional discontinuities prior to the (substantive) introduction of facsimile. This approach allows the decomposition of TFP growth estimates into continuous and discontinuous components, providing an opportunity to assess the impact of both predictable and radical technological change. While this analysis reflects Australian telecommunications sector activity, the installed technology is not dissimilar from that of most post-industrialized countries, nor was the introduction of competition and its regulation to that of most other OECD Member Countries. The paper is organized as follows. The following section presents data on network expansion and cost of the Australian telecommunications network. Next, the econometric model employed in estimation is outlined. That section is followed by a discussion of the estimation results, and reports cost elasticity and TFP estimates. A final section concludes.

Network Expansion

For the sample period, Australian telecommunications service was provided by a publicly mandated monopoly (Telstra). Justification for granting this monopoly is often based on natural monopoly arguments. Further, monopoly provision permitted a metropolitan-rural rate subsidy to fund rural network access. Clearly, short-run expansion costs vary by subscriber density, viz., extending the network into remote areas increases network costs. Table 7.1 provides indicative data on network expansion from 1922 through 1983, obtained from PMG Annual Reports for 1922 through 1975, and Australian Telecommunications Commission Annual Reports for 1976 through 1983. The number of telephone and TELEX services added per annum indicates the rate of network expansion. The sample average annual growth of telephone service is 4%, with the sub-period 1940-1983 averaging 10%. The 1940s and 1960s report above average growth rates of 22% and 11%, respectively. TELEX connections grew at 18% p.a., with the 1960s registering a sample high growth rate of 32%. Rural exchange growth is fastest in the 1930s. The next highest growth occurred in the 1950s. The most direct measure of geographic density available is the proportion of rural to total exchanges. As shown, the proportion of rural exchanges begins at less than 1% in 1937, and rises thereafter. For 1922 to 1958, new telephone subscription per new exchange statistics suggest proportionally increasing rural exchanges did not diminish increasing average subscriber density. Further, higher capacity exchanges in metropolitan areas offset the relatively inefficient rising proportion of rural exchanges. The average number of kilometers of cables laid per new subscriber reveal that rural expansion is greater for 1940-1944 (average 11.7 kilometers) and 1962-1976 (10.1 kilometers).

Table 7.1. Network Expansion Indicators

Year	New Telephony Service	New TELEX Service	Rural Exchanges (%)	Rural Exchange Growth (%)	New Exchange Additional Subscription	New Subscriber Additional Kilometers	Mainlines per Exchange
1922	12,523				44	4.3	76
1923	18,549				67	5.0	71
1924	28,166				31	4.2	71
1925	35,515				59	5.8	61
1926	31,090				74	3.1	61
1927	28,795				116	5.3	62
1928	25,075				100	5.2	64
1929	21,282		0.00		63	4.9	65
1930	11,454		0.00	0.00	2,446	9.2	65
1931	-19,569		0.00	0.00	745	-1.5	62
1932	-12,671		0.00	0.00	102	-1.2	60
1933	204		0.00	0.00	553	47.4	60
1934	8,845		0.00	4.00	780	0.9	61
1935	23,404		0.00	2.00	442	0.6	65
1936	21,636		0.00	0.00	326	1.4	68
1937	22,810		0.01	1.13	455	1.8	71
1938	25,027		0.01	0.56	525	3.4	74
1939	22,037		0.01	0.38	658	1.7	77
1940	19,077		0.02	0.49	9,468	4.0	80
1941	18,935		0.02	0.24	291	28.5	83
1942	5,528		0.02	0.13	-766	22.0	83
1943	9,186		0.02	0.02	-800	14.7	85
1944	16,810		0.02	0.01	690	5.9	88
1945	20,706		0.02	0.02	2,700	3.6	91
1946	29,698		0.02	0.01	1,443	3.0	95
1947	40,401		0.02	0.01	710	1.9	101
1948	40,472		0.02	0.00	427	2.5	106
1949	46,079		0.03	0.10	442	3.4	112
1950	60,167		0.04	0.42	530	3.8	118
1951	69,494		0.05	0.33	455	4.7	126
1952	63,185		0.06	0.37	828	6.2	133
1953	57,994		0.08	0.25	685	7.9	140
1954	64,359		0.10	0.28	984	6.1	147
1955	77,760		0.11	0.10	1,887	5.6	156
1956	79,239	76	0.12	0.12	8,027	5.1	166
1957	72,246	86	0.13	0.08	1,977	5.5	176
1958	83,053	102	0.15	0.15	n.a.	5.5	186
1959	129,393	143	0.17	0.11	-883	4.4	204
1960	70,651	182	0.18	0.08	-813	7.1	216
1961	69,116	292	0.19	0.06	-833	7.6	228
1962	87,485	239	0.21	0.05	-1,265	9.4	244
1963	93,612	224	0.22	0.04	-1,135	10.4	260
1964	106,699	376	0.26	0.18	-707	6.0	279
1965	91,244	364	0.28	0.07	-967	12.8	297
1966	110,254	265	0.31	0.08	-907	10.2	319
1967	114,325	710	0.34	0.08	-771	11.6	343

Table 7.1 (cont.)

1968	124,134	900	0.38	0.08	-1,191	11.5	371
1969	152,394	1,013	0.42	0.08	-1,887	11.1	403
1970	192,437	1,363	0.46	0.08	-2,255	7.5	441
1971	153,342	1,558	0.48	0.05	-1,491	8.9	471
1972	120,757	1,247	0.52	0.06	-2,687	11.7	498
1973	169,303	1,539	0.55	0.06	-3,689	9.3	532
1974	213,956	2,083	0.58	0.04	-2,070	7.0	574
1975	. 177,994	1,909	0.58	-0.02	-8,016	9.9	613
1976	128,250	2,281	0.60	0.03	-2,957	14.3	637
1977	206,959	2,554	0.62	0.03	-5,400	8.1	681
1978	259,178	3,123	0.64	0.02	-4,578	5.8	733
1979	283,813	3,177	0.67	0.03	-4,594	5.2	792
1980	289,412	3,830	0.70	0.03	-6,240	4.0	854
1981	330,734	4,244	0.72	0.02	-5,897	4.1	923
1982	288,974	3,827	0.74	0.02	-4,159	4.4	984
1983	241,225	1,586	0.76	0.02	44	7.4	1,040

Note. n.a. denotes not applicable. No new exchanges were installed in 1959.

Input and service measures summaries are contained in Table 7.2. Expenditure lists the main cost items (aggregated by the Törnqvist index). Capital, Labor and Materials comprise 90%-96% of annual expenditure. Buildings include Land expenditure. Fixed Capital expenditure is comprised of telecommunications plant that are mainly installed exchanges, cables, fixed machines and electrical generation plant. Source data for Materials is reported from 1948 and measures the average annual inventory. Detailed accounts, available for 1948-1959, indicate an average annual turnover of 40%. This turnover ratio is applied to available, less detailed, records of annual stock estimates for 1960-1983. Other Capital consists of engineers' movable plant, furniture and office equipment and a miscellaneous category. Motor Vehicles include cars, motorcycles, trucks and vans. Telecommunications output are classified as Data and Voice call numbers. Voice is an aggregate of local, long-distance and international calls, while TELEX and Telegraph subscribers and message volumes are combined to create Data. Table 7.2 highlights the relative magnitudes of Telegraph and Voice, with Voice output an average 50 times larger in magnitude than Telegraph. Voice cumulative output grew to more than 700 times 1915 output. Further output detail is provided in Fig. 7.1.

As shown in Fig. 7.1, the Voice series display a structural break at the end of World War I. After 1920 this series exhibits positive growth with the ten-year average growth rate increasing by 1% p.a. through to 1983. Casual observation of the Telegraph series suggests breaks at 1927 and 1943, with a highest value occurring during World War II.[1] Telegraph traffic levels began a decline that acceler-

---

[1] The possibility of structural breaks in Data and Voice series are examined by a Chow test for structural change within a univariate AR(1) model. The calculated F-statistic for the Voice series (that corresponds to a structural break in 1915) is 5.95 (DF=2, 60). The corresponding statistic for Data (with breaks at 1927 and 1943) is 6.84 (DF=1, 62). Both

ated with the widespread adoption of TELEX services. Clearly, there has been substantial quality advance through the sample period, making a unified examination of cost and TFP difficult. The model developed here, allows for such technological advance within identified periods. Shift parameters are introduced to allow for periodic drastic changes and are embedded within an economic framework that attempts to estimate common parameters representing optimizing behavior across the sample period of 1922-1983.

Table 7.2. Summary Statistics, 1915-1983

Variable	Minimum	Mean	Maximum	Std Deviation
Expenditure (A$ million)				
Buildings	0.11	8.94	34.09	9.40
Fixed capital	2.57	82.26	316.75	90.83
Labor	6.07	91.45	379.78	125.06
Materials	5.11	58.77	163.12	41.24
Motor Vehicles	0.00	2.76	13.15	3.15
Other Capital	0.08	2.77	13.80	3.62
Total Cost (A$ million)	9.60	188.13	710.34	226.09
Output (million)				
Voice (calls)	7.56	1,341.06	5,643.14	1,364.12
TELEX (calls)	0.01	11.62	37.42	12.82
Telegraph (messages)	4.05	26.77	72.65	14.14
Subscribers				
Telephone	160,000	2,090,000	7,150,000	1,850,000
Telex	95	10,780	39,388	12,351
Average Price (A$)				
Calls	0.21	0.73	1.50	0.38
Telegrams	0.11	0.57	2.23	0.54
Average Line Rental (A$)				
Telephone	16.10	24.67	41.23	6.12
TELEX	233.15	448.33	601.14	116.06
Total Revenue (A$ million)	15.60	173.82	679.65	192.83

Note. Real dollar values shown. Source: PMG (1915-1975) and Telecom (1976-1983)

tests support structural breaks at these dates. Given the small amount of data available prior to 1922, and the major discontinuity between 1921 and 1922, econometric analysis is restricted to 1922-1983.

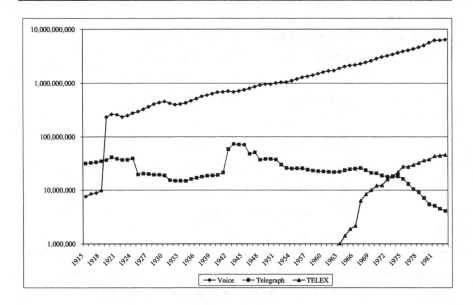

Fig. 7.1. Australian Telecommunications Output, 1915-1983

Econometric Cost Model

Close examination of these data reveal non-linear relationships between Data and Voice output, as well as between inputs and outputs. These observations coincide with tumultuous events such as the Great Depression and World War II. Additionally, the sample contains periods of radical technical change. Thus, to estimate a three-input, two-output translog cost function, this study specifies a discontinuous framework with non-linear functional forms embedded in the arguments and with discontinuity 'shifters' implicitly determined to ensure (approximate) continuity (at specified intervals) of key time varying economic parameters (such as output elasticities). The non-linear terms are a development of ideas introduced in Cooper et al. (2003), and are applied to capture the inherent non-linearity in the output and technology terms. The aim is to provide an approximation to the evolution of technology that arises from multiple regime shifts, and at times radical change in technology. In the process, the non-linear terms improve prospects for generation of economically meaningful results, such as positive marginal costs. The (modified) translog cost specification (with homogeneity and symmetry restrictions imposed) is:

$$
\begin{aligned}
\ln(C_t / p_{Mt}) = {} & \alpha_0 + \alpha_K \ln(p_{Kt} / p_{Mt}) + \alpha_L \ln(p_{Lt} / p_{Mt}) + \beta_D h^D (\ln q_{Dt}) \\
& + \beta_V h^V (\ln q_{Vt}) + \beta_T h^T (t) + \tfrac{1}{2} (\gamma_{KK} (\ln(p_{Kt} / p_{Mt}))^2 \\
& + 2\gamma_{KL} \ln(p_{Kt} / p_{Mt}) \ln(p_{Lt} / p_{Mt}) + \gamma_{LL} (\ln(p_{Lt} / p_{Mt}))^2) \\
& + \tfrac{1}{2} (\theta_{DD} h^{DD} (\ln q_{Dt}, \ln q_{Dt}) + 2\theta_{DV} h^{DV} (\ln q_{Dt}, \ln q_{Vt}) \\
& + \theta_{VV} h^{VV} (\ln q_{Vt}, \ln q_{Vt})) \\
& + \eta_{KD} \ln(p_{Kt} / p_{Mt}) h^{KD} (\ln q_{Dt}) + \eta_{KV} \ln(p_{Kt} / p_{Mt}) h^{KV} (\ln q_{Vt}) \\
& + \eta_{LD} \ln(p_{Lt} / p_{Mt}) h^{LD} (\ln q_{Dt}) + \eta_{LV} \ln(p_{Lt} / p_{Mt}) h^{LV} (\ln q_{Vt}) \\
& + \eta_{KT} \ln(p_{Kt} / p_{Mt}) h^{KT} (t) + \eta_{LT} \ln(p_{Lt} / p_{Mt}) h^{LT} (t) \\
& + \eta_{DT} h^{DT} (\ln q_{Dt}, t) + \eta_{VT} h^{VT} (\ln q_{Vt}, t)
\end{aligned}
\tag{7.1}
$$

where C_t is total cost, p are input prices, $\ln(\ )$ indicates the natural logarithm operator and $h^i (i = D, V, T)$, and $h^{ij} (i = D, V, K, L,\ j = D, V, T)$, denote non-linear functions. Let S_i denote the cost shares for the production inputs $(i = K, L, M)$. Since one share equation is satisfied by adding up, attention is directed to the structure of the capital and labor equations. These are appended to the cost function for efficiency in estimation. By Shephard's Theorem, the share equations for capital and labor are $S_{it} = \partial \ln C_t / \partial \ln p_{it}$, $(i = K, L)$, viz.:

$$
\begin{aligned}
S_{Kt} = {} & \alpha_K + \gamma_{KK} \ln(p_{Kt} / p_{Mt}) + \gamma_{KL} \ln(p_{Lt} / p_{Mt}) + \eta_{KD} h^{KD} (\ln q_{Dt}) \\
& + \eta_{KV} h^{KV} (\ln q_{Vt}) + \eta_{KT} h^{KT} (t)
\end{aligned}
\tag{7.2}
$$

and

$$
\begin{aligned}
S_{Lt} = {} & \alpha_L + \gamma_{KL} \ln(p_{Kt} / p_{Mt}) + \gamma_{LL} \ln(p_{Lt} / p_{Mt}) + \eta_{LD} h^{LD} (\ln q_{Dt}) \\
& + \eta_{LV} h^{LV} (\ln q_{Vt}) + \eta_{LT} h^{LT} (t) .
\end{aligned}
\tag{7.3}
$$

Notice that the precise structure of the share equations depends on the non-linear output functions $h^{ij} (i = K, L,\ j = D, V)$ and trend terms $h^{iT} (i = K, L)$. Also of importance in construction of the TFP measure is the cost elasticity for Data and Voice output $\varepsilon_{q_{it}} = \partial \ln C_t / \partial \ln q_{it}$ $(i = D, V)$. The elasticity estimates depend on the function forms $h^{ij} (i = D, V, K, L,\ j = D, V, T)$. These non-linear functions are constructed as functions of estimable parameters, so that characteristics of the time series determine the precise functional forms in econometric estimation, and with restrictions imposed that force them to equate to the types of functions employed in standard translog specifications at some base period t_0 (here, 1960). At this point the specification has the traditional characteristics of the flexible functional form associated with the translog. Let:

$$f^i(\ln q_{it}) = 2\ln q_{it_0}\ln q_{it}/(\ln q_{it_0} + \ln q_{it}), \quad i = D,V,$$

$$f^T(t) = 2t_0 t/(t_0 + t) - t_0,$$

$$f^{ij}(\ln q_{it},\ln q_{jt}) = f^i(\ln q_{it})f^j(\ln q_{jt}), \quad i,j = D,V,$$

$$f^{iT}(\ln q_{it},t) = f^i(\ln q_{it})f^T(t), \quad i = D,V$$

(7.4)

Note, by construction $f^i(\ln q_{it_0}) = \ln q_{it_0}$ $(i = D,V)$ and $f^T(t_0) = 0$. Next, introduce parameters μ_i $(i = D,V,T)$ and μ_{ij} $(i = D,V,K,L,\ j = D,V,T)$ to define the composite functions:

$$h^i(\ln q_{it}) = (\ln q_{it} + \mu_i f^i(\ln q_{it}))/(1+\mu_i), \quad i = D,V$$

$$h^T(t) = (t + \mu_T f^T(t))/(1+\mu_T),$$

$$h^{ij}(\ln q_{it},\ln q_{jt}) = (\ln q_{it}\ln q_{jt} + \mu_{ij}f^{ij}(\ln q_{it},\ln q_{jt}))/(1+\mu_{ij}), \quad i,j = D,V$$

$$h^{ij}(\ln q_{jt}) = (\ln q_{jt} + \mu_{ij}f^j(\ln q_{jt}))/(1+\mu_{ij}), \quad i = K,V, \quad j = D,V$$

(7.5)

$$h^{iT}(\ln q_{it},t) = t\ln q_{it} + \mu_{iT}f^{iT}(\ln q_{it},t), \quad i = D,V$$

$$h^{iT}(t) = (t + \mu_{iT}f^T(t))/(1+\mu_{iT}). \quad i = K,L$$

In this case, the functions $h^{iT}(\ln q_{it},t)$ $(i = D,V)$ and parameters μ_{iT} $(i = D,V)$ are not estimated freely but constrained to enforce a certain degree of continuity between time varying estimates of output elasticities across structural breaks due to technological change. In all other cases the parameters are structured so that the h functions are weighted averages of traditional translog-type variables and the f functions. For instance, with $\mu_{DV} = 0$, $h^{DV}(\ln q_{Dt},\ln q_{Vt}) = \ln q_{Dt}\ln q_{VT}$. As $\mu_{DV} \to \infty$, $h^{DV}(\ln q_{Dt},\ln q_{Vt}) \to f^D(\ln q_{Dt})f^V(\ln q_{Vt})$. The latter are functions with a different degree of curvature to the logarithmic functions. These forms are chosen with a view to improving the prospects for generating positive output elasticity estimates and marginal costs over a greater proportion of the sample than is common with cost function estimation using the translog specification. The weight given to the logarithmic versus alternative functions is determined by maximum likelihood estimation. Applying specifications Eq. 7.4 to the cost function, the cost elasticities for Data and Voice output are:

$$
\begin{aligned}
\varepsilon_{q_{Dt}} = &\ \beta_D (\partial h^D (\ln q_{Dt}) / \partial \ln q_{Dt}) + \tfrac{1}{2}\theta_{DD} (\partial h^{DD} (\ln q_{Dt}, \ln q_{Dt}) / \partial \ln q_{Dt}) \\
&+ \theta_{DV} (\partial h^{DV} (\ln q_{Dt}, \ln q_{Vt}) / \partial \ln q_{Dt}) \\
&+ \eta_{KD} \ln(p_{Kt} / p_{Mt})(\partial h^{KD} (\ln q_{Dt}) / \partial \ln q_{Dt}) \\
&+ \eta_{LD} \ln(p_{Lt} / p_{Mt})(\partial h^{LD} (\ln q_{Dt}) / \partial \ln q_{Dt}) \\
&+ \eta_{DT} (\partial h^{DT} (\ln q_{Dt}, t) / \partial \ln q_{Dt})
\end{aligned}
\tag{7.6}
$$

$$
\begin{aligned}
\varepsilon_{q_{Vt}} = &\ \beta_V (\partial h^V (\ln q_{Vt}) / \partial \ln q_{Vt}) + \tfrac{1}{2}\theta_{VV} (\partial h^{VV} (\ln q_{Vt}, \ln q_{Vt}) / \partial \ln q_{Vt}) \\
&+ \theta_{DV} \partial h^{DV} (\ln q_{Dt}, \ln q_{Vt}) / \partial \ln q_{Vt} \\
&+ \eta_{KV} \ln(p_{Kt} / p_{Mt})(\partial h^{KV} (\ln q_{Vt}) / \partial \ln q_{Vt}) \\
&+ \eta_{LV} \ln(p_{Lt} / p_{Mt})(\partial h^{LV} (\ln q_{Vt}) / \partial \ln q_{Vt}) \\
&+ \eta_{VT} (\partial h^{VT} (\ln q_{Vt}, t) / \partial \ln q_{Vt})
\end{aligned}
\tag{7.7}
$$

where the derivatives in Eq. 7.6 and Eq. 7.7 dependent on specifications Eq. 7.4 and Eq. 7.5. In view of constraints on β_D and β_V, which are imposed to capture major technological shifts, it is useful to note that these elasticities are freely estimated at t_0. At t_0, e.g., $\partial h^D (\ln q_{Dt}) / \partial \ln q_{Dt}$ reduces to $(1 + \mu_D / 2) / (1 + \mu_D)$, and parameter μ_D is freely estimated.[2]

In addition to the imposition of flexibility at the base point 1960, shifts in the cost function and in the share equations between ten year intervals (previously identified as representing different technological regimes) are allowed. Within these intervals a reference period is identified and intercepts in the estimated equations constrained so that the equations pass through the endogenous variable data points in the reference periods. Three parameters—α_K for the capital share equation, α_L for the labor share equation and α_0 for the cost function—are not freely estimated but are constrained to achieve this effect. This procedure imposes on the

---

[2] Of course, with a multiple-output specification in which the relative sizes of the Data and Voice outputs is important in determining costs, it is not desirable to scale both Data and Voice outputs simultaneously to unity at t_0 as is common in single output studies. In the multiple-output case other terms in Eq. 7.6 also influence ε_{q_D} at this base point. However, free estimation of μ_D ensures flexibility at t_0 with respect to the ultimate elasticity estimate no matter how complex are the additional terms in Eq. 7.6.

non-linear model a property which typically applies automatically at the mid-point of data in linear estimation. The reference periods in each interval are observation six. The approach in summary form is:

Table 7.3. Technology Intervals and Reference Periods

Technology Interval		Reference Period	
Identifier	Dates	Time	Identifier
1	1922-1931	1927	t_1
2	1932-1941	1937	t_2
3	1942-1951	1947	t_3
4	1952-1961	1957	t_4
5	1962-1971	1967	t_5
6	1972-1983	1977	t_6

The approach generates six values for each of the shift parameters for the technology intervals. First, Eq. 7.2 and Eq. 7.3 are solved for $\alpha_K(t_j)$ $j = 1,...6$ and $\alpha_L(t_j)$ $j = 1,...6$ respectively by evaluating all variables at the reference points. Substituting these parameter values in Eq. 7.1 and evaluating the remaining variables at the reference points, Eq. 7.1 is solved for $\alpha_0(t_j)$ $j = 1,...6$. From this point, the functions and reference points are substituted into the estimating equations Eq. 7.1 to Eq. 7.3 and the elasticity formulae Eq. 7.6 and Eq. 7.7.

Model Estimation: Issues and Results

Preliminary joint estimation of the cost and share equations, with shift terms allowing for technology regime shifts, produced parameter values which are compatible with the discontinuous equation systems in that the estimated equations perform well in terms of within sample prediction. However, without further modification many cost elasticity estimates are negative. Furthermore, discontinuities at the technology interval boundaries prove 'too sharp', and imply cost elasticity changes that are implausible. This may arise from imprecision in specifying the exact time at which drastic innovations become operative in terms of their impact on costs and optimal deployment of factors. While allowing for technology shifts is important in filtering these data prior to development of TFP measures, key components in the TFP measure in the multi-output case are the cost elasticities for outputs. For this reason the technology shift approach is overlaid with a procedure for smoothing time-series cost-output elasticity estimates. Effectively, the procedure takes account of information at the reference time periods to determine the values of the elasticities in the middle of a technology interval, but additionally utilizing the non-linear specifications to allow the elasticities to vary over time to smooth the transition from one mid-interval elasticity to another by interpolating the boundary elasticity (at the beginning of each technology interval) as the average of the two reference period elasticities in adjoining intervals. To initiate this process, information is utilized in the interval 1916-1921 to determine the first interval's boundary elasticity value.

Since this procedure requires some elasticity values to be imposed prior to estimation, arc elasticity estimates are initially constructed for each technology interval. In each case the measurement is over the six year period from the designated beginning of the interval to the reference period.[3] This procedure allows elasticities to vary through time within technology intervals in a manner prescribed by the model and subject to econometric estimation of a range of parameters which control the influence of the economic variables on the elasticities. However, the model characteristic of time varying elasticities creates some problems at the boundaries of the technology intervals, where the elasticities do not readily match. Although the elasticity estimates are expected to be different across technology intervals, the transition should arguably be smooth as adjustments are made to the technological change. Accordingly, to improve the economic meaning of elasticity estimates at the technological boundaries further smoothing on time variation in the elasticity estimates are imposed. The initial investigation for 1922-1982 data is conducted within the paradigm of an equilibrium model. In this context, the smoothing assumption is interpreted as an alternative to a disequilibrium adjustment approach. The elasticity smoothing procedure adopted requires the imposition of additional constraints in estimation, effectively interpolating the reference period elasticities to obtain elasticities at the boundaries of the technology intervals.

To sum, the approach, for both Data and Voice elasticity estimates, begins by examining the data and calculating six arc elasticity estimates at the technology interval reference points, $\varepsilon_D(t_j)$, $j = 1,...,6$ and $\varepsilon_V(t_j)$, $j = 1,...,6$. Elasticities at the beginning of technology intervals are interpolated by averaging adjacent values supplemented by an initial out-of-sample arc elasticity. These values provide another six elasticities for each case, $\varepsilon_D(\tau_j)$, $j = 1,...,6$ and $\varepsilon_V(\tau_j)$, $j = 1,...,6$,

---

[3] Arc elasticity estimates are inferior to econometric elasticity estimates as they represent cost changes due to a variety of factors, not merely output changes. It is also necessary to assign costs to the various outputs to construct arc elasticity estimates. For these reasons, six year periods are employed to average out extraneous influences. These values are used as constraints on elasticity estimates at all other points in time, so that the econometric estimation produces a time-series of elasticity estimates which are forced to be compatible with the arc elasticity estimates at the reference periods within technology intervals, but are also governed in their behavior over time by the evolving values of the economic variables and parameter estimates consistently with the specifications in Eq. 7.6 and Eq. 7.7.

where τ_j denotes the time period at the beginning of technology interval j. Using the first six sets of elasticities, technology shift variables $\beta_D(t_j)$ and $\beta_V(t_j)$ are obtained by rearranging Eq. 7.6 and Eq. 7.7 with variables evaluated at the t_j, $j = 1,...,6$. During the process, smoothness is imposed by constraining additional parameters in the functions $h^{iT}(\ln q_{it}, t)$, $i = D, V$ (see the specification of these functions given in Eq. 7.5, where the respective parameters μ_{DT} and μ_{VT} appear). Substituting $\beta_D(t_j)$ and $\beta_V(t_j)$ into Eq. 7.6 and Eq. 7.7 and evaluating Eq. 7.6 and Eq. 7.7 at τ_j, $j = 1,...,6$, produces values for ε_D and ε_V that are conditional on μ_{DT} and μ_{VT}, respectively. The 12 parameter values $\mu_{DT}(\tau_j)$ and $\mu_{VT}(\tau_j)$ for τ_j, $j = 1,...,6$ are then constructed to enforce the results $\varepsilon_D = \varepsilon_D(\tau_j)$ and $\varepsilon_V = \varepsilon_V(\tau_j)$ at points τ_j, $j = 1,...,6$ and the resultant functional forms for the shift parameters $\mu_{DT}(\tau_j)$ and $\mu_{VT}(\tau_j)$ are substituted back into the functions $h^{iT}(\ln q_{it}, t)$, $i = D, V$.

Maximum likelihood estimation results are presented in Table 7.4 and Table 7.5. The results are indicative only as there is evidence from the estimated residuals (not reported in detail) that disturbances are not serially independent.[4] This serial dependence may lead to some inefficiency but is unlikely to generate inconsistency.[5]

While the reported t-ratios are asymptotic, the values are quite large and are indicative of statistical significance in the majority of cases. Parameter estimates usually associated with disembodied technical change (β_T, μ_T) are insignificant, and indicate that technical progress is generally well accounted for in other aspects of the model. The technical change interaction terms for Capital and Labor inputs (η_{KT}, η_{LT}) are both insignificant. The technical change interaction terms on Data and Voice output (η_{DT}, η_{VT}) are significant, with opposite and roughly offsetting effects. Generally, the approach of estimating the cost function allowing for regime shifts seems to have successfully isolated the effect of measured technologi-

---

[4] A reaction is to consider the model within a partial adjustment context. This approach is too complex in the current case as the technological regime shift paradigm, which is important to the proposed methodology for examining TFP, may itself suggest that partial adjustment parameters vary across regimes. Further, the prospect introduces many parameters that reduce the degrees of freedom substantially. The current compromise is to recognize the issue and note that the technology shifters have the ability to absorb most non-stationarity in these data.

[5] Alternatively, atheoretical time series approaches to dealing with the serial dependence are likely to suffer from parameter instability in the context of the massive changes in technology which are being examined. The current study is seen as a first step in an attempt to find a resolution of the embedded theory/empirical dilemma.

cal change to the large shifts in regime. Estimates of the shift parameter values for the various technology intervals, constrained by the smoothing assumption on the cost elasticities, are given in Table 7.5.

Table 7.4. Estimation Results

Parameter	Estimate	t-ratio
γ_{KK}	0.135	11.895
γ_{KL}	-0.116	-11.602
γ_{LL}	0.157	16.399
θ_{DD}	-110.7	-9.151
θ_{DV}	47.83	9.269
θ_{VV}	-19.07	-7.403
η_{KD}	0.459	2.685
η_{LD}	0.214	2.469
η_{KV}	2.811	6.826
η_{LV}	-2.524	-6.266
η_{KT}	-0.004	-1.604
η_{LT}	-0.000	-0.154
η_{DT}	-0.405	-2.720
η_{VT}	0.308	5.045
β_{T}	-0.079	-0.159
μ_{D}	-1.114	-35.095
μ_{V}	-1.097	-48.396
μ_{DD}	-7.719	-12.350
μ_{DV}	-13.18	-10.024
μ_{VV}	-65,441	-6.184
μ_{KD}	-1.768	-10.478
μ_{KV}	-1.772	-80.057
μ_{LD}	-14,648	-2.334
μ_{LV}	-1.726	-115.66
μ_{KT}	-1.046	-35.635
μ_{LT}	-0.999	-24,454
μ_{T}	-0.918	-1.908
Observations	62	
Log-likelihood	221.11	

Table 7.5. Estimated Time Varying Technology Shift Parameters

Period	μ_{DT}	β_D	μ_{VT}	β_V	α_0	α_K	α_L
1922-1931	0.649	4.351	0.512	-1.723	-8.814	-21.538	18.229
1932-1941	-1.922	2.642	-1.825	0.003	-59.974	-21.506	18.228
1942-1951	-2.552	1.497	-2.631	0.790	-66.418	-21.567	18.292
1952-1961	6.571	3.681	3.153	-0.828	-43.366	-21.719	18.429
1962-1971	-12.611	4.321	-8.899	-1.032	-33.481	-21.810	18.504
1972-1983	-11.434	4.784	-6.272	-0.994	-56.459	-21.809	18.497

The α_0 estimate shows the extent to which total costs differ across technology regimes. Reduction in α_0 over time suggests either technological progress allowing the same output to be produced at lower cost and (almost certainly) include changes in output quality, so that the actual cost reduction for a given qualitative level of output is likely to be substantially greater. A qualitative shift in output appears evident in those cases where α_0 rises from 1942-1951 to 1952-1961, and again from 1952-1961 to 1962-1971. Apart from the very early period, movements in α_K and α_L suggest that there is a small but steady reduction in the capital share accompanied by an increase in the labor share across technology regimes. Again, this may be indicative of unmeasured qualitative change in the capital stock over the period. The elasticity shifters β_D and β_V allow for adjustments in the elasticities between reference periods in the different regimes, while the additional adjusters μ_{DT} and μ_{VT} help smooth the elasticity estimates across regimes. The reported substantial movements in μ_{DT} and μ_{VT} are interpreted with caution, since these are smoothing variations in elasticities which otherwise are associated with differences in a range of variables, but most notably with non-linear functions of different output levels across regimes. What is apparent is that change in μ_{DT} across the regime is similar to the change in μ_{VT}. Thus, the smoothing procedure is freely estimated as applying relatively equally to cost elasticity estimates for both Data and Voice output.

Post-estimation Analysis

Table 7.6 presents statistics derived post-estimation by calculation of elasticities, returns to scale, TFP growth estimates and economic regularity condition checks using the cost function parameter estimates from Table 7.4 and Table 7.5. The cost function satisfies concavity in input prices for 77% of the sample. Sustained violations occur in the 1930s, and at scattered points of time in the late-1920s and early-1940s. Apart from 1971, concavity is maintained throughout the post WWII period.[6]

Table 7.6. Elasticity, Returns to Scale, TFP Growth Estimates and Concavity Checks

Year	ε_D	ε_V	RTS	TFP	Det 1	Det 2	Det 3
1922	2.716	1.171	0.257	0.799	-0.090	-0.005	0.000
1923	2.793	1.034	0.261	-0.074	-0.082	0.000	0.000
1924	2.288	1.331	0.276	-0.058	-0.092	0.004	0.000
1925	1.372	2.011	0.296	0.035	-0.090	0.002	0.000
1926	1.493	2.011	0.285	0.019	-0.095	0.000	0.000
1927	2.737	1.209	0.253	0.002	-0.090	-0.002	0.000
1928	4.235	0.215	0.225	0.034	-0.080	0.000	0.000
1929	5.082	-0.229	0.206	-0.072	-0.069	0.000	0.000
1930	5.829	-0.537	0.189	-0.001	-0.054	-0.001	0.000
1931	10.170	-2.411	0.129	-0.085	0.047	0.002	0.000
1932	2.015	1.101	0.321	-0.030	0.054	0.003	0.000
1933	2.184	0.986	0.315	0.007	0.042	0.002	0.000
1934	3.248	0.340	0.279	-0.017	0.018	0.000	0.000
1935	1.043	1.424	0.405	0.085	-0.015	0.000	0.000
1936	1.059	1.302	0.424	0.026	-0.046	-0.001	0.000
1937	1.293	0.993	0.437	0.023	-0.067	-0.001	0.000
1938	1.173	1.021	0.456	-0.018	-0.083	-0.001	0.000
1939	1.529	0.611	0.467	-0.002	-0.084	0.001	0.000
1940	1.694	0.569	0.442	-0.128	-0.076	-0.002	0.000
1941	0.327	1.481	0.553	-0.017	-0.074	0.000	0.000
1942	1.636	1.859	0.286	0.000	-0.026	0.000	0.000
1943	0.530	4.251	0.209	-0.036	-0.001	0.000	0.000
1944	0.795	4.642	0.184	0.016	0.009	0.000	0.000
1945	1.156	4.508	0.177	0.010	-0.011	0.000	0.000
1946	1.515	3.990	0.182	0.019	-0.045	0.000	0.000
1947	1.980	2.725	0.213	-0.004	-0.071	0.000	0.000
1948	2.419	2.289	0.212	-0.011	-0.080	0.001	0.000
1949	2.580	2.090	0.214	-0.052	-0.099	0.001	0.000
1950	2.459	2.068	0.221	-0.077	-0.108	0.003	0.000
1951	2.721	1.469	0.239	-0.058	-0.112	0.003	0.000
1952	1.631	2.231	0.259	-0.124	-0.092	0.001	0.000
1953	2.412	1.603	0.249	-0.091	-0.085	0.003	0.000
1954	2.246	1.531	0.265	-0.038	-0.100	0.002	0.000
1955	1.495	1.787	0.305	-0.019	-0.108	0.001	0.000
1956	1.113	1.940	0.328	-0.056	-0.105	0.002	0.000
1957	1.282	1.738	0.331	-0.047	-0.112	0.001	0.000

[6] In Table 7.6, the columns Det 1, Det 2 and Det 3, represent estimated determinants of the matrix of second-order partial derivatives of the cost function with respect to input prices for 1922-1983. For concavity, Det 1 should be negative, Det 2 positive and Det 3 zero.

Table 7.6 (cont.)

1958	1.372	1.620	0.334	-0.033	-0.114	0.002	0.000
1959	1.121	1.680	0.357	-0.012	-0.115	0.002	0.000
1960	0.778	1.919	0.371	0.014	-0.114	0.001	0.000
1961	0.517	2.062	0.388	0.004	-0.115	0.000	0.000
1962	0.917	1.733	0.377	-0.030	-0.111	0.000	0.000
1963	2.048	1.176	0.310	0.013	-0.108	0.001	0.000
1964	1.840	1.179	0.331	0.039	-0.108	0.001	0.000
1965	1.875	1.118	0.334	0.001	-0.111	0.001	0.000
1966	2.084	0.972	0.327	-0.021	-0.114	0.002	0.000
1967	0.551	1.728	0.439	0.040	-0.114	0.000	0.000
1968	1.583	1.180	0.362	-0.026	-0.115	0.002	0.000
1969	3.106	0.815	0.255	-0.054	-0.111	0.000	0.000
1970	3.406	0.631	0.248	0.031	-0.110	0.000	0.000
1971	5.286	-0.095	0.193	-0.091	-0.109	-0.001	0.000
1972	1.657	1.886	0.282	0.039	-0.107	0.001	0.000
1973	1.368	2.122	0.287	0.016	-0.108	0.001	0.000
1974	0.932	2.408	0.299	0.030	-0.107	0.000	0.000
1975	0.538	2.815	0.298	0.034	-0.102	0.000	0.000
1976	2.077	2.240	0.232	0.168	-0.106	0.001	0.000
1977	2.763	2.045	0.208	0.003	-0.106	0.002	0.000
1978	2.999	2.068	0.197	0.026	-0.104	0.001	0.000
1979	3.140	2.168	0.188	0.037	-0.100	0.000	0.000
1980	4.016	1.899	0.169	0.025	-0.083	0.000	0.000
1981	3.371	2.203	0.179	0.085	-0.064	0.000	0.000
1982	3.533	2.759	0.159	0.016	-0.059	0.001	0.000
1983	3.695	2.940	0.151	0.008	-0.064	0.001	0.000

Concentrating on the post WWII period for which concavity is (largely) obtained, it is notable that there is a change in the relative importance of the elasticities, with Voice generally dominant in the early 1940s and 1950s, and Data beginning to dominate from the early-1960s. Both elasticity series show considerable variation as might be expected with a non-linear specification which has them dependent on both output levels and input prices. Interestingly, the estimates are compatible with decreasing returns to scale (see Column RTS in Table 7.6) throughout 1922-1983, and there is a trend decline in returns to scale based on these estimates.

The main purpose in estimating cost elasticities with respect to output is to make use of them in a measure of aggregate output growth for use in constructing a measure of TFP. Allowing technology regime shifts enables elasticity measures to be relevant to the determination of productivity of inputs, abstracting from qualitative changes in output where possible. Also, relevant, on the input side, are estimates of input shares that are associated with the production of current output. The impact of network construction lags on TFP measurement is also of concern. The use of estimated shares from an optimizing model is an attempt to allow for this, since actual shares can include variations in input levels associated with firm activity, i.e., better associated with investment rather than current production.

When inputs for construction are identified it is best to exclude them. However, when construction and operations accounts are not maintained separately other means of identification are required. That is, including non-operational inputs in an aggregate input index is problematic in that it suggests the factor is higher than required to produce the observed output. Accordingly, using actual input shares rather than predicted shares compatible with a cost-minimizing model, leads to incorrectly measuring relevant inputs for estimation of TFP. Estimation of Eq. 7.1 to Eq. 7.3 and subsequent calculation of Eq. 7.6 to Eq. 7.7 in a shifting regimes context permits appropriate index weights to be calculated by effectively netting out, e.g., construction influences in TFP on the input side and qualitative changes on the output side. The multi-input, -output Törnqvist index is,

$$\Delta \ln TFP = \Delta \ln Q - \Delta \ln X \qquad (7.8)$$

where Q and X are indices of outputs and inputs, respectively and Δ is the discrete first-difference operator. The log-change indexes Q and X are defined as

$$\Delta \ln X = \tfrac{1}{2} \sum (S_{it} + S_{it-1})(\ln X_{it} - \ln X_{it-1}) \qquad (7.9)$$

$$\Delta \ln Q = \tfrac{1}{2} \sum (M_{kt} + M_{kt-1})(\ln Q_{kt} - \ln Q_{kt-1}) \qquad (7.10)$$

where M and S represent output and input share weights, respectively. Following Fuss (1994) the output weights in Eq. 7.10 are the estimated output cost elasticity shares, constructed from Eq. 7.6 and Eq. 7.7. The input weights in Eq. 7.9 are the estimated input shares predicted by Eq. 7.2 and Eq. 7.3.

The column labeled TFP in Table 7.6 reports the difference in the percentage growth rates of aggregate output relative to aggregate input as defined in Eq. 7.8. Concentrating on the regular region from 1945 to 1983 (with the exception of 1971), it is observed that TFP is negative in the late-1940s and throughout the 1950s, but remains mostly positive thereafter. Given the earlier reported substantial increases in costs between the 1940s and 1950s, and again between the 1950s and 1960s (see Column α_0 in Table 7.5), and the fact that these suggest a concomitant qualitative improvement in output, these results need to be interpreted with caution. TFP, in the modeling context examined here, is interpreted as consisting of distinct components. α_0 in Table 7.5 reports a discontinuous component, associated with major change in technological regimes. The apparent increase in costs is likely associated with an increase in quality and a new type of output. To the extent that inputs remain steady, after controlling for input price and output level effects, is indicative of a (discontinuous) increase in TFP. The TFP estimates in Table 7.6 supplement this with a measure of (continuous) change in TFP, estimated under conditions that control for regime changes. In this paradigm, there are both discontinuous and continuous effects contributing to TFP.

Conclusion

This study examines long time series (1922-1983) of data on costs, output and inputs in the Australian telecommunications industry, in an effort to uncover estimates of TFP over a period associated with major technological changes as well as substantial network expansion. In doing so, the chapter estimates a cost-minimizing model augmented to control for inherent non-linearity in the output and trend terms, and allow for substantial changes in technological regimes. The resulting model is at this stage not fully well-specified econometrically. Although it satisfies the requirements of a proper cost function for 77% of the sample period, there is some evidence of serial correlation in the residuals which requires further examination. Despite this, cost elasticities and TFP reveal an industry accommodating substantial technological change over the sample period. In particular, the model provides the first econometric analysis of the joint production of Data and Voice outputs for the Australian telecommunications industry. Importantly, the method enables TFP series to be estimated, distinguishing between continuous improvements in productivity and occasional massive discontinuities which, unless controlled for, may contribute to error in the construction of traditional measures of TFP. As such, the approach identifies an issue of importance for the estimation of productivity using optimizing models when the data series cover periods of extensive technological change.

Appendix: Timeline of Technological Advance

1922-1931

Murray-Baudot machine printing telegraph equipment brought service between Melbourne and Sydney (1922: 19)

Carrier wave system of operating long-distance traffic between NSW and Victoria introduced (1925: 17)

Murray multiplex system of machine printing telegraph service installed between Melbourne and Adelaide (1927: 17)

Commercial telephone service established between Australia and UK. Later with the rest of Europe and the US (1930: 14)

Picturegram service available to the public between Melbourne and Sydney (1930: 16)

Small packet service introduced (1931: 14)

Direct radio telephone service between Australia and NZ—1st overseas call (1931: 18)

Inaugural long-distance telephony service between Adelaide and Perth. Perth in direct communication with Adelaide, Sydney via carrier-wave channel (1931: 17, 1931: 20)

1932-1941

Voice-frequency system installed between Sydney and Tamworth (1936: 17)

Direct telephone service with US. Tariff reduced considerably (1939: 15)

Direct telephone service to Canada (1939: 15)

1942-1951

Public radio telegram service in air-to-ground direction (1949: 17)
Mobile radio-telephone services in Sydney (1951: 21)

1952-1961

TELEX in Sydney and Melbourne (1953: 20)
Multiplex equipment replaced with teleprinter for major routes (1954: 28, 1955: 22)
Teleprinter exchange (TELEX) service in Sydney and Melbourne (1955: 21)
Code converter units introduced at Overseas Telecommunications Commission (1956: 17)
Automatic switching of trunk calls (STD) (1956:14). Initially between Dandenong and Melbourne
Contract for teleprinter reperforator switching scheme (TRESS) to the telegraph system (1957: 21-2)
Subscriber radiotelephone network in service at Broken Hill (April 1958)
International TELEX service countries in October 1958 (1958: 16, 1959: 20)
Radio-bearer system under the broadband system on Melbourne-Bendigo trunk route (1960: 18), i.e., broadband microwave system
Sydney-Canberra coaxial cable circuits (1961:3)
Large exchanges using crossbar automatic switching equipment. Initially in Toowoomba (1961: 3, 17, 1962: 16)

1962-1971

Nationwide transit switching introduced into TRESS system
Crossbar exchange equipment incorporating requirements for the National Telephone Plan in service at Petersham (1963: 11)
Completion of COMPAC cable (1964: 8)
Comprehensive data transmission service organized for Defence Force (1964: 8)
Brisbane-Cairns microwave system advance with completion of Brisbane-Maryborough segment. Part of SEACOM overseas cable link (1965: 9)
Sydney Data Processing Centre opened (1966: 9)
Brisbane-Cairns microwave radio link in service (1966: 8)
TELEX service cutover to automatic operates nationally (1966: 3)
SEACOM cable opened for traffic (1967: 9)
Introduction of Australian satellite circuits (1968: 3)
Time assignment speech interpolation (TASI) systems installed on COMPAC and CANTAT cables for international telephone service (1968: 9)
Online time-sharing computer in operation (1968: 10)
International fully subscriber-to-subscriber TELEX service (1969: 13)
Datel service to provide a data transmission facility over telephone circuits (1969: 13)
East-west microwave radio relay system (1970: 4, 1971:9)
Satellite circuits between Sydney and Perth (1970: 4, 6)
Contract for stored program controlled electronic telephone equipment for new trunk exchange in Pitt Street (1970: 4), i.e., computer-controlled telephone exchange

1972-1983

Computer installed in several departments to control costs (1972: 40)
Trials of closed circuit television (CCTV) conferencing (1973: 54)
Radio paging in Sydney and Melbourne (1973: 53, 1975: 18)

Mt Isa-Darwin microwave radio system. Provides telephone, telegraph and data communications, and additional broadcasting circuits (1974: 16)

Multi-purpose public telephone enabling dialing STD for local and trunk calls, and lodge telegrams via an operator (1974: 17)

Automatic data processing (ADP) systems for telephone accounting operating in Sydney and Melbourne. Introduced to Adelaide, Brisbane and Perth during the following three-year period (1974: 21)

Stored program controlled (SPC) exchange commissioned in Sydney (1975: 19). New electronic trunk exchange opened at Pitt Street, Sydney

Stored program controlled trunk exchange commissioned for Sydney (September). Similar trunk exchanges to be installed in major cities (1975: 19)

Push-button telephone (1975: 20).

Public telephones replaced with the CT3, offering full STD service. Housed in modern glass and aluminum cabinets

Pulse code modulation (PCM) systems, capable of simultaneous transmission of 30 telephone conversations over cable pairs. Placed in operation in Queensland (October) (1979: 78)

Inwards wide area telephone service (INWATS). Tenders let for a public automatic mobile telephone service (1980: 7)

Plans to introduce of a digital data network and packet switching service (1980: 87)

Call charge record (CCR) equipment for ISD calls on trial in NSW (March). Extension to other exchanges in Sydney and selected exchanges in capital cities 1980-81

Stored program controlled (SPC) PABXs available (1980: 57)

Plans for fully-automatic telephone service for vehicles in Melbourne and Sydney (1980: 67)

Optic fiber in link exchanges by Telstra

Mobile telephone service in Sydney and Melbourne

Stored program control telex exchange (AXB20) in Melbourne (1982: 48)

Section of Kimberley solar-powered microwave system completed (1983: 11)

Exhibition telephone exchange in Melbourne (1983: 11)

Digital data service (DDS) and AUSTPAC (public packet switching system) (1982: 13, 1983: 41)

References

Australian Telecommunications Commission (1976-1983), Annual Reports.

Bernstein JI (1989) An examination of the equilibrium specification and structure of production for Canadian telecommunications. Journal of Applied Econometrics 4: 265-82

Bloch H, Madden G, Coble-Neal G, Savage S (2001a) The cost structure of Australian telecommunications. Economic Record 77: 338-50

Bloch H, Madden G, Savage S (2001b) Economies of scale and scope in Australian telecommunications. Review of Industrial Organization 18: 219-27

Cooper R, Diewert WE, Wales TJ (2003) On the subadditivity of cost functions. Traditional Telecommunications Networks—The International Handbook of Telecommunications Economics, Volume I, ed. G Madden, Edward Elgar, Cheltenham

Fuss MA (1994) Productivity growth in Canadian telecommunications. Canadian Journal of Economics 27: 371-92

Madden G, Bloch H, Coble-Neal G (2002) Labor and capital saving technical change in telecommunications. Applied Economics 34: 1821-8

Nadiri MI, Prucha IR (1990) Dynamic factor demand models, productivity measurement, and rates of return: Theory and an empirical application to the US Bell System. Structural Change and Economic Dynamics 1: 263-89

Nadiri MI, Prucha IR (1999) Dynamic factor demand models and productivity analysis. NBER Working Paper 7079, pp 109

Schankerman M, Nadiri MI (1986) A test of static equilibrium models and rates of return to quasi-fixed factors, with an application to the Bell System. Journal of Econometrics 33: 97-118

Postmaster General's Department (1915-1975), Annual Reports

8 Dynamic Aspects of US Telecommunications Productivity Measurement

M. Ishaq Nadiri and Banani Nandi

Introduction

Information and communications technology (ICT) is an important source of economic growth and productivity improvement. Both the theoretical and empirical literature suggests a positive impact from ICT equipment investment and service production on firm, industry and national productivity.[1] Aggregate productivity growth is via improved interaction among advanced communications technology and computer hardware that leads to lower transaction costs. Such technological improvement enhances the speed and accuracy of analysis, storage and transmission of information. Also, improvement increases the availability of knowledge. This process necessarily generates spillover or network effects among firms (Katz and Shapiro 1985). In particular, recent analysis by Nadiri and Nandi (1999, 2003), Mun and Nadiri (2002a, 2002b) and Chun and Nadiri (2002) examine the role of the US ICT industry within the national economy.[2] These studies focus on the computer equipment and software, and telecommunication industry market segments as a growth industry and their impact on the economy more generally. Their reported results are based on econometric model estimates using data from US industry. The sample data for analysis of US telecommunications industry total factor productivity (TFP) growth trends is for 1935 through 1987. For analyzing the impact of the telecommunications industry on other US industry, they use data from 34 sectors of the economy for 1950 to 1991. These data are annual price, output and input time-series for industries that comprise the major economic sectors. A flexible functional form, translog cost function is employed.

This study discusses the contribution of the telecommunications industry to the economy by addressing the trend in TFP growth in the US telecommunications industry since the mid-1930s, and the changing sources of TFP growth. Dynamic aspects of the telecommunications sector and its impact on the production structure of other industries, i.e., changing factor ratio in different industries, and spillover or network effects on the output and productivity growth of other sectors and industry are also examined. The following section provides a brief discussion of pricing and investment in the sector. The next section considers substantial TFP

---

[1] Importantly, productivity growth by firms and industry engaged in ICT equipment and service production leads to substantial declines in their product prices. Such price falls led to wider adoption of ICT in the US, with this investment gathering momentum through 1980s and 1990s.

[2] Some are published and others are in various stages of completion.

growth and its sources within US telecommunications industry. A discussion of dynamic aspects of the sector and their impact on the production structure of other industries follows. The spillover or network effects of telecommunications infrastructure are then explored. Total benefit derived from telecommunications infrastructure by US industry is then presented. A final section concludes.

Prices and Investment

ICT is used both as a productive input by firms and for the communication network of the economy more generally. Cronin et al. (1997), Jorgenson and Stiroh (2000) and Jorgenson (2001) emphasize that the rate of productivity growth and price decline in the ICT sector has led to substitution of ICT and telecommunications services for traditional inputs.[3] Table 8.1 shows the ICT equipment sector, particularly the computer industry, experienced substantial decline in output and input prices. Similarly, price declines are observed for telecommunications services since the mid-1980s. This price decline is reflected in the downward movement of both the CPI (consumer price index) and PPI (producer price index) for long-distance (toll) telephony service during 1985-1989—accounting for approximately 26% of reduction in the CPI and 18% of reduction in PPI (FCC 2001/2002).[4] During 1990-2000, the CPI toll price index experienced volatile trend decline. The PPI toll service index had a corresponding 17% price reduction. This decline stimulated demand for telecommunications services by business customers. Table 8.2 presents the growth in the toll service PPI index, and business line and special access line demand for 1985-2000. Based on Table 8.1 and Table 8.2, it appears that the price trend of ICT inputs is different to that for labor and non-information technology (non-ICT) capital inputs. These substantial relative price declines, in both the computer and telecommunications industry, led to a shift in the contribution of computer and telecommunications to productivity and GDP growth, and a movement toward more ICT-intensive production. Some of this change is reported by Jorgenson (2001) (see Table 8.3). Clearly the contributions to productivity growth and GDP, particularly by the computer industry, are substantial for 1990-1995, and accelerated further during 1995-1999. Further, Table 8.3 provides evidence as to the contribution of the ICT industry and the components to national productivity growth.

---

[3] ICT prices declined almost at 18% p.a. from 1959 through 1995. Since 1995, prices have declined at an average rate of 32% p.a.

[4] CPI indices for interstate and intrastate toll services are obtained from Table 5.5 of *Statistics of Communications Common Carriers*, 2001/2002 edition. These indexes are combined into a single toll service index using weights created using the distribution of calls between interstate and intrastate call volume records from the FCC report.

Table 8.1. ICT Sector Output and Input Price and Quantity Growth Rates

	1990-1995		1995-1999	
	Prices	Quantities	Prices	Quantities
Outputs				
Gross domestic product	1.99	2.36	1.62	4.08
Information technology	−4.42	12.15	−9.74	20.75
Computers	−15.77	21.71	−32.09	38.87
Software	−1.62	11.86	−2.43	20.80
Communications equipment	−1.77	7.01	−2.90	11.42
Information technology services	−2.95	12.19	−11.76	18.24
Non-information technology investment	2.15	1.22	2.20	4.21
Non-information technology consumption	2.35	2.06	2.31	2.79
Inputs				
Gross domestic income	2.23	2.13	2.36	3.33
Information technology capital services	−2.70	11.51	−10.46	19.41
Computer capital services	−11.71	20.27	−24.81	36.36
Software capital services	−1.83	12.67	−2.04	16.30
Communications equipment capital services	2.18	5.45	−5.90	8.07
Non-information technology capital services	1.53	1.72	2.48	2.94
Labor services	3.02	1.70	3.39	2.18

Note. Average annual percentage growth rates. Source: Jorgenson (2001)

Table 8.2. US Telecommunications Business Service Prices and Demand Growth

	1985-1989	1990-1994	1995-2000
PPI Index for toll service	−18.10	−2.90	−14.10
Business access line	17.61	19.14	26.80
Special access line	118.00	271.00	301.00

Note. Growth in sub-periods is shown. Source: Statistics of Communications Carriers, FCC, (2001/2002), Table 4.10 and Table 5.5.

TFP Trends and Sources

Telecommunications investments' direct contribution to economic growth arises from rapid productivity growth within the sector. Investment in the sector, as a percentage of national investment, and its industry share in GDP has risen in both developed and developing nations (ITU 1999). Also, in most countries telecommunication sector productivity growth is relatively high. For the US economy, Cronin et al. (1993) report an average sector productivity growth of 3% p.a. from

1965 to 1991.[5] Nadiri and Nandi (1999) estimate US telecommunications industry productivity growth magnitudes and sources using a multiple-output, multiple-input cost function (Denny et al. 1981; Fuss and Waverman 1981) with time-series data for 1935 to 1987. The series encompasses both the pre- and post-AT&T divestiture period. This study formulates a structural model to account for both changes in cost and demand for communications service. Conventionally measured TFP growth is decomposed into contributions from aggregate demand, price-cost margin, relative factor price, direct and indirect technological progress and R&D investment to TFP growth.

Table 8.3. US Gross Domestic Product Growth Sources

	Average 1948-99	1948-73	1973-90	1990-95	1995-99
	Outputs				
Gross domestic product	3.46	3.99	2.86	2.36	4.08
Contribution of information technology	0.40	0.20	0.46	0.57	1.18
Computers	0.12	0.04	0.16	0.18	0.36
Software	0.08	0.02	0.09	0.15	0.39
Communications equipment	0.10	0.08	0.10	0.10	0.17
Information technology services	0.10	0.06	0.10	0.15	0.25
Contribution of non-ICT	3.06	3.79	2.40	1.79	2.91
Contribution of non-ICT investment	0.72	1.06	0.34	0.23	0.83
Contribution of non-ICT consumption	2.34	2.73	2.06	1.56	2.08
	Inputs				
Gross domestic income	2.84	3.07	2.61	2.13	3.33
Contribution of ICT capital services	0.34	0.16	0.40	0.48	0.99
Computers	0.15	0.04	0.20	0.22	0.55
Software	0.07	0.02	0.08	0.16	0.29
Communications equipment	0.11	0.10	0.12	0.10	0.14
Contribution of non-ICT capital services	1.36	1.77	1.05	0.61	1.07
Contribution of labor services	1.14	1.13	1.16	1.03	1.27
Total factor productivity	0.61	0.92	0.25	0.24	0.75

Note. Average annual growth rates. The contribution of an output or input is the rate of growth, multiplied by the value share. Special access lines are used by business customers. Source: Jorgenson (2001)

Table 8.4 shows positive telecommunications industry TFP growth since 1938. However, from 1975 to 1983 TFP growth is higher than the preceding period, with an average annual TFP growth of 5.3% p.a. TFP growth declined post-divestiture (1985-1987). Following the method of Nadiri and Prucha (1990), TFP growth is decomposed into the component effects of direct technical change, marginal cost pricing departure and scale. The scale effect is further decomposed into: demand,

---

[5] Jorgenson et al. (1987), Crandall (1991) and Nadiri and Nandi (1999) report similar results.

factor price, quasi-fixed factor, (indirect) technology on scale, and cost elasticity and markup effects. Decomposed average elasticity magnitudes are summarized in Row 3 through Row 11. The estimates clearly indicate the rate of technological change is critical in accounting for approximately 50% of TFP growth. However, this contribution declines substantially post-divestiture. The mark-up contribution to TFP growth arises from the non-marginal cost pricing of telecommunications services. This effect is insubstantial prior to the mid-1970s, but subsequently increased and represents more than 50% of measured TFP growth since 1975. The primary source of this increased impact is through an increased mark-up in toll service that prevailed until end-1980s. Although the introduction of competition to toll service markets substantially reduced prices since the early-1980s, a more rapid decline in the incremental cost of toll services due to technological progress caused the toll service mark-up to increase post-divestiture.

Table 8.4. US Telecommunication Industry Decomposition of TFP Growth 1938-1987

	1938-44	1945-54	1955-64	1965-74	1975-83	1985-87	Average
TFP	3.809	2.201	3.845	4.736	5.351	2.401	3.874
Direct technical change effect	1.813	2.381	2.438	2.471	1.745	0.382	2.061
Markup effect	0.911	−0.444	0.607	1.336	2.500	1.495	0.987
Scale effect	1.085	0.264	0.799	0.929	1.105	0.524	0.827
Source of Scale Effect							
Exogenous demand	1.071	0.190	0.943	0.705	0.763	0.728	0.713
Factor price	−0.095	−0.067	−0.133	−0.108	−0.032	0.130	−0.074
Capital adjustment	0.046	0.024	0.046	0.050	0.048	0.009	0.040
R&D capital adjustment	0.087	0.001	−0.001	0.004	0.021	0.061	0.021
Indirect technical change	0.015	0.076	0.208	0.190	0.201	0.040	0.158
Residual scale effect	−0.179	0.040	−0.263	0.088	0.104	−0.444	−0.061

Note. TFP is average annual growth. Source: Nadiri and Nandi (1999)

The contribution of scale economies to TFP growth for the sample period and selected sub-periods are reported in Row 5 of Table 8.4. The scale effect accounts for approximately 26% of measured TFP growth for 1938-1944, followed by a decline during the period in 1945-1954. However, since 1955 the effect accounts for 25% of TFP. The scale effect component magnitudes are also reported in Table 8.4 (Row 7 through Row 12). Sector TFP growth is principally stimulated due to a shift in production function that result from the rapid growth of ICT stocks, and an indirect technological progress effect. The indirect effect arises from productivity growth leading to reduction in telecommunications service prices that, in turn, stimulate demand and generate scale effects that increases the (traditionally measured) productivity growth. Even though telecommunications use is usually price inelastic, price decline does increase network subscription, and so aggregate demand. As the model allows for both cost and demand impacts on TFP growth, it is capable of capturing any change in aggregate demand as a part of the scale effect. Telecommunications demand may also increase due to macroeconomic variables such as population and income. On average, changes in aggregate demand accounts for 8%-30% of TFP growth (Row 7). For the entire sample, the average per annum contribution of exogenous demand to TFP growth is 0.71%. Further, the reported scale elasticity effects indicate that US telecommunications experienced increasing returns to scale, except for the early sample period.

Dynamic Aspects

Rapid innovation in ICT, increasing telecommunications sector productivity growth and subsequent ICT equipment and service price decline led to substantial change in the structure of production across the US economy. Modern telecommunications play a dual role in influencing economy-wide production processes. For instance, telecommunication equipment and services are production inputs, and are endogenous to firms. Alternatively, the telecommunications network is part of the national infrastructure, i.e., exogenously available to industry, and increases firm productive efficiency. As such, telecommunications infrastructure assists industry to increase operational efficiency by facilitating coordination among producers and by allowing intra- and inter-corporate information flows to occur in a cost-efficient manner. Telecommunications networks are also a support infrastructure to access and utilize new ICT. This exogenously provided network infrastructure is developed on the basis of private and public firm investment decisions.[6]

As production inputs, telecommunications services are potentially substitutes for labor, materials and non-ICT capital. For instance, within the services sector telecommunication services are close substitutes for information-handling labor. As such, their intensity of use of services within the sector has increased substantially. However, with advanced telecommunication networks and efficient computers, manufacturers are also using more telecommunications services in production. That is, telecommunication equipment and services together with other ICT capital (e.g., computers) are facilitating improved inventory and human resource management, decision-making based on instant market data and supply chain control. Antonelli (1990), Cronin et al. (1991), Cronin et al. (1993) and Cronin et al. (1997) examine the substitution and complementarity of telecommunications inputs among other inputs in production by industry. They find telecommunications service is often a substitute for labor and material inputs, and complementary with capital. Antonelli (1990) observed a rising demand for telecommunications services in Italian manufacturing and examined the reasons behind this growth in demand. By differentiating capital input according to their information and non-information components, he finds that telecommunication services are highly complementary with the use of ICT capital.

Cronin et al. (1993) and Cronin et al. (1997) estimate sector average Allen elasticity of substitution among capital, labor, materials and telecommunications for the US economy. They report a positive (13.01) telecommunications substitution

---

[6] US telecommunications infrastructure is primarily private sector funded and owned.

elasticity with capital, a negative elasticity (-0.15) for labor, and a positive elasticity (0.07) for materials. They also observe falling telecommunications price and a high substitution elasticity are associated with substantial change to the input mix of other industry, *viz.*, increased telecommunications service use. Based on US input-output tables (Bureau of Economic Analysis, Department of Commerce), Cronin et al. (1997) infer telecommunications use intensity by industry, and show for 1963 to 1991, that telecommunications service consumption per dollar output increased at 4.3% p.a., whereas labor, capital, material grew by -0.5% p.a., 1.3% p.a. and -0.1% p.a., respectively. They also report that the average industry used US\$ 0.62 of telecommunication service to produce US\$ 100 of output in 1987 compared to US\$ 0.27 to produce the same output in 1965.[7] Further, they report some sectors are more intense in their telecommunications service use, e.g., finance, insurance, and wholesale and retail.[8] As these industries are relatively information-intensive the marginal benefit from increasing telecommunications use is high. Using an input-output approach, Cronin et al. (1993) and Cronin et al. (1997) also estimate the telecommunications-induced resource savings for 1963 to 1991, and report a benefit of US\$ 134.4 billion for the economy due to modernization of the telecommunications system.[9] Further, this study finds the average elasticity of demand of labor for telecommunications infrastructure is negative (-0.0147), while that for capital is positive (0.002).[10]

Spillover or Network Effect Measurement

Rapid telecommunications sector technological progress impacts on firm productivity through improved service quality and service price reduction. Service price reductions encourage substitution of telecommunication services for other inputs and in doing so reduce production costs. Also, a spillover or network effect arises as the telecommunications network comprises part of the national infrastructure. Network effects allow producers to increase output with the same input levels. Such effects are more important with increases in the information intensity of production. Leff (1984) argues network expansion generates cost saving by lowering search costs (and so increasing arbitrage opportunity), improving price distribution information and in diffusing of key technological innovations. Stiroh (2001) examines US manufacturing TFP growth in considering the TFP growth-ICT consumption nexus. The study reports a positive correlation between accelerated labor productivity across industry with ICT use (which includes both ICT and services).

---

[7] Projecting these findings, an average industry is currently using more than US\$ 1 worth of telecommunications services to produce US\$ 100 of output.

[8] According to Cronin et al. (1997) these industries on average spend at least US\$ 1 on telecommunications input for production of US\$ 100 worth of output.

[9] Savings are of primary inputs only.

[10] Since the share of telecommunications expenditure in total cost is small for any industry, the measured elasticity is small.

However, the study is unable to identify whether this outcome reflects higher TFP growth or ICT equipment deepening.[11]

New growth theory explicitly recognizes the contribution of human capital to economic growth (Mankiw et al. 1992). In particular, it is emphasized that human capital is more productive when interacting with the stock of knowledge. Röller and Waverman (1996) note that expansion and modernization of communications infrastructure has a similar impact to increased innovation in that it is formally represented by a production function shift. Further, telecommunications networks generate a consumption effect through the value subscribers place on the size of the subscriber base connected to the network. That is, as subscribers have access to larger a subscriber network the value of subscription is increased. As network subscription benefits are common to subscribers, access exhibits public good characteristics. Also, network use by a subscriber does not diminish such benefit (in the absence of network congestion). Thus, while the network is a public good its use by an individual is as a private good. Additionally, with expansion and modernization, the introduction of new products is facilitated. Accordingly, the utility received from telecommunications service consumption increases (with prices fixed) as the ability to exchange information and knowledge among users is enhanced.

To sum, similar to service provided by public infrastructure, telecommunications infrastructure can be treated as an unpaid input in an industry production process (Aschauer 1989; Munell 1992). That is, telecommunications infrastructure is exogenously given to firms and is independent of industry outputs. Conversely, individual telecommunications service use (with private good characteristics) is considered a utility increasing argument. Thus, in general equilibrium telecommunications network infrastructure reduces cost to production and increases consumer demand for services.

Measurement of Infrastructure Investment Benefit

Nadiri and Nandi (2003) model the effects of telecommunications network infrastructure and the publicly-financed infrastructure capital on the cost structure of 34 US industries. The model is estimated using time-series data for the period 1950-1990. Such infrastructure benefits are in addition to direct benefit from service use. Write industry cost functions as:

$$C = C(q, Y, S_1, S_2, T) , \qquad (8.1)$$

where q denotes the vector of input prices for labor (L), private capital (K) and material (M). Y is output, T is an index representing disembodied technical change, S_1 is the service flow from telecommunications infrastructure and S_2 the

---

[11] Importantly, Stiroh (2001) studies the role of ICT as an input and ignores the role of the ICT network as information infrastructure influencing TFP growth.

public infrastructure service flow.[12] Fees paid for telecommunications services by industry are included in materials payments.[13] While there are no market prices for the externality effect of infrastructure their shadow prices (or willingness-to-pay) can be determined. The shadow values for S_1 and S_2 are measured by the potential cost savings from a decline in variable costs and indirectly capture the network effect of telecommunications infrastructure.

The marginal benefit of infrastructure by industry are:

$$U_{S_1} = -\partial C / \partial S_i \, , \qquad\qquad i = 1, 2 \qquad\qquad (8.2)$$

and show an increase in any infrastructure result in a saving of U_{S_1} of total production cost. These cost reductions are the marginal benefit derived from infrastructure S_1 and S_2 in an industry. Thus, aggregate marginal benefit is obtained by summing the marginal benefits received by industry ($SMBS_1$). Similar estimates of the aggregate marginal benefit from public infrastructure are denoted $SMBS_2$.[14] Further, the average cost elasticity (η_{CS1}) for S_1, excluding telecommunications industry, are shown in Table 8.5. Column 2 shows an increase in telecommunications capital reduces manufacturing and non-manufacturing industry costs. The cost elasticity (η_{CS1}) magnitudes vary considerably by industry. With these estimates and Eq. 8.2, the marginal benefit from telecommunications infrastructure capital by industry is calculated. The calculations show the cost reduction experienced or average marginal benefit (MB) in real terms, i.e., in terms of the change in materials price, by industry in Table 8.5 (Column 6). The benefits indicate an industry willingness to pay for an additional unit of telecommunications infrastructure services. The estimated marginal benefits by industry are positive however, their magnitudes vary considerably. Low marginal benefits are observed, e.g., in Metal Mining, Coal Mining and Non-metallic Mining. Magnitudes are higher for Trade, and Finance, Insurance and Real Estate. These high MB values reflect the relatively high industry information intensity. Column 3 through Column 5 of Table 8.5 report the average input demand elasticity for labor, private capital stock and material inputs. The signs of labor and material input elasticity are negative suggesting substitution with telecommunications infrastructure. Alternatively, the physical capital elasticity suggests a complementary relationship between S_1 and traditional capital. In general, telecommunications investment is

---

[12] S_1 and S_2 are derived by multiplying an infrastructure stock by its utilization. Due to available data, industry utilization rates capture utilization of public and telecommunication infrastructure even though input-specific utilization rates are the more appropriate.

[13] It is implicitly assumed that demand for private inputs fully adjusts to their cost-minimizing levels within a period.

[14] Parameter estimates are available from the authors.

labor- and materials-saving and capital using in most industries.[15]

An increase in the telecommunications infrastructure raises the production efficiency of other industry, and so national productivity. To estimate the contribution
of telecommunications infrastructure to the economy industry level estimates are
aggregated. The average cost elasticity, social marginal benefits and the net rate of
return to telecommunications infrastructure are shown in Table 8.6. The average
aggregate elasticity of cost with respect to the telecommunications infrastructure
(η_{CS1}) is -0.0091. This estimate suggests a percent increase in telecommunications
capital reduces aggregate costs by approximately 0.01%. The net rate of return to
the telecommunications infrastructure capital is calculated by:

$$NRRS_1 = \sum_f (\partial C_f / \partial S_1)/P_{S_1} - \delta \tag{8.3}$$

where $\sum_f \partial C_f / \partial S_1$ is the sum of industry marginal benefit, P_{S1} is the acquisition
price and δ is the rate of telecommunications infrastructure depreciation. Applying Eq. 8.3, the net rate of return to this capital is estimated. Table 8.6 also shows
the values of the sum of industry marginal benefit ($SMBS_1$). The magnitude of
$SMBS_1$ suggests an average marginal benefit from telecommunications infrastructure to the national economy of 0.15 % p.a.. $SMBS_1$ values rose since the 1950s,
and had a value of 0.35 at 1991. The average gross value of S_1 ($GRRS_1$) is 0.26%,
while the average net return ($NRRS_1$) is 0.17. Both $GRRS_1$ and $NRRS_1$ consistently
increased through time. In 1991 they are 0.33 and 0.24, respectively, and suggest a
substantial social rate of return to the telecommunications infrastructure. The total
rate of return, consisting of direct payments received by the telecommunications
firms from industry and the social rate of return on telecommunications infrastructure, is even greater.

Finally, the aggregate elasticity of labor, capital and materials for telecommunications infrastructure are shown in Table 8.6 (Panel B). The average elasticity of
labor (η_{LS1}) is -0.0057, while that for material (η_{MS1}) is negative but larger, i.e., -
0.0147. The capital elasticity (η_{KS1}) is positive but smaller at 0.002. These elasticity estimates suggest telecommunications infrastructure is mostly material saving.
On average, a percent increase in S_1 generates a saving of 0.6% in labor, but increases demand for capital by 0.2% p.a.. Allowing the level of output to vary in
response to reduced costs from increases in S_1, the demand for other production
inputs is affected, since the more output generates demand for other inputs. This
output expansion effect on input demand is addressed. Input elasticity estimates,
including the output expansion and substitution (or complementary) effects, are
shown in Table 8.6 (Panel C). The estimates indicate telecommunications infrastructure increases labor and traditional capital demand, while it is material sav-

---

[15] As firm-specific telecommunications equipment and service expenditure is incorporated
in material cost, the elasticity of substitution between telecommunications as traditional
input (for which industry pays) and other inputs cannot be estimated.

ing. That is, the net effect on the production structure of the economy is biased against material and increases demand for labor and capital.

Table 8.5. US Telecommunications Infrastructure Demand Elasticities by Industry

Industry	η_{CSI}	η_{LSI}	η_{KSI}	η_{MSI}	MBS_1 (%)
Agriculture, Forestry and Fisheries	−0.0104	−0.0062	0.0005	−0.0152	0.0072
Metal Mining	−0.0116	−0.0074	−0.0047	−0.0171	0.0003
Coal Mining	−0.0108	−0.0080	0.0001	−0.0170	0.0007
Crude Petroleum and Natural Gas	−0.0093	−0.0031	−0.0056	−0.0165	0.0028
Nonmetallic Mineral Mining	−0.0119	−0.0081	−0.0044	−0.0175	0.0004
Construction	−0.0091	−0.0061	0.0128	−0.0141	0.0120
Food and Kindred Products	−0.0100	−0.0027	0.0149	−0.0135	0.0088
Tobacco Manufactures	−0.0119	−0.0032	−0.0038	−0.0162	0.0006
Textile Mill products	−0.0111	−0.0045	0.0135	−0.0147	0.0023
Apparel and Other Textile Products	−0.0108	−0.0072	0.0241	−0.0150	0.0022
Lumber and Wood Products	−0.0122	−0.0084	0.0166	−0.0165	0.0021
Furniture and Fixtures	−0.0115	−0.0082	0.0105	−0.0162	0.0010
Paper and Allied Products	−0.0103	−0.0059	0.0022	−0.0148	0.0027
Printing and Publishing	−0.0102	−0.0073	0.0050	−0.0155	0.0029
Chemicals and Allied Products	−0.0100	−0.0045	−0.0016	−0.0145	0.0045
Petroleum Refining	−0.0090	0.0110	0.0155	−0.0122	0.0036
Rubber and Plastic Products	−0.0102	−0.0067	0.0217	−0.0146	0.0028
Leather and Leather Products	−0.0115	−0.0080	−0.0007	−0.0160	0.0005
Stone, Clay and Glass Products	−0.0102	−0.0070	0.0030	−0.0155	0.0016
Primary Metals	−0.0081	−0.0022	0.0132	−0.0119	0.0042
Fabricated Metal Products	−0.0101	−0.0066	0.0058	−0.0149	0.0030
Machinery, Except Electrical	−0.0085	−0.0052	0.0078	−0.0135	0.0047
Electrical Machinery	−0.0097	−0.0065	0.0067	−0.0147	0.0038
Motor Vehicles	−0.0095	−0.0038	0.0103	−0.0134	0.0046
Other Transportation Equipment	−0.0102	−0.0073	0.1605	−0.0150	0.0034
Instruments	−0.0088	−0.0060	0.0049	−0.0146	0.0019
Miscellaneous Manufacturing	−0.0114	−0.0079	0.0069	−0.0160	0.0010
Transportation and Warehousing	−0.0099	−0.0073	0.0012	−0.0165	0.0067
Electric Utilities	−0.0104	−0.0049	−0.0063	−0.0177	0.0033
Gas Utilities	−0.0098	−0.0006	−0.0019	−0.0141	0.0019
Trade	−0.0078	−0.0057	0.0033	−0.0167	0.0181
Finance, Insurance, and Real Estate	−0.0087	−0.0057	−0.0029	−0.0169	0.0136
Other Services	−0.0085	−0.0061	−0.0009	−0.0168	0.0196
Government Enterprises	−0.0107	−0.0078	−0.0030	−0.0177	0.0023

Note. Mean values: 1950-1991. Source: Nadiri and Nandi (2003)

Table 8.6. Telecommunications Capital Elasticity and Averages

Panel A			
η_{CSI}	$SMBS_I$	$GNRRS_I$	$NRRS_I$
−0.0091	0.1514	0.2571	0.1713

Panel B		
η_{LSI}	η_{KSI}	η_{MSI}
−0.0057	0.0020	−0.0147

Panel C		
$\bar{\eta}_{LSI}$	$\bar{\eta}_{KSI}$	$\bar{\eta}_{MSI}$
0.0019	0.0062	−0.0031

Source: Nadiri and Nandi (2003)

Conclusion

ICT contributes to output and productivity growth. There is also evidence that with the recent revolution in telecommunications technology and service innovation that modern telecommunications can contribute to the economic growth of developing countries. This study examines dynamic factors behind the increasing contribution of telecommunications. With industry data, the trend in productivity growth for the US telecommunications industry, sources of this growth and how it contributes economy-wide through spillovers are analyzed. Even though the existence of spillover effects for telecommunications infrastructure is accepted it is measured here. This study develops a method to evaluate telecommunications network effects for US industry. The results suggest industry benefits from network effects, especially service industries that are more information intensive. At an aggregate level, the estimated annual marginal benefit to the national economy is 15%. The estimate suggests a high social rate of return to the telecommunications infrastructure. The impact of the increasing use of telecommunications networks on the demand for traditional inputs is also calculated. Without allowing for output expansion effects, negative impacts on demand for labor and material inputs, and a positive effect on capital demand are found. When output expansion effect is allowed the net effect on demand for both labor and capital is positive, suggesting that on the whole expansion and improvement in telecommunications network do not lead to technological unemployment.

Acknowledgement

The authors thank NBER, Treena Wu and Jin Liu for their assistance.

References

Antonelli C (1990) Information technology and the demand for telecommunications service in the manufacturing industry. Information Economics and Policy 4: 45–55

Aschauer DA (1989) Is public expenditure productive? Journal of Monetary Economics 23: 177–200

Chun H, Nadiri MI (2002) Decomposing productivity growth in the US computer industry. New York University, mimeo

Crandall RW (1991) Efficiency and productivity. In: Cole BG (ed) After the breakup: Assessing the new post-AT&T divestiture era. Columbia University Press, New York, pp 409–18

Cronin FJ, Colleran E, Gold M (1997) Telecommunications, factor substitution and economic growth. Contemporary Economic Policy 15: 21–31

Cronin FJ, Colleran E, Herbert PL, Lewitzky S (1993) Telecommunications and economic growth. Telecommunications Policy 17: 677–90

Cronin FJ, Parker E, Colleran E, Gold M (1991) Telecommunications infrastructure and economic growth: An analysis of causality. Telecommunications Policy 15: 529–35

Denny M, Fuss M, Waverman L (1981) The measurement and interpretation of total factor productivity in regulated industries, with an application to Canadian telecommunications. In: Cowing T, Stevenson M (eds) Productivity measurement in regulated industries. Academic Press, New York, pp 179–218

Federal Communications Commission (2001/2002) Statistics of communications common carriers. FCC, Washington

Fuss MA, Waverman L (1981) Regulation and multi-product firm: The case of telecommunications in Canada. In: Fromm G (ed) Studies in public regulation. MIT, Cambridge

ITU (1999) Telecommunications and economic growth. ITU, Geneva

Jorgenson DW (2001) Information technology and the US economy. American Economic Review 91: 1–32

Jorgenson DW, Gollop FM, Fraumeni BM (1987) Total factor productivity and US economic growth. Harvard University Press, Cambridge

Jorgenson DW, Stiroh KJ (2000) Raising the speed limit: US economic growth in the information age. Brookings Papers on Economic Activity 1: 125–211

Katz ML, Shapiro C (1985) Network externalities, competition and compatibility. American Economic Review 75: 424–40

Leff NH (1984) Externalities information costs and social benefit-cost analysis for economic development: An example from telecommunications. Economic Development and Cultural Change 32: 255–76

Mankiw NG, Romer D, Weil D (1992) A contribution to the empirics of economic growth. Quarterly Journal of Economics 107: 407–37

Mun S-B, Nadiri MI (2002a) Information technology externalities: Empirical evidence from 42 US industries. NBER working paper # 9272

Mun S-B, Nadiri MI (2002b) Computer adjustment costs: Is quality improvement important? New York University, mimeo

Munnell AH (1992) Policy watch: Infrastructure investment and economic growth. Journal of Economic Perspectives 6: 189–98

Nadiri MI, Nandi B (1999) Technical change, markup, divestiture and productivity growth in the US telecommunications industry. Review of Economics and Statistics 81: 488–98

Nadiri MI, Nandi B (2003) Telecommunications infrastructure and economic development. In: Madden G (ed) Traditional telecommunications networks. The international handbook of telecommunications economics volume I. Edward Elgar, Cheltenham, pp 293–314

Nadiri MI, Prucha IR (1990) Dynamic factor demand models, productivity and measurement, and rates of return: Theory and an empirical application to the US Bell System. Structural Change and Economic Dynamics 1: 263–89

Röller LH, Waverman L (1996) The impact of telecommunications infrastructure on economic development. In: Howitt P (ed) The implications of knowledge-based growth for microeconomic policies. University of Calgary Press, Calgary

Stiroh KJ (2001) Are ICT spillovers driving the new economy? Mimeo

Part III: Demand and Pricing

9 Korean Wireless Data Communication Markets and Consumer Technology

Sang-Kyu Byun, Jongsu Lee, Jeong-Dong Lee and Jiwoon Ahn

Introduction

At the end of the 20th century the focus of communication service markets shifted from voice to data. Simultaneously, wireless communication became mobile, spreading globally, and produced an immense economic impact, with external effects impacting on related industry. However, voice service provider revenues declined due to a slowdown in subscription growth and falls in calling charges. Service providers have offered data services based on the mobile communication networks as an alternative (mobile Internet). However, the traditional economic framework is not applicable to this mobile Internet environment since both the volume of data traffic and revenues are slight, as is the rate of increase. Moreover, new wireless technology such as wireless LAN and Bluetooth are developed to further threaten incumbents. Accordingly, uncertainty about the returns to enormous investments in the communications facilities has increased. Further, within the mobile communications industry, the relative importance of data communication as opposed to voice is growing.

This study forecasts the pattern of wireless data communication (WDC) services and proposes effective strategy for service providers and efficient national economic policy. For this purpose the main characteristics of WDC technology and services are identified, and used to estimate consumer preference for attributes, and so enable prediction of likely communication service market evolution. The chapter describes mobile Internet and wireless LAN services, and discusses recent market trends. Common and competing technology used by these services is also discussed. Conjoint analysis design and empirical application are then detailed, and estimation results are presented and analyzed. Finally, conclusions are drawn and policy implications made.

Wireless Data Communication

Several WDC technologies, including the mobile Internet, use mobile communication networks. Wireless LAN offers last-mile service, in the last step of asymmetric digital subscriber line (ADSL) service, while Bluetooth is designed for use for very short-range wireless data exchange. Other WDC technologies are high-speed mobile Internet, broadband wireless local loop, trunked radio system and telephaser. Of these technologies, service providers regard mobile Internet, wireless phaser. Of these technologies, service providers regard mobile Internet, wireless LAN and Bluetooth as the most likely to succeed. This assessment is based on market demand prospects, size of supporting equipment base, spillover to other industry and data transmission efficiency. Mobile communication networks provide the basis for mobile Internet service as it has both a wide customer base and industry penetration. In the early stage of analog technology diffusion, service providers offered voice calling and basic data communication service, such as e-mail and SMS. Introduction of WAP/ME and virtual machine began the era of mobile Internet. Since 2002, service providers have offered 2.5G (GPRS and CDMA2000 1x) services. Recently, commercial 3G services have been launched, to introduce packet systems and improve transmission speed. 3G systems include 1xEV-DO and W-CDMA. Some regions are now provided with color-multimedia WDC service. Meanwhile, wireless LAN transmits and receives data with network interface and LAN cards by setting up ADSL access points (APs). The standardization of wireless LAN has progressed rapidly, leading to rapid wireless-LAN market growth. The IEEE 802 series, in particular, is in a superior position to the European HyperLAN in the global market through its Wi-Fi recognition interoperability between wireless-LAN products. 802.11b leads the market in 2.4GHz ISM (information, science and medical) band, and competition between 802.11a and HiperLAN2 is predicted to unveil itself in the 5 GHz ISM band.

Mobile Internet and Wireless LAN Market Trends

Mobile communication subscribers provide the base market demand for mobile Internet as mobile Internet service is provided through mobile communication networks and handsets. At end-2002, global mobile telephony subscription attained 1,136 million subscribers and 494 million of these subscribers use mobile Internet. Predictions are that by 2007, when service providers are actively offering 3G service, 1,203 million subscribers or 63 % of mobile telephone subscribers will use mobile Internet. However, current mobile Internet services revenue is only US$ 22 billion or 6.7% of mobile communication revenue. Average revenue per user (ARPU) is US$ 53 or 18.3% of mobile service ARPU, and is low compared to registered customers.

Wireless LAN service started as private networks, but soon commercial public wireless LAN service became available through the increased use of ADSL. Increased compatibility with other systems, equipment price reduction and standardization also contributed to rapid wireless LAN diffusion. Public service wireless LAN systems prevail in the US (T-mobile Broadband, Wayport and WiFi Metro), Japan (MIS, NTT Communications and Soft-Bank), and the European Union (Telia in Sweden, Sonera in Finland and British Telecom in the UK). These systems focus on the expansion of hot spots[1]. Gartner Dataquest forecast that the

---

[1] A hot spot is an area where a service provider has set up an AP, typically crowded places such as fast-food restaurants, cafes, railway stations and universities.

wireless LAN equipment sales will increase from 15.47 million units (US$ 2,120 million) to 73.32 million units (US$ 3.940 million) (Gartner Dataquest 2002).

Korean mobile Internet was introduced in 1998. 29.085 million persons have since subscribed, and diffusion reached 89.8% of cellular subscription at end-2002. However, only 45.4% of mobile Internet subscribers use the service once per month, and generate only 8.4% of mobile communication revenue. Public wireless LAN commercial service commenced operation in September 2001. At end-2002, 100,000 customers used the service, which is small compared to that of mobile Internet subscription. KT (Korea Telecom) is the main provider with an 86% market penetration. Other service providers entered the market to compete with KT. KT has constructed 8000 hot spots at end-2002, and is planning to set up 930,000 APs by a further investment of 301,300 million won. In general, an increase of 58.3% per annum is predicted, with an expected 1,570,000 customers by 2007.

Table 9.1. Forecast Subscription to Korean Public Wireless LAN Service (thousand)

Year	2002	2003	2004	2005	2006	2007	Compound Growth
Subscribe	100	322	631	993	1,358	1,572	58.3% p.a.

Source. Data: Wireless Industry Research Team, ETRI (2002)

WDC Service Business Structure

Demand for public wireless LAN service increased rapidly due to the expansion of the ADSL user base and the market for WDC, making the relationship between wireless LAN and existing mobile Internet an issue of policy concern. Early in wireless LAN development it was believed that mobile Internet and wireless LAN would constitute separate markets because of the technologies used. Public wireless LAN service subscription remains much smaller than that of mobile Internet, and little use is made of dual handsets that provide access to WDC services, resulting in little competition. Currently, 73% of PCs are equipped with a wireless LAN card, and 78% of notebook PC's and PDAs are forecast to have similar cards by 2007. At that time it is expected that competition with mobile Internet will start. As wireless LAN service and dual handsets diffuse, competition between 3G mobile Internet and wireless LAN will intensify. Such competition will fall heavily on mobile Internet service providers.

Characteristics of mobile Internet and wireless LAN services, however, might create complementarities. That is, a dual handset owner might use a location based service to watch news when outside a hot spot and switch to acquire this content through wireless LAN when inside. Wireless LAN users obtain service without restriction by location, and mobile Internet users obtain reduced charges. Therefore, subscribers to both services gain utility due to service fusion. Such fusion is likely to continue through provider cooperation since the required investment is enormous at early stages of network deployment. Actually, Korean service providers attempt to offer synthetic services, though the form of synthesis between mobile Internet and wireless LAN is rather elementary. KT, the wire-line leader and second provider of mobile telecommunications, is purchasing synthetic public wireless LAN products and CDMA2000 1xEV-DO. Further, SKT the leading mobile telecommunication provider, and Hanaro the ADSL service provider also have plans to synthesize services with LG Telecom (largest 3G provider).

Competitive WDC Factors

Mobile Internet and wireless LAN have a large potential consumer base. This customer base is large because of the diffusion of handsets containing a browser based on mobile Internet such as WAP/ME/Brew, and functions for short message service and multimedia messaging service.[2] Marginal user costs are small since providers offer mobile Internet as a voice complementary service, and most customers purchase handsets for voice communication. Wireless LAN service potential is large because ADSL subscribers are potential wireless LAN customers.[3] ADSL is readily available and additional customer utility is obtained by accessing the Internet at hot spots.[4] Customer entry costs are small as only a LAN card (or rental fee) and service base rate need to be paid for.

Additionally, both classes of service providers are confronted with rapid change in their market environment. ADSL providers offer wireless LAN service and mobile carriers provide mobile Internet service. Typically, wireless LAN providers face a stagnant market for the existing ADSL, and decreased profit due to mobile Internet carrier entry. Mobile Internet carriers, in turn face gradually declining ARPU from voice service.[5] Accordingly, carriers in seeking new profit sources consider WDC a potential source, and so are investing in mobile Internet and wireless LAN. Finally, both services have common technology characteristics, such as relatively high speed, improved compatibility due to common technology standards, completion of equipment systems, continuous R&D investment and demand creation for WDC services. At least 400 million handsets, the main component of mobile Internet equipment, are made annually, and wireless LAN equipment provides another 15 million units. Completion of supply systems serves as a principal factor behind service diffusion.

---

[2] WAP and ME are mobile Internet browsers, and Brew is a mobile Internet platform.
[3] At end-2002, 26.27 million Korean subscribers use ADSL service—55.1% of the population.
[4] Providers of high-speed Internet offer service by ADSL, VDSL, cable modem and satellite LAN.
[5] Voice ARPU decreased globally due to reduction in calling charges and market stagnation. Korean ARPU fell from 34,952 won to 31,733 won, a 9% fall from 2001 to 2002.

Table 9.2. Mobile-Internet and Wireless LAN Competitive Features

	W-LAN	M-Internet	Common factors
Consumer	ADSL	Mobile	Large base
Provider	ADSL / wire-line provider	Carrier	
Market	Voice by ADSL	Mobile voice	Declining profit
Advantage	Content / speed / price system	Mobility	
Equipment	Notebook PC / PDA	Handset / PDA	Sufficient diffusion
Service	Early stage	Early stage	
Investment	Small	Large	
Technology	802.11b→802.11g→802.11a	2G→2.5G / 3G	
Entry cost	LAN card	Browser /Base rate	Small

Competition by Service Characteristics

A source of competitive advantage mobile Internet provides is mobility. This technology allows seamless transmission of data irrespective of location due to extensive network coverage. Mobile networks consist of hexagonal cells with a base station at the center. Networks have a handoff function that makes continuous connection possible by transferring information to other base stations as users move between cells. Conversely, wireless LAN does not have a handoff function, and users obtain LAN service only at APs. Further, users cannot access LAN services hot spots when movement is rapid. Quality is another source of competitive advantage for mobile Internet. Mobile Internet providers exclusively use frequency bands that Government assigns. Wireless LAN uses the ISM band gratis, leading to service QoS deterioration, e.g., jamming, noise and breaks due to radio interference between wireless LAN or other equipment. However, wireless LAN has an advantage related to transmission speed. The maximum speed of 802.11b is 11 Mbps within 2.4GHz band and 802.11g is over 20 Mbps. Further, the maximum speed 802.11a can obtain in the 5GHz band in the future reaches 54 Mbps. Mobile Internet has a maximum speed of 2.4 Mbps only for latest technology CDMA2000 1xEV-DO, and the expected maximum speed of 1xEV-DO, the technology of the future, is at most 2.4 Mbps to 5.2 Mbps. Further, mobile Internet users with large displays can assess many Internet features as wireless LAN complements ADSL. Mobile Internet users face restrictions including small display and an awkward input apparatus. Currently, 95% of mobile Internet subscribers use handsets that limit content access. Another mobile Internet advantage is low price and a flexible pricing system. Wireless LAN pricing systems are based on a monthly flat system, while mobile Internet pricing is usually metered. In Korea, existing ADSL users can access wireless LAN services by making an additional fixed monthly payment of 16,000 won to 20,000 won. Should consumers not be ADSL subscribers, they can obtain service by paying an additional 25,000 won / month. Alternatively, mobile Internet operators have a circuit pricing system based on time, and a packet pricing system based on data download. The charge is 16 won to 17 won / 10 seconds for the circuit pricing system, while the charge in the packet pricing system is 1.3 won to 6.5 won / packet.[6] When a user downloads a video clip of 2.3Mbs through mobile Internet, 5,980 won must be paid at the minimum price, making mobile Internet access the more expensive. Moreover, consumers gain higher utility from the monthly flat system as they obtain additional utility from unlimited use without extra payment. Such pricing strengthens the competitive position of wireless LAN.

Conjoint Design and WDC Competitive Advantage

This study analyzes patterns of WDC development and changes to market structure through service provider market power. An outcome of this analysis is identification of market strategy and estimates of macroeconomic impact. Collecting revealed preference data for analysis is not feasible since the WDC services market is too early in its development to allow sufficient time series for estimation. Further, service providers are reluctant to reveal commercially sensitive information. The WDC service market is distinguished by the rapid development of content in an intensely competitive environment. As such revealed preference sales data readily indemnifies most preferred products, however, information on the part-worth product attributes is not so readily obtained. Accordingly, a stated preference approach is undertaken. The conjoint approach of Green and Srinivasan (1978) has been applied to communications (Batt and Katz 1997), transportation (Hensher 1994; Calfee et al. 2001), environmental studies (Roe et al. 1996; Alvarez-Farizo and Hanley 2002) and medical care (Slothuus et al. 2002).

Sample Design

The sample is drawn based on the distribution of mobile communication subscription by gender, age and residence. Table 9.3 presents the sample design. 252 respondents are male (50.3%). The age distribution is: 30.1% (151) are 15-19 years, 30.5% (153) are 20-24, 29.3% (147) are 25-29 and 10% (50) are aged over 30 years. Questionnaires are completed by residents of Seoul and its satellites. Seoul is allocated 400 respondents, and both satellites are allocated 50 respondents. 374 respondents (74.7%) have experience using mobile Internet, and 430 respondents (85.8%) use the ADSL service. 313 respondents (62.5%) use wireless LAN service. Clearly, this group of respondents is interested in communication service.

---

[6] In Korea, a packet is 512kbytes.

Table 9.3. Sample Design

Design feature	Details
Population	Seoul, Ilsan and Bundang mobile telephone subscribers
Period	August 26 to September 30, 2002
Sample size	501 respondents
Sampling method	Residential location, gender and age
Data collection method	Interview by questionnaire
Sample error	±4.4% of 95% confidence level

However, as WDC services are only emerging some bias may arise due to respondent lack of familiarity.

Conjoint Questionnaire

To elicit information from respondents the conjoint method uses cards listing product attributes and so provides information on consumer preferences (see the Appendix for sample cards). This information allows calculation of attribute part worth values and willingness-to-pay (WTP) for attribute value changes can be obtained from parameter estimates. Estimation also allows the derivation of consumer valuation of alternative WDC services. Table 9.4 shows that cards contain six attributes: mobility, content, transmission speed, price, price system and quality of communication. Each attribute has levels for mobile Internet and for wireless LAN.

MOBILE is the ability to access mobile Internet whilst in transit. *FIXED* is a communication service, such as wireless LAN that can be used only in a specified zone. *UNLIMITED* means no limit on access to Internet content by the mobile Internet users, viz., an equivalent content as provided by fixed wire Internet is available. *LIMITED* is the limit on accessible Internet content due to equipment speed. Data transmission speeds considered are 2 Mbps (IMT-2000) and 10 Mbps (existing wireless LAN). To aid understanding, respondents are provided with information on download times for different speeds, e.g., 43 minutes and 9 minutes, to download a movie clip (650 Mbytes). Pricing is by metered and monthly flat systems. *METERED* refers to pricing based on time taken or amount of data transfer—the case for mobile Internet. *MONTHLY FLAT* is pricing by a fixed sum irrespective of amount of time taken or amount of data transferred—applies to wireless LAN and ADSL. Price levels are 15 won / 10 seconds for mobile Internet and 1.5 won / 10 seconds—obtained by dividing the additional payment made by an ADSL user for access to the wireless LAN services by the average access time per month (the time unit is 10 seconds). The base rate or entrance fee and a charge for special contents excluded. Finally, *GOOD* and *BAD* quality levels are identified by transmission speed, viz., *GOOD* seldom slows down. An orthogonal plan provides 10 alternatives for the choice experiment.

Table 9.4. Quality Attributes

Attribute	Level
Mobility	Mobile / Fixed
Content	Limited / Unlimited
Speed	2 Mbps / 10 Mbps
Pricing system	Monthly flat / Metered
Communication quality	Good / Bad
Price	1.5 (Won/10s) / 15 (Won/10s)

Empirical Model

To analyze these survey product rank data a type-I Tobit model in which data are censored from 10 above and 1 from below (Tobin 1958; Quester and Green 1982; Amemiya 1985; Roe et al. 1996). When data are not censored the rate on alternative j given by respondent i y_{ij}^* has a monotonic relationship with utility:

$$y_{ij}^* = T(U_{ij})$$

where $T(\cdot)$ is a monotone function, such that $y_{ij}^* > U_{ij} \Leftrightarrow y_{il}^* > U_{il}$ for any j and l. Let vector x_{ij} consist of attributes of alternative j, individual characteristics of respondent i and their interaction terms. Also let y_{ij} be the censored rate, i.e., the rate given by the respondent i to alternative j. The model is:

$$y_{ij}^* = \beta' x_{ij} + \varepsilon_{ij} \qquad \varepsilon_{ij} \sim N(0, \sigma^2) \tag{9.1}$$

and

$$
y_{ij} = \begin{cases} 1 & \text{if} & y_{ij}^* \leq 1 \\ y_{ij}^* & \text{if} & 1 < y_{ij}^* < 10 \\ 10 & \text{if} & y_{ij}^* \geq 10 \end{cases}
\tag{9.2}
$$

where ε_{ij} is an error term. The corresponding log-likelihood function is:

$$
\begin{aligned}
\log L = & \sum_{1 < y_{ij} < 10} -\frac{1}{2}(\log(2\pi) + \log\sigma^2 + \frac{(y_{ij} - \beta' x_{ij})^2}{\sigma^2}) \\
& + \sum_{y_{ij}=1} \log(\Phi(\frac{1 - \beta' x_{ij}}{\sigma})) + \sum_{y_{ij}=10} \log(1 - \Phi(\frac{10 - \beta' x_{ij}}{\sigma}))
\end{aligned}
\tag{9.3}
$$

where Φ and σ are the cumulative density function of the standard normal distribution and standard deviation of ε_{ij}, respectively. The first term on the right-hand side of Eq. 9.3 is the log-likelihood function of uncensored data. The remaining right-hand side terms are the log-likelihood for the area censored by lower bound and upper bound, respectively. These latter bracketed terms are the point-weight probabilities that condense the probability y_{ij}^* belonging to the censored areas. Assuming respondent preferences are transitive allows ratings to be transformed into rankings, and the ordered probit model can be applied to these data (Zavoina and McElvey 1975). Only y_{ij} and not y_{ij}^* is observed. Thus, numbering the ranks from zero through nine, implies:

$$
y_{ij}^* = \beta' x_{ij} + \varepsilon_{ij} \qquad \varepsilon_{ij} \ \square \ N(0,1)
\tag{9.4}
$$

$$
\begin{aligned}
y_{ij} = 0 & \quad \text{if} \quad y_{ij}^* \leq 0 \\
y_{ij} = 1 & \quad \text{if} \quad 0 < y_{ij}^* \leq \mu_1 \\
y_{ij} = 2 & \quad \text{if} \quad \mu_1 < y_{ij}^* \leq \mu_2 \\
& \quad \vdots \\
y_{ij} = 9 & \quad \text{if} \quad y_{ij}^* \geq \mu_8
\end{aligned}
\tag{9.5}
$$

where $0 < \mu_1 < \mu_2 < \cdots < \mu_8$.

For simplicity, assume $\mu_0 = 0$, so that the probability of receiving a particular rank for an alternative is:

$$\text{Prob}(y_{ij} = 0) = \Phi(-\beta' x_{ij})$$
$$\text{Prob}(y_{ij} = 1) = \Phi(\mu_1 - \beta' x_{ij}) - \Phi(-\beta' x_{ij})$$
$$\text{Prob}(y_{ij} = 2) = \Phi(\mu_2 - \beta' x_{ij}) - \Phi(\mu_1 - \beta' x_{ij}) .$$
$$\vdots$$
$$\text{Prob}(y_{ij} = 9) = 1 - \Phi(\mu_8 - \beta' x_{ij})$$

The corresponding log-likelihood function for these probabilities is:

$$\log L = \sum_{y_{ij}=0} \log \Phi(-\beta' x_{ij}) + \sum_{k=1}^{7} (\sum_{y_{ij}=k} \log(\Phi(\mu_k - \beta' x_{ij}) - \Phi(\mu_{k-1} - \beta' x_{ij})))$$
$$+ \sum_{y_{ij}=9} \log(1 - \Phi(\mu_8 - \beta' x_{ij})) .$$

From both type-I tobit estimation and the ordered probit estimation, WTP for attribute k is provided by:

$$\text{WTP}_k = -\frac{\partial U / \partial k}{\partial U / \partial p} = -\frac{\partial T^{-1} / \partial y}{\partial T^{-1} / \partial y} \cdot \frac{\partial y / \partial k}{\partial y / \partial p} = -\frac{\partial y / \partial k}{\partial y / \partial p} . \tag{9.6}$$

Estimation Results

Variables contained in the model are described in Table 9.5. Independent variables and the interaction terms are elements of x_{ij}. Dependent variables *RATE* and *RANK* are the y_{ij} in Eqs. 9.1 and 9.4, respectively. Equation 9.1 and Eq. 9.2, and Eq. 9.3 through Eq. 9.5 are estimated employing the empirical specification:

$$y_{ij} = \beta_M MOBILE_j + \beta_C CONTENT_j + \beta_S SPEED_j + \beta_{PS} PRICES_j$$
$$+ \beta_{C30} CONA30_{ij} + \beta_Q QUALITY_j + \beta_P PRICE_j$$
$$+ \beta_{M30} MOBA30_{ij} + \beta_{S30} SPEA30_{ij} + \beta_{PS30} PA30_{ij}$$
$$+ \beta_{Q30} QUAA30_{ij} + \beta_{P30} PRIA30_{ij} + \varepsilon_{ij} \tag{9.7}$$

where y_{ij} is ratings data *RATE* for type-I tobit estimation and ranking data *RANK* for ordered probit estimation. Type-I tobit and ordered probit estimation results are provided in Table 9.6.

Table 9.5. Variables

	Description	Value
RATE	Respondent rating	= Integer, 1-10
RANK	Respondent ranking	= Integer, 0-9 [a]
MOBILE	WDC service mobility	= 1, if mobile; = 0, otherwise
CONTENT	WDC service content	= 1, if PC Internet; = 0, otherwise
SPEED	Maximum data speed	= 2 Mbps or 10 Mbps
PRICES	Pricing scheme	= 1, if monthly flat; = 0, if metered
QUALITY	Data transmission quality	= 1, if no disconnection; = 0, otherwise
PRICE	Price level	= 1.5 Won / 10 sec or 15 Won / 10 sec [c]
MOBA30 [b]	Mobility and age interaction	= 1, if interaction; = 0, otherwise
CONA30	Content and age interaction	= 1, if interaction; = 0, otherwise
SPEA30	Data speed and age interaction	= 1, if interaction; = 0, otherwise
PSA30	Price scheme and age interaction	= 1, if interaction; = 0, otherwise
QUAA30	Data quality and age interaction	= 1, if interaction; = 0, otherwise
PRIA30	Price level and age interaction	= 1, if interaction; = 0, otherwise

Note. *a*: 0 is most preferable choice and 9 the least. *b*: *A30* is the respondent aged at least 30 years. *c*: US$=1,201 Won at May 28, 2003.

WTP for attributes is calculated by combining Eq. 9.6 with parameter estimates from Eq. 9.7, e.g., sample WTP for *MOBILE* is:

$$WTP_{i,MOBILE} = -\frac{\partial y_{ij}/\partial MOBILE}{\partial y_{ij}/\partial PRICE} = -\frac{\beta_M + \beta_{M30}A30_i}{\beta_P + \beta_{P30}A30_i}. \qquad (9.8)$$

Average WTP for an attribute is the average value of sample WTP. The part worth of an attribute, e.g., *MOBILE*, in terms of respondent *i* utility is:

$$\text{Part Worth}_{MOBILE} = \left| \beta_M(1-0) + \beta_{M30}A30(1-0) \right|. \qquad (9.9)$$

Table 9.6. Estimation Results

Variable	Type-I tobit model			Ordered probit model		
	Estimate	Std. Error	t-value	Estimate	Std. Error	t-value
Constant	2.84*	0.14	20.67	2.34*	0.06	41.40
MOBILE	2.18*	0.09	23.69	-0.77*	0.03	-24.07
CONTENT	1.07*	0.09	11.62	-0.38*	0.03	-11.78
SPEED	0.07*	0.01	6.56	-0.03*	0.00	-6.79
PRICES	0.32*	0.09	3.46	-0.12*	0.03	-3.76
QUALITY	1.84*	0.09	19.92	-0.66*	0.03	-20.83
PRICE	-0.07*	0.01	-10.81	0.03*	0.00	11.19
MOBA30	0.56*	0.27	2.07	-0.23*	0.10	-2.36
CONA30	-0.26	0.27	-0.94	0.09	0.10	0.97
SPEA30	-0.04	0.03	-1.50	0.02	0.01	1.52
PSA30	0.33	0.27	1.22	-0.11	0.10	-1.15
QUAA30	-0.04	0.27	-0.14	0.01	0.10	0.07
PRIA30	-0.01	0.02	-0.378	0.00	0.01	0.42
	Log Likelihood -11240			Log Likelihood -10906		

Business Strategy for Service Providers

Table 9.7 reports attribute part worth, importance and WTP estimates. Importance is the share of attribute part worth to the summed attribute part worth. Both tobit and ordered probit model estimation results reveal similar distributions on attribute importance. *MOBILE* (32%) is most important followed by *QUALITY* (26%), *CONTENT* (15%), *PRICE* (14%), *SPEED* (8%) and *PRICES* (5%). The WTP ordering is the same. Mobile Internet is characterized by almost perfect mobility due to base station placement nationally and hand-off technology. However, this technology is impossible to implement for wireless LAN and will remain so when AP and network coverage increase. This situation is likely to encourage the further development of mobile Internet, thereby increasing its relative importance. However, it remains unclear when mobile Internet will dominate wireless LAN since this technology has a comparative advantage in content variety *CONTENT*, data transmission speed *SPEED* and price level *PRICE*.

QUALITY is the next most important attribute. While improved *MOBILE* is constrained, *QUALITY* can be improved for mobile Internet and wireless LAN through network investment. Wireless LAN service faces QoS issues, e.g., signal and radio interference, and decline in transmission speed as it uses the designated ISM band. To address these wireless LAN problems, the Korean Ministry of Information and Communication issued the *Recommendation for Operation of Wireless LAN* that requires service providers to give service set ID recognition of APs and nominate an arbitrator for problems of frequency interference.[7] Further, another argument considers the 2.3 GHz band, currently allocated to wireless local loop should be reallocated for exclusive use by public wireless LAN. Conversely, mobile communications service providers use a dedicated band that provides the basis for superior QoS. Korean providers have focused investment on network speed and not quality of communication. Low quality is a major source of complaint.

Table 9.7. WDC Service Part Worth, Importance and WTP Estimates

Attribute	Type-I tobit model			Ordered probit model		
	Part worth	Importance (percent)	WTP (Won/10s)	Part worth	Importance (percent)	WTP (Won/10s)
MOBILE	2.24	31.9	30.37	0.79	31.6	30.37
CONTENT	1.04	14.8	14.17	0.37	14.9	14.34
SPEED	0.55	7.9	7.57	0.20	8.0	7.79
PRICES	0.35	5.0	4.71	0.13	5.1	4.91
QUALITY	1.84	26.2	24.95	0.65	26.3	25.31
PRICE	0.99	14.2	-	0.35	14.0	-

---

[7] SSID is an identifier added to a packet header and identifies certain wireless LAN services.

With regard to *CONTENT* respondents appear willing to pay 14 won / 10 seconds for access to unlimited content provided by fixed wire Internet. Thus it is expected that mobile Internet service is inferior to wireless LAN service in terms of content. However, given that the *MOBILE* part worth is more than double that for *CONTENT*, development of content based on mobility appears an effective strategy. In other words, demand for Internet content while mobile is much larger than that for content variety, i.e., WTP for *MOBILE* is much larger than for *CONTENT*. Accordingly, wireless LAN service providers should enlarge their support for content providers to develop mobile Internet content. The high *PRICE* part worth suggests a relatively higher growth of wireless LAN systems when coupled with their relatively low prices. Apart from these attributes, *SPEED* and *PRICES* are advantages for wireless LAN, but are effectively unimportant because of their small WTP. Finally, WDC equipment includes handsets, PDAs and notebook PCs. Currently handsets are used for Internet access, notebook PCs for wireless LAN and PDAs for both. However, dual-mode equipment containing both wireless LAN card and mobile modem is likely become popular when competition or cooperation emerges. The conjoint analysis indicates mobility is more important than content variety. Accordingly, it is expected that handsets and PDAs will gain superiority over notebook PCs.[8] Subscribers access to content with a notebook PC gained popularity because of its large display and calculating power, but notebook PCs cannot satisfy the mobility need. Conversely, handsets and PDAs provide access to limited content due to display size and input device constraints, but have excellent mobility properties.

Economic Impact

When mobile communications infrastructure is upgraded from the CDMA 2000 1x (2.5G) to early 3G, e.g., CDMA2000 1xEV-DO or W-CDMA, maximum data transmission speed will increase from 144 kbps to 2 Mbps. Assuming linearity between speed and WTP, the *SPEED* WTP for an increase from 144 kbps to 2 Mbps is 1.76 won / 10 seconds.[9] For 2002, Korean mobile Internet timed use is 86,096 million seconds and should this volume be maintained, a 15.1 billion won (US$ 12.6 million) revenue increase will result.[10] However, this value is only 1.54% of the 2002 Korean mobile Internet revenue of 984 billion won (US$ 819 million). Thus, based on the *SPEED* WTP estimates, system upgrade investment is

---

[8] Wireless LAN services have advantages in pricing and speed, but mobile Internet service providers can improve these attributes. Conversely, it is difficult for wireless LAN providers to improve their mobility due to equipment and system characteristics.

[9] According to Table 9.7, WTP for speed increase from 2 Mbps to 10 Mbps is 7.57 won / 10 seconds for the tobit model.

[10] Mobile Internet time consists of circuit traffic (84,382 million seconds) and packet traffic (61,679 million packets). Since packet and circuit traffic are summed assuming maximum speed, total time use is calculated at minimum levels. Timed Internet use and its impact will change with price and network improvements.

unlikely to be justified in terms of generated revenue.[11] When the *QUALITY* WTP are summed, approximately 25 won / 10 seconds, for 2002 mobile Internet, consumer welfare increases by 215 billion won (US\$ 179 million) per annum or approximately 22% of mobile Internet revenue. Finally, at present a metered system operates for mobile Internet service, and a switch to a monthly flat system, based on a WTP of 4.7 won / 10 seconds, translates to 40.5 billion won (US\$ 33.7 million) per annum. This gain is additional consumer welfare derived from the free use of mobile Internet under a flat payment system, and serves as a benchmark for firm pricing decisions and Government regulatory policy.

Conclusion

Mobile telecommunications service providers and equipment manufacturers have increased their global revenue by virtue of an explosion in subscription and handset penetration. Currently, mobile Internet is identified as a new business model that has the potential to compensate for the recent decline in voice markets. Owing to rapid IT evolution, several wireless data communications technologies are emerging and they have the potential to compete with mobile Internet, an important factor behind delays in 3G and 4G investment. In this study, the core competency of WDC is examined, especially for mobile Internet and wireless LAN. Econometric estimates of the importance of technology attributes and consumer WTP are provided. Estimation results indicate improved mobility, especially for mobile Internet service is important to consumers. However, mobility alone cannot provide competitive advantage, as the summed WTP (or part worth) for attributes other than mobility for wireless LAN exceed those for mobility. Investment to improve QoS is revealed a more effective strategy. The analysis developed here can be used to examine the WTP of improved wireless data transfer technology that will overcome existing limitations. The model identifies the competitive advantage of potential subscribers' WTP, business prospects and spillover effects to regional economies by using attribute valuations has the potential to direct technology roll out by providing the engineers, entrepreneurs and Government with more detailed information.

---

[11] Mobile network upgrade, however, provides other benefits, e.g., improved efficiency of voice transmission and more content variety—available at higher speed. As such benefits are difficult to quantify, this study focuses on estimated WTP.

Appendix: Sample of Cards Used in the Survey

Card 1	
MOBILITY	Mobile
CONTENT	Limited
SPEED	2 Mbps
PRICES	Metered
QUALITY	Bad
PRICE	1.5(won/10s)

Card 2	
MOBILITY	Fixed
CONTENT	Limited
SPEED	2 Mbps
PRICES	Flat sum
QUALITY	Good
PRICE	15(won/10s)

References

Alvarez-Farizo B, Hanley N (2002) Using conjoint analysis to quantify public preferences over the environmental impacts of wind farms: An example from Spain. Energy Policy 30: 107–16

Amemiya T (1985) Advanced econometrics. Basil Blackwell, Oxford

Batt C, Katz J (1997) A conjoint model of enhanced voice mail services: Implications for new service development and forecasting. Telecommunication Policy 21: 743–60

Calfee J, Winston C, Stempski R (2001) Econometric issues in estimating consumer preferences from stated preference data: A case study of the value of automobile travel time. Review of Economics and Statistics 83: 699–707

Gartner Dataquest (2002) Wireless LAN equipment: Worldwide 2001-2007

Green P, Srinivasan V (1978) Conjoint analysis in marketing: New developments with implications for research and practice. Journal of Marketing 54: 3–19

Hensher DA (1994) Stated preference analysis of travel choices: The state of the practice. Transportation 21: 107–33

OVUM (2002) Mobile @Ovum: Forecasts, OVUM

Quester A, Green W (1982) Divorce risk and wives' labor supply behavior. Social Science Quarterly 63: 730–8

Roe B, Boyle KJ, Teisl MF (1996) Using conjoint analysis to derive estimates of compensating variation. Journal of Environmental Economics and Management 31: 145–59

Slothuus U, Larsen ML, Junker P (2002) The contingent ranking method: A feasible and valid method when eliciting preference for health care? Social Science and Medicine 54: 1601–9

Tobin J (1958) Estimation of relationships for limited dependent variables. Econometrica 26: 314–21

Wireless Industry Research Team (2002) The analysis of the mutual relation between Public Wireless LAN and mobile communication: Competitive or complementary? Electronics and Telecommunications Research Institute, Korea

Zavoina R, McElvey W (1975) A statistical model for the analysis of ordinal level dependent variables. Journal of Mathematical Sociology, Summer: 103–20

10 WTP Analysis of Mobile Internet Demand

Paul Rappoport, Lester D. Taylor and James Alleman

Introduction

Pricing new services is a mostly trial and error process, and wireless Internet access is no exception. Additionally, judging by recent publication, wireless Internet pricing and consumer willingness to pay (WTP) is the subject of much interest. In particular, Maier (2002) indicates wireless Internet provider expectations have not been met by realized market demand. Maier further argues that this disappointing consumer response is caused by complicated service provider pricing schemes. However, CNET News.com considers that while complex pricing is an impediment to market growth, it is not the only problem, viz.

> Mobile phone carriers have sold phones and services for several years that allow subscribers to check sport scores or purchase merchandise online. But not many have. According to various estimates, fewer than two million of the nation's 140 million cell phone owners do any phone-based web surfing. ... Small monochrome screens and the chunky interfaces that make typing a web address into a keypad challenging are partly to blame, but pricing has also been a deterrent. Carriers generally charge by the minute or by the amount of data that is downloaded, significant hang-ups for people used to all-you-can-eat Web access via their home and work PCs [CNET News.com 2002].

Currently, most wireless Internet provider plans price by the volume of data transferred. The wireless industry has adopted a tiered pricing strategy, typically consisting of a fixed monthly access price, including threshold minutes or bytes 'freely' transferred, and additional charges for transfers beyond the threshold.[1] From the consumers' perspective, plans with an endogenous pricing component, either by byte or timed use makes expenditure monitoring difficult. Not surprisingly, consumers favor fixed rate over usage sensitive pricing plans. Accordingly, wireless operators have responded with plans containing both limited and unlimited minute buckets. Plan price ranges from US$ 10 per month for several hundred minutes to US$ 300 for unlimited access. Industry observers believe that when mobile Web access becomes a commodity, competition will force carriers to offer very similar plans (News.com 2003).

Within this context, this study examines consumers' WTP for wireless Internet access.[2] In particular, the chapter provides estimates of the price elasticity of demand for mobile (wireless) Internet access. To do so, the analysis focuses on wireless Internet services that are accessible via a cellular telephone or personal digital

---

[1] See, e.g., Verizon Wireless' Express Network Plans, http://www.verizonwireless.com.

[2] It is assumed here that the price observed by consumers is a fixed price for an unspecified bundle of bytes or minutes.

assistant device (PDA).[3] Model estimation employs data obtained from an omnibus survey, conducted in March 2003 that contains respondent WTP for wireless Internet access information.[4] Apart from the elasticity estimates, the study also finds: a threshold effect at the monthly US$ 50 price; younger household heads are more likely to subscribe to wireless Internet; households with Internet access—especially with broadband access—have a higher WTP for wireless Internet access; households that have recently relocated have a higher WTP; and households with a personal communications services (PCS) telephone have a higher WTP. The following section compares the relationship between age and income and the market penetration of PCS (cellular) telephony, any Internet access, broadband Internet access and PCS Internet ready telephones. A description of the access/use framework, is given, and followed by the presentation a two-stage probit regression model that attempts to explain consumer WTP for wireless Internet access. Next, price elasticity estimates for wireless Internet access are calculated from kernel-smoothed cumulative WTP distributions. Conclusions are then given.

Descriptive Analysis

Wireless Internet access demand is the union of broadband Internet access and PCS (or cellular) service demand. To obtain a preliminary subscriber profile for this emerging market, the age and income distributions for subscribers are examined. Figure 10.1 displays the sample relationship between subscriber income and PCS telephone, any Internet access, broadband Internet access and wireless Internet access penetrations.[5] Not surprisingly, service penetration rates monotonically increase with income. Further, this group of services has above mean penetration for subscribers with an income greater than US$ 50,000 p.a. Casual observation suggests that the broadband market has the strongest income effect. Corresponding service penetration rates, as a function of age, are presented in Fig. 10.2. As expected, penetration rates for PCS telephony and the Internet decline with age. Wireless Internet access subscription is highest for the 21-30 age group (a point noted by providers), and highest in the 31-40 age group for the broadband market.

---

[3] Wireless WiFi Internet packages for access to the Internet in designated 'hot spots' are not considered.

[4] The authors acknowledge the Management Systems Group for access to their omnibus survey. Available at: www.m-s-g.com.

[5] Series are normalized at their means and expressed in terms of an index. An index value of 100 is the average across income groups. A value of 150 means this groups penetration is 50% higher than market penetration.

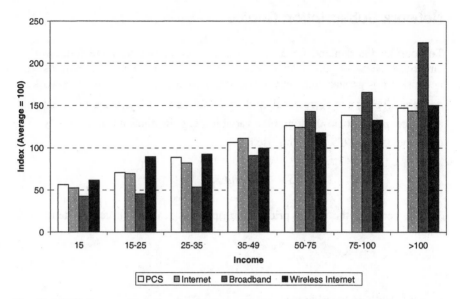

Fig. 10.1. PCS, Internet, Broadband and Wireless Internet Penetration by Income

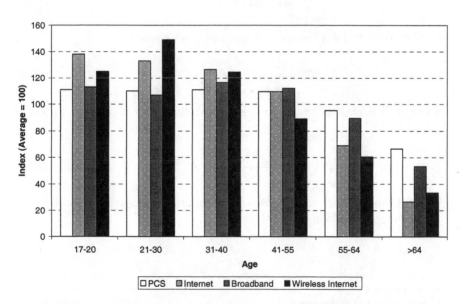

Fig. 10.2. PCS, Internet, Broadband and Wireless Internet Penetration by Respondent Age

Network Subscription Choice

To consider the demand for access to a network an access/usage framework is employed. Network subscription occurs when subscriber consumer surplus from network use is greater than network subscription price.[6] Let q denote network usage, and $q(p, y)$ the corresponding demand for network use, conditional on a usage price p and socio-demographic variables, e.g., income and education, y. The consumer surplus CS from this use is

$$CS = \int q(z, y)dz .$$
(10.1)

With π denoting the access price, subscription is realized (access is demanded) when

$$CS \geq \pi ,$$
(10.2)

or equivalently in logarithms if

$$\ln CS \geq \ln \pi .$$
(10.3)

Perl (1983) applied this framework empirically, by assuming:[7]

$$CS = \int_p^\infty A e^{-az} y^\beta e^u dz$$
(10.4)

where u is a random error term with distribution $g(u)$. Consumer surplus is given by

$$CS = \frac{A e^{-\alpha p} y^\beta e^u}{\alpha} .$$
(10.5)

With net benefits from network use and access price expressed in logarithms, the condition for subscribing to the network is:

---

[6] See Chapter 2 of Taylor (1994).

[7] Since its introduction by Perl in 1978 in an earlier version of his 1983 paper, this function has been used extensively in the analysis of telecommunications access demand (see, e.g., Kridel 1988 and Taylor and Kridel 1990). The great attraction of this demand function is its non-linearity in income, and an ability to handle both zero and non-zero usage prices.

$$P(\ln CS \geq \ln \pi) = P(u \geq \ln \pi - a + \alpha p - \beta \ln y) \qquad (10.6)$$

where $a = \ln(A/\alpha)$. To specify a probability law for CS, which, in view of Eq. 10.6, is reduced to the specification for the distribution of u in the use demand function. Assuming u is distributed normally leads to a simple probit model. Alternatively, assuming u is logistic implies to a logit model. Empirical studies employing both approaches abound in the literature.[8]

To estimate access demand usually requires that consumer surplus from network use is obtained by estimating a use demand function, and then performing integration beneath it. In the present context the procedure is reversed as information obtained from respondents in the survey are statements indicating their maximum WTP for wireless Internet access. Such statements represent, at least in principle, the maximum price at which a respondent will subscribe to the network. Thus, access is realized for a WTP value that is equal or greater than the network access price π^*. Conversely, no subscription will occur for any WTP less than π^*. Hence, implicit in the WTP distribution is an aggregate demand function, or more precisely, penetration function, for wireless Internet access. In particular, the function is

$$\begin{aligned} D(\pi) &= P(WTP \geq \pi) \\ &= 1 - CDF(\pi) \;, \end{aligned} \qquad (10.7)$$

where $CDF(\pi)$ denotes the WTP cumulative distribution function. Once the WTP CDFs are constructed, price elasticity estimates are obtained, without intervention of the demand function, via:

$$\text{Elasticity}(\pi) = \frac{d \ln CDF(\pi)}{d \ln \pi} \;. \qquad (10.8)$$

Modeling WTP for Wireless Internet Access

Information on WTP for wireless Internet access is collected from an omnibus national survey of 3000 US households in early-March, 2003. The omnibus survey is

---

[8] Perl (1983) and Taylor and Kridel (1990) are among econometric studies that employ a probit specification. Studies using the logit framework include Train et al. (1987) and Bodnar et al. (1988). Most empirical studies of telecommunications access demand that employ the consumer-surplus framework focus on local use, and so ignore the net benefit from toll use. Hausman et al. (1993) and Erikson et al. (1998) are exceptions.

a continuous random telephone survey.[9] Participants are asked either:

(a) What is the most you would be willing to pay on a monthly basis for wireless access to the Internet? or

(b) What is the highest monthly price at which you would consider purchasing wireless access to the Internet?

A total of 3,014 households responded to the survey. However, after eliminating for question non-response, together with refusal to provide information on income or other socio-demographic variables, reduced the number of usable responses to 1,487. Exclusion of responses of 'zero' for WTP further reduces the data set to 1,174 observations from which to construct the WTP CDFs. The resulting WTP frequency distribution for wireless access is given by Appendix Table 1. WTP values range from US\$ 1 through to US\$ 184, with a sample mean of US\$ 28.17 and standard deviation of US\$ 15.92. Not surprisingly, responses are concentrated at WTP values divisible by US\$ 5, viz., US\$ 5, US\$ 10 and US\$ 15. Moreover, for several respondents, the values elicited appear to be prices currently paid for PCS access or Internet service, e.g., US\$ 39.95, US\$ 39.99 and US\$ 45.95.

An interesting question is the extent to which WTP, that in principle represents areas beneath demand network use curves, can be explained by the determinants of demand, i.e., prices and income. To explore this question, reconsider the expression for consumer surplus of Eq. 10.6, in logarithmic form:

$$\ln CS = f(p, y, x, u) \qquad (10.9)$$

where p, y, x and u denote the usage price, income, socio-demographic variables and an error term, respectively. However, Eq. 10.9 cannot be estimated as the dependent variable is not defined for zero WTP values. [10] To define the dependent variable, all observations with zero WTP values are removed from the sample. In doing so a two-stage procedure, in which a discrete-choice probit model is estimated to explain zero and non-zero WTP values. The corresponding inverse Mills ratio is constructed and used as a correction term in a second-stage model, with ln *WTP* regressed on income and other socio-demographic variables.[11] Variables are presented in Table 10.1 and results of first-stage probit estimation are contained in Table 10.2.

---

[9] www.Centris.com

[10] Since we are considering a fixed price for wireless Internet access, implying usage is bundled with access, hence WTP and consumer surplus coincide.

[11] Treating zero and non-zero WTP values equally in building a CDF implies 100% penetration at zero access prices. To correct, specify a model to explain WTP non-zero likelihood, and add this correction term in to explain non-zero WTP values, penetration is determined only by households that value wireless Internet access.

Table 10.1. Variables

	Description	Value
GENDER	Respondent gender	=1, if male; =0, otherwise
NOT HIGH	Graduated from high school	=1, if not graduated; =0, otherwise
COLLEGE	Graduated from college	=1, if graduated; =0, otherwise
TECH	Attended technical school	=1, if attended; =0, otherwise
AGE	Respondent age	=Integer
AGE20	Respondent age less than 20	=1, if Age<20; =0, otherwise
AGE29	Respondent age 20 to 29	=1, if 20<Age<29; =0, otherwise
AGE39	Respondent age 30 to 39	=1, if 30<Age<39; =0, otherwise
AGE49	Respondent age 40 to 49	=1, if 40<Age<49; =0, otherwise
AGE64	Respondent age 50 to 64	=1, if 50<Age<64; =0, otherwise
AGE64+	Respondent age over 64	=1, if Age>64; =0, otherwise
INC15	Income less than US$ 15K [a]	=1, if Income<15K; =0, otherwise
INC25	Income US$ 15K to US$ 25K	=1, if 15<Income<25K; =0, otherwise
INC35	Income US$ 25K to US$ 35K	=1, if 25<Income<35K; =0, otherwise
INC50	Income US$ 35K to US$ 50K	=1, if 35<Income<50K; =0, otherwise
INC75	Income US$ 50K to US$ 75K	=1, if 50<Income<75K; =0, otherwise
INC100	Income US$ 75K to US$ 100K	=1, if 75<Income<100K; =0, otherwise
INC100+	Income greater than US$ 100K	=1, if Income>100K; =0, otherwise
HOUSE	Persons reside in household	=Integer, 1-7 [b]
ADULTS	Adults reside in household	=Integer
CHILD11	Children aged 6-11	=Integer
CHILD17	Children aged 12-17	=Integer
INTERNET	Household Internet access	=1, if access; =0, otherwise
BROAD	Household broadband access	=1, if access; =0, otherwise
PCS	Household PCS telephone	=1, if phone; =0, otherwise
COMP	Household computer	=1, if computer; =0, otherwise
PAGER	Household pager	=1, if pager; =0, otherwise
PURCH	On-line purchase last 30 days	=1, if purchase; =0, otherwise
MOVED	Moved in last 6 months	=1, if moved; =0, otherwise
MILLS	Mill's ratio	=Continuous

Note. *a:* US$ K denotes thousand US dollars. *b:* Household>7, recorded as 7.

Table 10.2. First-stage Probit Estimation Results

Variable	Wald Statistic	Prob > χ^2
GENDER	2.58	0.11
NOT HIGH	0.24	0.62
COLLEGE	0.62	0.43
TECH	0.86	0.35
AGE	0.58	0.45
AGE20	1.52	0.22
AGE29	0.49	0.48
AGE39	0.90	0.34
AGE49	1.63	0.20
AGE64	1.21	0.27
INC25	0.04	0.84
INC35	1.75	0.19
INC50	1.30	0.25
INC75	3.24	0.07
INC100	1.12	0.29
INC100+	1.14	0.29
HOUSE	5.42	0.02
ADULTS	1.46	0.23
CHILD11	0.05	0.83
CHILD17	0.32	0.57
INTERNET	2.23	0.14
BROAD	1.26	0.26
PCS	3.54	0.06
COMP	4.32	0.04
PAGER	0.26	0.61
PURCH	0.11	0.74
MOVED	2.05	0.15

From Table 10.2, the importance of possessing a computer, cellular telephone (PCS telephone) and having a high income are apparent. Aside from household size, other demographic variables such as age, education and gender are relatively unimportant in identifying households with non-zero WTP values for wireless Internet access. The estimated coefficients, standard errors, t-ratios and p-statistics for the second-stage WTP models for wireless Internet access are displayed in Table 10.3. The dependent variable in this model is ln*WTP*. The independent variables include those that appearing in the first-stage model, plus the first-stage Mills ratios.[12]

The results for second-stage estimation suggest that WTP is increasing in household income and decreasing in respondent age, viz., WTP is higher for younger respondents. Households with Internet access have higher WTP values for wireless Internet. Not surprisingly, households with broadband also have higher WTP values. The Mills ratio is significant, and indicates that the two-stage estimation procedure is appropriate.

Table 10.3. WTP Model Estimates

Variable	Coefficient	Std Error	t-value	P-value
MILLS	0.49	0.19	2.61	0.01
INTERCEPT	2.83	0.07	43.2	0.00
AGE20	0.81	0.15	5.26	0.00
AGE29	0.39	0.07	5.16	0.00
AGE39	0.26	0.06	3.36	0.01
AGE49	0.21	0.06	3.36	0.01
INC25	-0.01	0.07	-0.10	0.92
INC35	0.19	0.07	2.78	0.01
INC50	0.12	0.06	1.87	0.06
INC75	0.20	0.07	2.83	0.01
INC100	0.19	0.07	2.66	0.01
INC100+	0.18	0.07	2.48	0.01
HOUSE	0.23	0.07	3.55	0.00
ADULTS	-0.17	0.06	-2.63	0.01
INTERNET	0.29	0.06	4.66	0.00
BROAD	0.19	0.05	3.80	0.00
PCS	0.18	0.05	3.89	0.00
PURCH	0.08	0.00	1.86	0.06
MOVED	0.19	0.05	3.69	0.00

[12] Mills ratios correct for the fact that, as the second-stage model is interpreted as the conditional expectation of WTP given WTP is positive, the error-term is 'drawn' from a truncated distribution, and so does not have a zero mean. Mills ratios are calculated by the rule $n(\pi_i)/N(\pi_i)$, where π_i is the predicted value of the probability respondent WTP value is non-zero, and $n(\pi_i)$ and $N(\pi_i)$ represent the standard normal density and cumulative distribution functions (Maddala 1986: Chapter 6).

Price Elasticity Calculations

WTP CDFs provide measures of market penetration, and their price elasticity estimates are obtained through Eq. 10.8. The elasticity estimates are derived from a non-parametric approximation to the empirical CDFs shown in Fig 10.3.

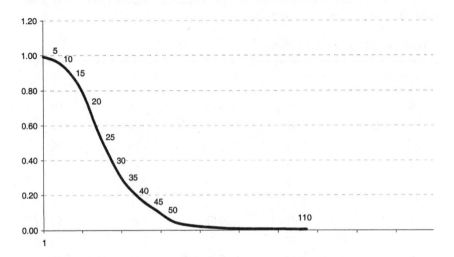

Fig. 10.3. WTP CDF

To approximate the CDF, kernel density functions (using the unit normal density function as the kernel weighting factor) are estimated as approximations to the empirical WTP frequency distributions. These empirical approximations are used to calculate elasticity estimates at several points. The goal of kernel estimation is to develop a continuous approximation to an empirical frequency distribution that can be used to assign density, in a statistically valid manner, in a small neighborhood for an observed frequency point. For example, from these data, there are 90 observations at a WTP of US$ 50 (treating US$ 49.95 as US$ 50), but none at US$ 51, six at US$ 55 and four at US$ 49. Such an uneven observation distribution makes calculation of arc elasticity estimates in a neighborhood of US$ 50 problematic.

Further, there is no compelling reason to believe, in such a large population, that concentrations of WTP values divisible by US$ 5 reflects anything other than 'convenient' rounding by respondents. Accordingly, there is no reason to accept that true WTP density values of US$ 49 or US$ 51 are much different than the density at US$ 50. An intuitive way to deal with this contingency, i.e., concentrations of observations at particular discrete points, is to tabulate frequencies within intervals and calculate density as frequency within an interval divided by the length of the interval, i.e., as an average within an interval. However, in doing so the 'density' within an interval is calculated using only observations within that interval. That is, assuming, e.g., an interval is from US$ 40 to US$ 45, then a WTP of US$ 46 (which is as 'close' to US$ 45 as is US$ 44) will not be given weight in calculating the density for that interval. Kernel density estimation allows every observation to have weight in the calculation of intervals. However, this weight varies inversely with the distance the observations lies from the center of the interval.

Let $\hat{g}(x)$ denote the density function to be constructed for the random variable x. x can take any value from x_1 through to x_n. For wireless Internet WTP values, e.g., the range [x_1, x_n] corresponds to [US$ 1, US$ 184]. Next, divide this interval or support into k sub-intervals. $\hat{g}(x)$ is then constructed as:

$$\hat{g}(x_i) = \sum_{j=1}^{N} K(\frac{x_i - x_j}{h}) \frac{1}{Nh} . \qquad i = 1, ..., k \qquad (10.10)$$

where K denotes the kernel-weighting function, h a smoothing parameter and N the number of observations. The density function in Eq. 10.10 is constructed using the unit normal density as the kernel weighting function and a support of $k = 1000$ intervals.[13]

With estimation of the kernel density functions, wireless Internet WTP price elasticity estimates are calculated from the associated CDFs using a numerical counterpart to Eq. 10.8. The resulting elasticity estimates, calculated for WTPs of US$ 10, US$ 20, US$ 30, US$ 40, US$ 50, US$ 60 and US$ 70, are presented in Table 10.4.[14] Table 10.4 reports elasticity estimates that range mostly from -0.2 to -3.0. However, since half market penetration occurs for WTP values of US$ 30 or less, the most relevant elasticity estimates range in value from -1.0 to -2.0. This range of elasticity values is similar to those reported for broadband Internet access demand.[15]

---

[13] Silverman's (1986) rule, $\hat{h}$ = 0.9 minimum of [standard deviation, inter-quartile range/1.34]($N^{-0.2}$), is used to smooth the parameter h. References for kernel density estimation are Silverman (1986) and Wand and Jones (1995). Ker and Goodwin (2000) provide an application to estimate crop insurance rates.

[14] The values for the cumulative densities at US$ 10, US$ 20 and so on, are taken from the CDFs for the empirical frequency distributions in Table 10.2 and Appendix Table 1. The elasticity estimates are calculated using corresponding points on the kernel CDFs via:

$$Elasticity(x) = \frac{\Delta CDF(x)/CDF(x)}{\Delta WTP(x)/WTP(x)},$$

with x the smallest value on the kernel CDF greater than x, for x = US$ 10, ..., US$ 70.

[15] See, e.g., Rappoport et al. (1998, 2002, 2003a, b) and Kridel et al. (1999, 2002).

Table 10.4. Wireless Internet Price Elasticity Estimates

Price (US$)	Elasticity Value
10	-0.20
20	-1.07
30	-2.13
40	-2.77
50	-3.05
60	-3.17
70	-3.28

Conclusions

This chapter analyzes wireless Internet access demand using information on WTP collected from an early-2003 omnibus survey of 3,000 US households. A theoretical framework is utilized to identify WTP for consumer surplus from network use. The approach allows WTP values to be modeled as a function of socio-demographic variables. Results suggest that demand is elastic for prices currently charged by wireless service providers. Determinants of demand are consistent with those for Internet and broadband access. Age of the household head and household income are important determinants in the WTP function. Indicators of Internet use include the presence of a PCS telephone and personal computer within the household. Additionally, WTP is strongly related to whether the household has been recently relocated. Further, the WTP frequency distribution, displayed in Appendix Table 1, indicates many respondents are willing to pay at least US$ 50 for wireless access. Finally, concentrations of WTP values at US$ 20, US$ 25 and US$ 30 suggest that wireless providers should vary their plans to better meet the needs of wireless customers. The study does not examine consumer WTP for services provided on a usage sensitive basis, e.g., by the number of messages or bytes transferred. Low penetration for wireless Internet access suggests that plans that price by usage are not readily accepted by consumers. In addition, this study is silent on issues related to the WTP for combined wireless voice and data access. Finally, by modifying Cramer's (1969) procedure, it has been possible to obtain non-parametric price elasticity estimates for wireless Internet access from the frequency distributions of the WTP. The elasticity estimates obtained range from -0.2 (inelastic) to −3.0 (elastic). However, the estimated elasticity values, for the price range of US$ 20 through US$ 35, accord with received estimates for broadband access demand. A caveat to the ready acceptance of these elasticity estimates at face value is that they are effectively obtained from contingent valuation data. [16] However, several aspects of the estimates add credence to the results.

---

[16] The critical literature on the contingent valuation method is well represented by the NOAA Panel Report (1993), Smith (1993), Diamond and Hausman (1994), Hanneman

First, the reported elasticity magnitudes are plausible. Second, the services being dealt with are, more or less, familiar to respondents—unlike those for circumstances where there is no meaningful market-based valuation. Finally, the results reported are encouraging that, in that the question eliciting WTP are posed in different forms, no apparent framing differences are found. To conclude, this study shows that demand analysis involving contingent valuation data, in circumstances in which the products are familiar and well defined, is a fruitful approach.

Appendix

Appendix Table 1. WTP (US$) Frequency Distribution

WTP	Respondents	Percent	WTP	Respondents	Percent
1.00	1174	100.00	24.00	649	55.28
2.00	1161	98.89	24.99	648	55.20
3.00	1160	98.81	25.00	646	55.03
5.00	1157	98.55	28.00	526	44.80
5.95	1144	97.44	29.00	524	44.63
6.00	1143	97.36	29.95	513	43.70
8.00	1142	97.27	29.99	510	43.44
9.00	1140	97.10	30.00	507	43.19
9.95	1139	97.02	32.00	323	27.51
9.99	1127	96.00	34.95	322	27.43
10.00	1124	95.74	35.00	320	27.26
10.99	1053	89.69	36.00	265	22.57
12.00	1052	89.61	39.00	264	22.49
12.50	1047	89.18	39.95	256	21.81
13.00	1046	89.10	39.99	255	21.72
14.00	1044	88.93	40.00	253	21.55
14.95	1041	88.67	42.00	163	13.88
14.99	1040	88.59	45.00	162	13.80
15.00	1039	88.50	48.00	141	12.01
15.99	970	82.62	49.00	140	11.93
16.00	968	82.45	49.95	136	11.58
17.00	966	82.28	50.00	135	11.50
18.00	965	82.20	55.00	46	3.92
19.00	964	82.11	59.00	41	3.49
19.95	955	81.35	60.00	40	3.41
19.98	943	80.32	65.00	24	2.04
19.99	942	80.24	67.00	22	1.87
20.00	936	79.73	70.00	21	1.79
20.99	667	56.81	75.00	14	1.19
21.00	666	56.73	80.00	10	0.85
21.95	661	56.30	89.99	8	0.68
22.00	659	56.13	100.00	7	0.60
22.99	654	55.71	110.00	5	0.43
23.00	653	55.62	150.00	4	0.34
23.95	650	55.37	184.00	1	0.09

(1994), McFadden (1994) and Portnoy (1994). Alternatively, particularly successful uses of contingent valuation data include Hammitt (1986) and Kridel (1988).

References

Bodnar J, Dilworth P, Iacono S (1988) Cross-section analysis of residential telephone subscription in Canada. Information Economics and Policy 3(4): 311–31

CNET News.com (2002) At: http://zdnet.com.com/2100-1105-954614.html

Cramer JS (1969) Empirical econometrics. Elsevier, New York

Diamond PA, Hausman JA (1994) Contingent valuation: Is some number better than no number? Journal of Economic Perspectives 8(4): 45–64

Erikson RC, Kaserman DL, Mayo JW (1998) Targeted and untargeted subsidy schemes: Evidence from post-divestiture efforts to promote universal service. Journal of Law and Economics 41(2): 477–502

Hammitt JK (1986) Estimating consumer willingness to pay to reduce food borne risk. Report # R-3447-EPA, RAND Corporation, Santa Monica

Hannemann WM (1994) Valuing the environment though contingent valuation. Journal of Economic Perspectives 8(4): 19–44

Hausman JA, Tardiff TJ, Bellinfonte A (1993) The effects of the breakup of AT&T on telephone penetration in the United States. American Economic Review 83(2): 178–84

Ker AP, Goodwin BK (2000) Non-parametric estimation of crop insurance and rates revisited. American Journal of Agricultural Economics 83(2): 463–78

Kridel DJ (1988) A consumer surplus approach to predicting extended area service (EAS) development and stimulation rates. Information Economics and Policy 3(4): 379–90

Kridel DJ, Rappoport PN, Taylor LD (1999) An econometric study of the demand for access to the Internet. In: Loomis DG, Taylor LD (eds) The future of the telecommunications industry: Forecasting and demand analysis. Kluwer Academic Publishers, Boston

Kridel DJ, Rappoport PN, Taylor LD (2002) The demand for high-speed access to the Internet: The case of cable modems. In: Loomis DG, Taylor LD (eds) Forecasting the Internet: Understanding the explosive growth of data communications. Kluwer Academic Publishers, Boston

Maddala GS (1986) Limited-dependent and qualitative variables in econometrics. Cambridge University Press, Cambridge

Maier G (2002) At: http://www.business2.com/articles/web/0,1653,45377,00.html.

McFadden D (1994) Contingent valuation and social choice. American Journal of Agricultural Economics 76: 689–708

National Oceanographic and Atmospheric Administration (NOAA) (1993) 58, Federal Register, 4601, January 15

News.com (2003) At: http://news.com.com/2100-1039-9971178.html?part=dht&tag=ntop, March 5

Perl LJ (1978) Economic and demographic determinants for basic telephone service. National Economic Research Associates, White Plains, March 28

Perl LJ (1983) Residential demand for telephone service 1983. Prepared for the Central Service Organization of the Bell Operating Companies, Inc., National Economic Research Associates, White Plains, December

Portnoy PR (1994) The contingent valuation debate: Why economists should care. Journal of Economic Perspectives 8(4): 3–18

Rappoport PN, Kridel DJ, Taylor LD, Alleman J, Duffy-Deno K (2003a) Residential demand for access to the Internet. In: Madden G (ed) Emerging telecommunications networks: The international handbook of telecommunications economics, vol II. Edward Elgar, Cheltenham

Rappoport PN, Taylor LD, Kridel DJ (2003b) Willingness-to-pay and the demand for broadband access. In: Shampine A (ed) Down to the wire: Studies in the diffusion and regulation of telecommunications technologies. Nova Science Publishers, New York

Rappoport PN, Taylor LD, Kridel DJ (2002) The demand for broadband: Access, content and the value of time. In: Crandall RW, Alleman JH (eds) Broadband: Should we regulate high-speed Internet access? AEI-Brookings Joint Center for Regulatory Studies, Washington

Rappoport PN, Taylor LD, Kridel DJ, Serad W (1998) The demand for Internet and on-line access. In: Bohlin E, Levin S (eds) Telecommunications transformation: Technology, strategy and policy. IOS Press, Amsterdam

Silverman BW (1986) Density estimation for statistics and data analysis. Monographs on Statistics and Applied Probability 26, Chapman and Hall, London

Smith VK (1993) Non-market valuation of natural resources: An interpretive appraisal. Land Economics 69(1): 1–26

Taylor LD (1994) Telecommunications demand in theory and practice. Kluwer Academic Publishers, Dordrecht

Taylor LD, Kridel DJ (1990) Residential demand for access to the telephone network. In: de Fontenay A, Shugard MH, Sibley DS (eds) Telecommunications demand modeling: An integrated view. North-Holland, Amsterdam

Train KE, McFadden DL, Ben-Akiva M (1987) The demand for local telephone service: A fully discrete model of residential calling patterns and service choices. Rand Journal of Economics 18(1): 109–23

Wand MP, Jones MC (1995) Kernel smoothing. Monographs on Statistics and Applied Probability 60. Chapman and Hall, London

11 Asymmetry in Pricing Information Goods

Yong-Yeop Sohn

Introduction

Firms producing information goods that exhibit a network externality often adopt an introductory pricing strategy. The argument is that to secure a critical mass or installed base of customers, a firm sets price lower than marginal cost at the time of introducing an information good. A salient example is the web browser, Navigator, which could be freely downloaded in the early-1990s. During this period, economists identified the existence of introductory pricing strategies by firms within information good markets.[1] For example, Bensaid and Lense (1996) develop a monopoly information good producer model to consider profit maximizing price setting behavior. Low initial prices are set, with price increasing later when a learning-by-doing network externality adds further benefit attracting new consumers to the market. Additionally, Cabral et al. (1999), using a dynamic model, show that duopoly firms may choose an introductory pricing strategy.

As powerful as such demonstrations are, many examples of information goods whose prices fall through time are documented. For instance, the price of software programs such as word processing packages are initially set high but typically fall as the upgrade version release nears. Further, the Hangul word processor was available for free in the early-1990s when Hangul entered the Korean market, but the upgraded version HWP3.0 sold at a substantial price and was protected from illegal copying. As such, it seems reasonable to argue not all information goods are priced according to introductory pricing strategies. Accordingly, this study investigates the pricing strategy of an information good producing firm, and examines factors affecting its pricing strategy. Next, salient features of information good demand are examined. The following section identifies price discrimination by a monopolist which produces both an information good, and an upgraded version continuously. Introductory pricing and later price reductions depend on the size of the installed base of customers. The following section deals with actual pricing strategy by a monopolist. A conclusion and research direction suggestions follow.

---

[1] Rohlfs (1974) shows that a communication firm should charge higher prices to new customers, as new customers receive utility from the installed base of customers.

Information Good Demand Characteristics

Monopolistic firms produce information goods provided there is no threat of immediate entry by competitors. While the development of a new information good is costly, post development marginal reproduction costs ($c \geq 0$) are trivial. The monopolist is aware that a consumer gains more satisfaction from the good the larger is the installed base of consumers. In recognizing this network externality, the firm attempts to secure a critical mass or customer base to ensure its survival. The firm is assumed to develop and sell both an initial version of an information good and an upgraded version to be made available to the market at a later date.

Assume a population of η potential customers. Consumers are heterogeneous in their preference for the good and are distributed uniformly in the interval $x \in [0,1]$. Consumers located close to zero derive greater utility from consumption of the good than consumers with an index value close to one, who receive little utility from consumption of the good. A consumer is assumed to purchase a unit of both versions of the good. Let q denote quantity demanded and p its price. The utility function of a consumer of type x is

$$U_x = (1-x)q^e - p \tag{11.1}$$

where q^e is expected demand. In Eq. 11.1, utility is proportional to q^e and reflects the presence of a network externality. Since consumers with a greater utility for the good are the more likely to purchase the good, the last purchaser must be indifferent toward purchasing the good. This consumer $\hat{x}$ must satisfy

$$\hat{x} = \frac{q^e - p}{q^e}. \tag{11.2}$$

All consumers located $x \leq \hat{x}$ purchase the good, and consumers located in the area $x > \hat{x}$ do not. Consumers who purchase the good number $q = \eta\hat{x}$. Assuming perfect information, then $q^e = q = \eta\hat{x}$. Substituting into Eq. 11.2, the derived demand function is

$$p = (1-\hat{x})\eta\hat{x}. \tag{11.3}$$

The demand curve Eq. 11.3 is drawn in Fig. 11.1 as an inverted hyperbola, whose intercepts with x-axis are $\hat{x} = \{0,1\}$.[2] Fig. 11.1 shows that consumers are willing to pay more for the good as the consumer base expands until it reaches the population midpoint. In this range, the network externality dominates the price ef-

---

[2] Economides and Himmelberg (1995) proved, with a fulfilled expectation assumption, that this demand function can also exist in a perfectly competitive market.

fect on demand for the good. However, the willingness to pay for the good falls beyond this point as the price effect outweighs the network effect. Fig. 11.1 indicates that demand does not increase indefinitely as the impact of the network externality diminishes in importance beyond some threshold.

If the monopolist sets price at $p_0 = c$, the quantities that satisfy Eq. 11.3, assuming $\eta = 1$, are

$$\hat{x}_0^L = \frac{1-\sqrt{(1-4c)}}{2}, \quad \hat{x}_0^H = \frac{1+\sqrt{(1-4c)}}{2}. \tag{11.4}$$

The point $\hat{x}_0^L$ is an unstable equilibrium. That is, if the customer base is less than $\hat{x}_0^L$, then this demand dissipates. However, when demand is greater than $\hat{x}_0^L$, then demand increases exogenously through the externality to $\hat{x}_0^H$. In this sense, $\hat{x}_0^L$ is a critical mass of consumers required to guarantee the firm's survival. For a realized market demand of less than $\hat{x}_0^L$, the firm should subsidize consumer purchase of the good until realized market demand exceeds $\hat{x}_0^L$.

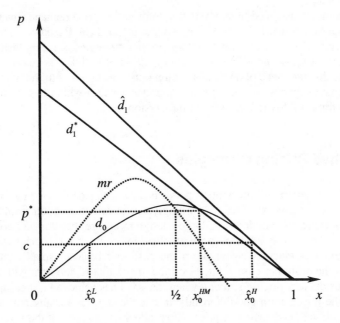

Fig. 11.1. Market Demand Curve for an Information Good

When the monopolistic firm induces a demand for the good $x > \hat{x}_0^L$, demand increases to $\hat{x}_0^H$. During the period when demand expands from $\hat{x}_0^L$ to $\hat{x}_0^H$, the firm can charge the profit-maximizing price. The final demand for the initial version of the good is $\hat{x}_0^H$.

Now, with an installed base of customers $\hat{x}_0^H$, the demand function for the upgraded version, which is backward compatible with the initial version, with $\eta = 1$, is

$$p = (1-x)\hat{x}_0^H = \hat{x}_0^H - x \cdot \hat{x}_0^H \tag{11.5}$$

which is linear and decreasing in x. The demand curve for the upgraded product needs to be compatible with the original version since backward compatibility is a common feature of information goods since the late-1980s. Another feature of Eq. 11.5 deserving attention is that it remains negatively sloped in the presence of a network externality. With the installed base secure, a consumer's willingness to pay is proportional to his (her) strength of preference (utility) for the good. For this negatively sloped demand curve, the network externality affects the slope and vertical intercept.

It is commonly assumed by analysts that information good demand functions in the presence of a network externality are similar to Eq. 11.3. Based on such specifications, critical values and introductory pricing strategies are demonstrated to operate in information good markets (Rohlfs 1974; Shy 2001). A reason for this approach is that the focus of analysis is commonly concerned with the demand for new information goods, and not for an upgraded version which has an installed base of customers inherited from the initial version.

Alternative Pricing Strategies

A salient characteristic of information goods is the existence of economies of scale in demand or a network effect. Accordingly, an information good producer attempts to achieve an installed base large enough to ensure that demand continues to grow exogenously. For this purpose, the firm may charge a low price (not covering marginal production costs) when the customer base is 'small', and raise price when the customer base achieves critical mass, that is, the firm will not employ an introductory pricing strategy when the good has achieved critical mass. In this case, the information good demand curve is the same as for any non-network normal good, viz., decreasing in price. Next, allow an upgrade of the information good to be produced. With this in mind, consider a monopolist that develops a new information product and a subsequent upgraded version, and who wishes to maximize long-run profit. In particular, when the good exhibits a strong network effect, consumers are less willing to purchase the good and the result is a smaller installed base. Clearly, the monopolist's effort to maximize profit is further complicated by the negative impact on demand by the introduction of an upgraded version.

To consider both issues suppose a monopolist places a new information good on the market every second period. Also assume that consumers purchase at most a unit of both the initial and upgraded good. The firm sets prices every period. Even though the new good is sold for only two periods, the size of its consumer base is important for next version demand as the upgrade is backward compatible. The higher is new good demand the greater is demand for an upgraded version and consumer willingness to pay caused by the inherited installed base. Hence, the upgraded version is capable of producing substantial returns. Further, for simplicity, assume that the Coase (1972) conjecture does not hold, and neither consumers nor the firm discount future utility or revenue streams, respectively. Also, the development cost of a version is sunk at the time of reproducing the good as a commodity, and so is not taken into account when making pricing decisions. Finally, the marginal unit production cost of a version is assumed equal to c.

In the initial period, the monopolist attempts to secure the necessary installed base for the new information good by employing an introductory pricing strategy. Accordingly, the period-1 price is subsidized, and related to the price chosen by the firm in the period-2 as the critical mass $\hat{x}_0^L$ is directly bound to period-2 price. The higher is period-2 price, the greater $\hat{x}_0^L$ and the subsidy. For a period-2 price of t the corresponding consumer subsidy is $t \cdot \hat{x}_0^L$, an increasing function of t. Further, in selecting a period-2 price the firm also needs to consider the impact on period-2 profit, since the quantity demanded of the new information good in period-2 provides the installed base for the upgraded version. While there are many ways to price these goods, the cases investigated here include when the firm sets price equal to marginal cost to ensure a greater demand for the upgraded version of the good, viz., enhance the installed base for the upgraded version. Alternatively, the firm selects the profit-maximizing price to gain period-2 monopoly profits for the new good and accepts a smaller installed base for the upgraded version.

Marginal Cost Pricing

The marginal cost of reproducing of the new information good is constant at c. Suppose the firm sets a price equal to this marginal cost. That is, the firm attempts to secure an installed base of,

$$x_1^{L*} = \frac{1 - \sqrt{(1-4c)}}{2}$$

(11.4a)

by offering a subsidy to consumers with a relatively high utility value for the new information good. Should the good be given away then the total subsidy value is

$s_1 = x_1^{L*} \cdot c$. After gaining a critical mass of consumers x_1^{L*} , demand for the good rises exogenously, since consumer willingness to pay is higher than period-1 price when the installed base is x_1^{L*} , and so market demand of the new information good reaches,

$$x_1^{H*} = \frac{1 + \sqrt{(1-4c)}}{2} .$$
(11.4b)

The firm's net revenue, excluding the subsidy, is $NR = c(x_1^{H*} - x_1^{L*})$. When the firm adopts a marginal cost pricing strategy, it must bear the cost of subsidizing the new information good in period-1.

When the firm introduces an upgraded version of the good to the market the economic life of the initial version ends. The upgraded version is engineered backward compatible so that customers of the initial version are hopefully 're-tained' by the firm as the installed base for the upgraded version. Accordingly, demand for the upgraded version is

$$p = k_2 (1-x) x_1^{H*}$$
(11.6)

where k_2 represents technical advance embodied in the upgraded version. Usually $k_2 > 1$, however, to simplify notation in the arguments that follow it is set at $k_2 = 1$. Next, the firm exerts monopoly power and selects output and price, respectively, as

$$x_2^{1*} = \frac{x_1^{H*} - c}{2x_1^{H*}}, \qquad p_2^{1*} = \frac{x_1^{H*} + c}{2}$$
(11.7)

to provide a profit of

$$\pi_2^{1*} = \frac{(x_1^{H*} - c)^2}{4x_1^{H*}} .$$
(11.8)

When marginal cost is zero, profit of Eq. 11.8 is ¼.

In the period following the upgraded versions market release, taking the residual demand into consideration, demand for an upgraded version of the good is

$$p = \begin{cases} (1-x)(x_1^{H^*} - x_2^{1M}), & x_2^{1M} \leq x \leq x_1^{H^*} \\ (1-x)x, & x_1^{H^*} < x \leq 1 \end{cases}. \qquad (11.9)$$

At this point, the firm decides whether to employ a profit-maximizing or marginal cost pricing strategy. When a third or second-upgraded version of the information good is introduced, it faces a demand function similar to Eq. 11.6. The firm's decision-making is similar to that for the initial upgrade of the good.

Monopolistic Pricing

Assume the firm sets price as a monopolist. Since the demand function exhibits an inverted hyperbolic shape the profit function, without fixed costs, has the form

$$\pi = (p - c) = ((1-x)x - c)x. \qquad (11.10)$$

The corresponding profit maximizing output and price, respectively, are

$$x_1^{HM} = \frac{1 + \sqrt{1 - 3c}}{3}, \quad p_1^M = (1 - x_1^{HM})x_1^{HM}. \qquad (11.11)$$

Profit for Eq. 11.11 is $\pi_1^{2M} = (p_1^M - c)(x_1^{HM} - x_1^{LM})$, where $x_1^{LM} = \frac{1 - \sqrt{1 - 3c}}{3}$. The firm's profit is $\frac{4}{27}$ in period-2 for the new information good when marginal cost is zero. But this calculation ignores the subsidy transferred to ensure a critical mass or installed customer base is achieved in period-1 of $s_1^M = p_1^M \cdot x_1^{LM}$. Hence, the net profit of the new information good is $\pi_1^M = \pi_1^{2M} - s_1^M$, and converges to $\frac{4}{27}$ as c approaches zero, since the subsidy also approaches zero with c. However, π_1^M declines with increases in c.

When the upgraded version is introduced to the market, the firm sets price monopolistically for the installed base. The monopolist could price at marginal cost to encourage further expansion of the customer base, however, that strategy does not benefit the firm as the residual demand for the upgraded version in period-2 plummets. To demonstrate this point, consider an installed base x_1^{HM} of the initial version of the information good, the demand function for the upgraded version in period-1 is

$$p = k_2(1-x)x_1^{HM} \qquad (11.12)$$

With $k_2 = 1$, profit maximizing output and price, respectively, are:

$$x_2^{1M} = \frac{1}{2} - \frac{c}{2x_1^{HM}},$$ (11.13)

$$p_2^{1M} = (1 - x_2^{1M})x_1^{HM} = \frac{1 + 3c + \sqrt{1 - 3c}}{6}.$$ (11.14)

Profit, $\pi_2^{1M} = (p_2^{1M} - c)x_1^{HM}$ approaches $\frac{1}{6}$ as c approaches zero.

Next, consider the period following the introduction of the upgraded version. Demand for upgraded good in this period is comprised of the residual upgraded good demand from period-1, is given by the inverted hyperbolic shape of Fig. 11.2.

$$p = \begin{cases} (1-x)x_1^{HM}, & x_2^{1M} \le x \le x_1^{HM} \\ (1-x)x, & x_1^{HM} < x \le 1 \end{cases}$$ (11.15)

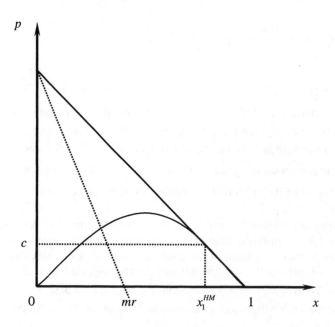

Fig. 11.2. Market Demand Curve for the Upgraded Version of an Information Good

If the firm maximizes profit on this residual demand base, then the period-2 demand is:

$$x_2^{2M} = \frac{3}{4} - \frac{3c}{4x_1^{HM}},$$

(11.16)

with a corresponding price of

$$p_2^{2M} = (1 - x_2^{2M})x_1^{HM} = \frac{1 + 9c + \sqrt{1 - 3c}}{12}.$$

(11.17)

In this period, $\pi_2^{2M} = (p_2^{2M} - c)(x_2^{2M} - x_1^{1M})$ converges to $\frac{1}{24}$ as c approaches zero.

If the firm sets price so as to maximize profit on this inverted hyperbolic shaped demand curve, then $p_2^M = p_1^M$ and period-2 profit for the upgraded version $\pi_2^{HM} = (p_1^M - c)(x_1^{HM} - x_2^{1M})$, approaches $\frac{1}{27}$ as c converges to zero. However, when the firm intends to market a third version of the information good, the further upgraded version faces the problem that profit on an installed base of x_2^{2M} is smaller than that for a base of x_1^{HM}, as $x_2^{2M} < x_1^{HM}$. Therefore, the firm is best served by setting price at p_1^M and securing a customer base of x_1^{HM}.[3]

Firm Survival and Pricing in Information Goods Markets

Firm Industry Exit with the Initial Version

With a monopoly selling only an initial version of the information good, the firm maximizes profit by supporting consumer adoption through a low period-1 price and setting a monopoly period-2 price. The firm sets a monopolistic price in period-2 as it cannot recover the period-1 subsidy and earn positive net profit by setting price at c. Should, in period-1, the firm set the subsidy independently of period-2 price, equilibrium quantity in that period is x_1^{HM}, and the corresponding profit π_1^{HM} is $\frac{4}{27}$ at $c = 0$. However, this approach is suboptimal as subsidy and price is related, viz., the higher price in period-2 the greater is the period-1 subsidy.

Suppose the firm supports a consumer by price minus his (her) willingness to pay. Then, from Fig. 11.3, marginal cost pricing results in a variable cost of

---

[3] To gain the highest profit from this third version, the firm should price at c. Profit is greater because of the enhanced installed base. For additional versions, the firm acts as a monopolist in the period-1 and as a price-taker in the period-2.

A+B+C+D+E, which is naturally the same as revenue. Further, this subsidy to consumers is equal to A. Conversely, when the firm employs a monopolistic pricing strategy, it earns revenue A+B+C+D+F+G+H and incurs costs of A+B+C+D as cost and a subsidy F to high utility consumers. The monopoly can earn positive net profit when G+H > A+F. A loss can also arise, depending on the marginal cost of reproduction.

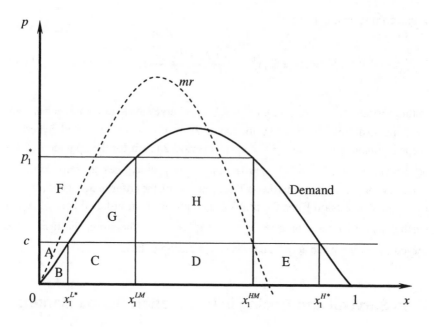

Fig. 11.3. Alternative Pricing Regimes for a New Information Good

Firm Industry Exit with the Upgraded Version

Irrespective of the success of the initial version, the firm will continue to develop and market an upgraded version of the good. If the improvement of the upgraded good is considerable then its price should be higher. However, the price of the upgraded version is typically, more or less, equal to the price of the initial version. Accordingly, assume that the demand function for the upgraded version coincides with that of the initial version, i.e., $k_i = 1$ and $i = 2, 3, ..., t$. In the last period of the firm's life, the firm charges the monopolistic price p_2^{2M}.[4] Hence the firm sets a profit-maximizing price based on the installed base from the initial version, which

---

[4] It is important to note that the firm would not employ marginal cost pricing in period-1 for the upgraded version, a period-1 profit is low, and this holds for period-2 with its small residual demand.

relies on the strategy adopted by the firm in period-2 for the initial version. The higher the profit the firm makes from monopolistic pricing of the initial version, the smaller is profit next period because of a low installed base, and vice versa. For example, with $c = 0$, the sum of profits from monopolistic pricing setting in period-2 for the initial version and period-1 for the upgraded version, respectively, is $\pi_1^{MH} + \pi_2^{1M} = \frac{4}{27} + \frac{1}{6}$ and $\pi_1^{2*} + \pi_2^{1*} = 0 + \frac{1}{4}$. Finally, the firms decision on which pricing strategy to adopt in period-2 for the initial version depends on whether $(\pi_2^{1*} - s_1) - (\pi_2^{1M} + \pi_1^{HM} - s^M)$ is greater than zero. If the difference is positive, the firm will set price equal to marginal cost.

Firm Survives Indefinitely

If the firm produces subsequent versions, attention must be paid to the firm's decision-making for period-2 for the existing version and period-1 for the newly upgraded version. Again there is a trade-off between period-2 profit for the previous version and that for the following period for the upgraded version. If the firm decides to make a higher profit in period-2 for the updated version, then Eq. 11.16 and the following, show the sum of profits in the successive periods is $\pi_2^{HM} + \pi_3^{1M}$ or $\frac{1}{27} + \frac{1}{6}$ with zero marginal cost. Conversely, marginal cost pricing in period-2 for the upgraded version, and monopolistic pricing in the following period makes profit sum $\pi_2^{2*} + \pi_3^{1*}$ or $0 + \frac{1}{4}$ with $c = 0$. Therefore, the firm wants to price at marginal cost in period-2 for the upgraded version, and then maximize profit in the next period on a larger installed base. This pricing strategy over period-2 for any version and period-1 for the following upgraded version is likely to be employed by a monopoly for subsequent versions.

Conclusion

A firm producing an information good characterized by a network externality is usually shown to adopt an introductory pricing strategy. To secure an installed base of customers, the firm sets price below marginal cost in marketing a new information good. However, in real-world markets introductory pricing is not exclusively practiced. In some cases, firms initially set a high price when marketing a specific version of an information good, and then lower price prior to introducing an upgraded version. This study investigates the pricing of an information good, and analyzes factors that determine the pricing strategy adopted. A simple model posits an information good demand function with an inverted hyperbolic shape, due to a network externality. However, the upgraded version demand function is negatively sloped and the firm produces upgraded versions continuously. A finding is that firms producing information goods do not always employ an introductory pricing strategy. Only when a firm invents a new good is it likely to use this strategy. When the product is an upgrade of an existing version, and compatible with it, then it is not necessary for the firm to charge a lower price in period-1. In this situation, the firm is better off charging a profit-maximizing price initially to the installed base of established customers, and then lower price to help secure the customer base for the next version. Another finding is that an information good with network externality is likely to provide less profit than non-information goods, since information good producers must ensure they retain a substantial customer base. The models considered in this paper need to be extended to consider the distribution of consumers, potential entry and switching costs.

References

Bensaid B, Lense JP (1996) Dynamic monopoly pricing with network externalities. International Journal of Industrial Organization 14: 837–55

Economides N, Himmelberg C (1995) Critical mass and network evolution in telecommunications. In: Broch GW (ed) Toward a competitive telecommunications industry: Selected papers from the 1994 Telecommunications Policy Research Conference. University of Maryland, College Park, pp 31–42

Coase RH (1972) Durability and monopoly. Journal of Law and Economics 15: 143–9

Cabral LMB, Salant DJ, Woroch GA (1999) Monopoly pricing with network externalities. International Journal of Industrial Organization 17: 199–214

Katz M, Shapiro C (1985) Network externalities, competition, and compatibility. American Economic Review 75(3): 424–40

Rohlfs J (1974) A theory of interdependent demand for communications service. Bell Journal of Economics and Management Science 5(1): 16–37

Shapiro C, Varian HR (1999) Information rules: A strategic guide to network economy. Harvard Business School Press, Boston

Shy O (2001) The economics of network industries. Cambridge University Press, Cambridge

Part IV: Market Growth, Regulation and Investment

12 Measuring Telecommunication System Network Effects

Gary Madden, Aniruddha Banerjee and Grant Coble-Neal

Introduction

The idea that adding new subscribers to a telecommunications network increases the value of subscription to individual subscribers, or network effect, has attracted much interest in the network economics literature. Artle and Averous (1973), Rohlfs (1974) and Littlechild (1975) pioneered formal theoretical analyses of the welfare implications of such network effects. In particular, Artle and Averous (1973) demonstrated the public good dimension of networks. Rohlfs (1974) extends the analysis to derive a demand function containing telephone price as an argument. However, difficulty in formulating structural demand models that incorporate network effects meant that few empirical studies provide evidence of the magnitude of network effects. However, Perl (1983) estimates the value of the telephone service for different sized local calling areas and density of telephone main stations per square mile. Further, Taylor and Kridel (1990) report evidence of a network effect using arguments similar to Perl (1983). Following Economides and Himmelberg (1995), Madden and Coble-Neal (2002, 2004) and Madden et al. (2004) estimate a network effect based on a structural model consistent with optimizing consumer behavior for individual telecommunications services.

This paper analyses consumer demand for several network services. The subscription model is adapted from Becker et al. (1994), and allows interaction among the network effects for competing networks. The model also examines whether consumers consider the combined network size of compatible fixed-line and mobile telephone systems when deciding to subscribe. Additionally, the model allows the testing of whether consumers' subscription decisions are based on forward-looking expectations of future network size. Model estimates of the network effect magnitudes for fixed-line telephony, mobile telephony and the Internet are derived. Interactions among network effects are also reported. The chapter is organized as follows. The following section derives estimating equations, adapted from Becker et al. that are consistent with intertemporal utility maximization. A discussion relating to OECD Member Country data follows. The next section presents the estimation results and further analysis. A final section provides concluding remarks.

Modeling Telecommunications Demand

Following Becker et al., consider a model with three telecommunications services and utility in period t given by the concave utility function,

$$U(N_{Ft}, N_{Ft-1}, N_{Mt}, N_{Mt-1}, N_{It}, N_{It-1}, Y_t, e_t), \tag{12.1}$$

where F, M and I denote fixed-line telephone, mobile telephone and the Internet, respectively. N_F is the size of the fixed-line telephone network, N_M is the size of the mobile telephone network and N_I is the size of the Internet network. Network size is indexed by time. Y_t is a quantity index of all other goods and e_t captures the impact of any unmeasured life-cycle variables on utility.

The consumer's objective is to maximize lifetime utility,

$$\max \sum_{t=1}^{\infty} \beta^{t-1} U(N_{Ft}, N_{Ft-1}, N_{Mt}, N_{Mt-1}, N_{It}, N_{It-1}, Y_t, e_t) \tag{12.2}$$

subject to

$$\sum_{t=1}^{\infty} \beta^{t-1} (P_{Ft} C_{Ft} + P_{Mt} C_{Mt} + P_{It} C_{It} + Y_t) = A^0 \tag{12.3}$$

and

$$C_{F0} = C_F^0, C_{M0} = C_M^0, C_{I0} = C_I^0, \tag{12.4}$$

$$C_F^0, C_M^0, C_I^0 \geq 0, \tag{12.5}$$

where $\beta = (1+r)^{-1}$, r is the market rate of interest, P_F, P_M and P_I denote access price for fixed-line telephony, mobile telephony and the Internet, respectively.[1] A^0 is the present value of the consumer's wealth and C_F, C_M and C_I are consumption of fixed-line telephony, mobile telephony and the Internet, respectively. Zero subscripts in Eq. 12.4 denote the initial consumption of services. Eq. 12.5 contains initial consumption within non-negative quantities. Further, telecommunication service consumption requires more than one subscriber. Assuming consumption is proportional to network size, Eq. 12.3 is restated as

---

[1] Note that the price of the composite good Y_t is normalized to 1.

$$\sum_{t=1}^{\infty} \beta^{t-1}(P_{Ft}\delta_F N_{Ft} + P_{Mt}\delta_M N_{Mt} + P_{It}\delta_I N_{It} + Y_t) = A^0 \,, \tag{12.6}$$

where δ_F, δ_M and δ_I are constant.[2] Eq. 12.2 and Eq. 12.6 state that consumers maximize wealth-constrained utility as

$$\begin{aligned}
\max V &= U(N_{Ft}, N_{Ft-1}, N_{Mt}, N_{Mt-1}, N_{It}, N_{It-1}, Y_t, e_t) \\
&+ \beta U(N_{Ft+1}, N_{Ft}, N_{Mt+1}, N_{Mt}, N_{It+1}, N_{It}, Y_{t+1}, e_{t+1}) \\
&+ \sum_{j=2}^{\infty} \beta^j U(N_{Ft+j}, N_{Ft+j-1}, N_{Mt+j}, N_{Mt+j-1}, N_{It+j}, N_{It+j-1}, Y_{t+j}, e_{t+j}) \\
&+ \lambda(A^0 - \sum_{t=1}^{\infty} \beta^{t-1}(P_{Ft}\delta_F N_{Ft} + P_{Mt}\delta_M N_{Mt} + P_{It}\delta_I N_{It} + Y_t))\,,
\end{aligned} \tag{12.7}$$

where the Lagrange multiplier, λ, represents the marginal value of wealth. The associated first-order conditions (current period) are

$$U_{Y_t}(N_{Ft}, N_{Ft-1}, N_{Mt}, N_{Mt-1}, N_{It}, N_{It-1}, Y_t, e_t) = \lambda\,, \tag{12.8}$$

$$\begin{aligned}
&U_{N_{Ft}}(N_{Ft}, N_{Ft-1}, N_{Mt}, N_{Mt-1}, N_{It}, N_{It-1}, Y_t, e_t) \\
&+ \beta U_{N_{Ft}}(N_{Ft}, N_{Ft-1}, N_{Mt}, N_{Mt-1}, N_{It}, N_{It-1}, Y_t, e_t) = \lambda P_{Ft}\delta_F\,,
\end{aligned} \tag{12.9}$$

$$\begin{aligned}
&U_{N_{Mt}}(N_{Ft}, N_{Ft-1}, N_{Mt}, N_{Mt-1}, N_{It}, N_{It-1}, Y_t, e_t) \\
&+ \beta U_{N_{Mt}}(N_{Ft}, N_{Ft-1}, N_{Mt}, N_{Mt-1}, N_{It}, N_{It-1}, Y_t, e_t) = \lambda P_{Mt}\delta_M
\end{aligned} \tag{12.10}$$

and

$$\begin{aligned}
&U_{N_{It}}(N_{Ft}, N_{Ft-1}, N_{Mt}, N_{Mt-1}, N_{It}, N_{It-1}, Y_t, e_t) \\
&+ \beta U_{N_{It}}(N_{Ft}, N_{Ft-1}, N_{Mt}, N_{Mt-1}, N_{It}, N_{It-1}, Y_t, e_t) = \lambda P_{It}\delta_I\,.
\end{aligned} \tag{12.11}$$

---

[2] It is straightforward to relax the constant proportionality assumption, e.g., consumption may be assumed proportional to connections among telephone subscribers, which grow geometrically with network size. This relationship is more likely for fixed-line and mobile telephony.

Eq. 12.8 states that current period marginal utility derived from the consumption of other goods Y_t is equal to the marginal utility of wealth. Eq. 12.9 through Eq. 12.11 imply that the current period marginal utility plus the present value of next period utility with respect to each of the telecommunications services is equal to the marginal utility of wealth multiplied by the marginal price of use, scaled by the constant of proportionality. Since the marginal utility of wealth is constant through time, variations in telecommunications service prices define marginal utility of wealth-constant demand curves for telecommunications services and all other goods.

To make the model operational, following Becker et al., consider a generalized quadratic approximation to the general current period utility function in Eq. 12.1[3]

$$
\begin{aligned}
U(N_{Ft}, & N_{Ft-1}, N_{Mt}, N_{Mt-1}, N_{It}, N_{It-1}, Y_t, e_t) = \\
& u_{F1}N_{Ft} + u_{F2}N_{Ft-1} + u_{M1}N_{Mt} + u_{M2}N_{Mt-1} + u_{I1}N_{It} + u_{I2}N_{It-1} + u_Y Y_t + u_e e_t \\
& + \tfrac{1}{2}u_{FF1}N_{Ft}^2 + \tfrac{1}{2}u_{FF2}N_{Ft-1}^2 + \tfrac{1}{2}u_{MM1}N_{Mt}^2 + \tfrac{1}{2}u_{MM2}N_{Mt-1}^2 + \tfrac{1}{2}u_{II1}N_{It}^2 \\
& + \tfrac{1}{2}u_{II2}N_{It-1}^2 + \tfrac{1}{2}u_{YY}Y_t^2 + \tfrac{1}{2}e_t^2 + u_{F12}N_{Ft}N_{Ft-1} + u_{F1M1}N_{Ft}N_{Mt} \\
& + u_{F1M2}N_{Ft}N_{Mt-1} + u_{F1I1}N_{Ft}N_{It} + u_{F1I2}N_{Ft}N_{It-1} + u_{F1Y}N_{Ft}Y_t + u_{F1e}N_{Ft}e_t \\
& + u_{F2M1}N_{Ft-1}N_{Mt} + u_{F2M2}N_{Ft-1}N_{Mt-1} + u_{F2I1}N_{Ft-1}N_{It} + u_{F2I2}N_{Ft-1}N_{It-1} \\
& + u_{F2Y}N_{Ft-1}Y_t + u_{F2e}N_{Ft-1}e_t + u_{M1M2}N_{Mt}N_{Mt-1} + u_{M1I1}N_{Mt}N_{It} \\
& + u_{M1I2}N_{Mt}N_{It-1} + u_{M1Y}N_{Mt}Y_t + u_{M1e}N_{Mt}e_t + u_{M2I1}N_{Mt-1}N_{It} + u_{M2I2}N_{Mt-1}N_{It-1} \\
& + u_{M2Y}N_{Mt-1}Y_t + u_{M2e}N_{Mt-1}e_t + u_{I1I2}N_{It}N_{It-1} + u_{I1Y}N_{It}Y_t + u_{I1e}N_{It}e_t \\
& + u_{I2Y}N_{It-1}Y_t + u_{I2e}N_{It-1}e_t + u_{Ye}Y_t e_t .
\end{aligned}
\tag{12.12}
$$

Similarly, next period utility is defined as

---

[3] Positive but diminishing marginal utility—the concavity property—likely implies, e.g., that the coefficients for the first partials, such as $u_{F1}, u_{F2}, ..., u_e$, are positive. Similarly, one could expect the coefficients for the second-order direct partials, e.g., $u_{FF1}, u_{FF2}, ..., u_{YY}$, are negative.

$$u_{F1}N_{Ft+1} + u_{F2}N_{Ft} + u_{M1}N_{Mt+1} + u_{M2}N_{Mt} + u_{I1}N_{It+1} + u_{I2}N_{It} + u_Y Y_{t+1} + u_e e_{t+1}$$
$$+\tfrac{1}{2}u_{FF1}N^2_{Ft+1} + \tfrac{1}{2}u_{FF2}N^2_{Ft} + \tfrac{1}{2}u_{MM1}N^2_{Mt+1} + \tfrac{1}{2}u_{MM2}N^2_{Mt} + \tfrac{1}{2}u_{I1}N^2_{It+1}$$
$$+\tfrac{1}{2}u_{I2}N^2_{It} + \tfrac{1}{2}u_{YY}Y^2_{t+1} + \tfrac{1}{2}e^2_{t+1} + u_{F12}N_{Ft+1}N_{Ft} + u_{F1M1}N_{Ft+1}N_{Mt+1}$$
$$+u_{F1M2}N_{Ft+1}N_{Mt} + u_{F1I1}N_{Ft+1}N_{It+1} + u_{F1I2}N_{Ft+1}N_{It} + u_{F1Y}N_{Ft+1}Y_{t+1}$$
$$+u_{F1e}N_{Ft+1}e_{t+1} + u_{F2M1}N_{Ft}N_{Mt+1} + u_{F2M2}N_{Ft}N_{Mt} + u_{F2I1}N_{Ft}N_{It+1}$$
$$+u_{F2I2}N_{Ft}N_{It} + u_{F2Y}N_{Ft}Y_{t+1} + u_{F2e}N_{Ft}e_{t+1} + u_{M1M2}N_{Mt+1}N_{Mt} \qquad (12.13)$$
$$+u_{M1I1}N_{Mt+1}N_{It+1} + u_{M1I2}N_{Mt+1}N_{It} + u_{M1Y}N_{Mt+1}Y_{t+1} + u_{M1e}N_{Mt+1}e_{t+1}$$
$$+u_{M2I1}N_{Mt}N_{It+1} + u_{M2I2}N_{Mt}N_{It} + u_{M2Y}N_{Mt}Y_{t+1} + u_{M2e}N_{Mt}e_{t+1}$$
$$+u_{I1I2}N_{It+1}N_{It} + u_{I1Y}N_{It+1}Y_{t+1} + u_{I1e}N_{It+1}e_{t+1}$$
$$+u_{I2Y}N_{It}Y_{t+1} + u_{I2e}N_{It}e_{t+1}$$
$$+u_{Ye}Y_{t+1}e_{t+1} \; .$$

The first-order condition (Eq. 12.8) becomes

$$u_Y + u_{YY}Y_t + u_{F1Y}N_{Ft} + u_{F2Y}N_{Ft-1} + u_{M1Y}N_{Mt} + u_{M2Y}N_{Mt-1}$$
$$+u_{I1Y}N_{It} + u_{I2Y}N_{It-1} + u_{Ye}e_t = \lambda \; . \qquad (12.14)$$

First-order conditions Eq. 12.9 through Eq 12.11 are accordingly defined:

$$u_{F1} + u_{FF1}N_{Ft} + u_{F12}N_{Ft-1} + u_{F1M1}N_{Mt} + u_{F1M2}N_{Mt-1} + u_{F1I1}N_{It} + u_{F1I2}N_{It-1}$$
$$+u_{F1Y}Y_t + u_{F1e}e_t + \beta(u_{F2} + u_{FF2}N_{Ft} + u_{F12}N_{Ft+1} + u_{F2M1}N_{Mt+1} \qquad (12.15)$$
$$+u_{F2M2}N_{Mt} + u_{F2I1}N_{It+1} + u_{F2I2}N_{It} + u_{F2Y}Y_{t+1} + u_{F2e}e_{t+1}) = \lambda P_{Ft}\delta_F \; ,$$

$$u_{M1} + u_{MM1}N_{Mt} + u_{F1M1}N_{Ft} + u_{F2M1}N_{Ft-1} + u_{M1M2}N_{Mt-1} + u_{M1I1}N_{It} + u_{M1I2}N_{It-1}$$
$$+u_{M1Y}Y_t + u_{M1e}e_t + \beta(u_{M2} + u_{MM2}N_{Mt} + u_{F1M2}N_{Ft+1} + u_{F2M2}N_{Ft}$$
$$+u_{M1M2}N_{Mt+1} + u_{M2I1}N_{It+1} + u_{M2I2}N_{It} + u_{M2Y}Y_{t+1} + u_{M2e}e_{t+1}) = \lambda P_{Mt}\delta_M \qquad (12.16)$$

and

$$u_{I1} + u_{I1I1}N_{It} + u_{F1I1}N_{Ft} + u_{F2I1}N_{Ft-1} + u_{M1I1}N_{Mt} + u_{M2I1}N_{Mt-1} + u_{I1I2}N_{It-1}$$
$$+u_{I1Y}Y_t + u_{I1e}e_t + \beta(u_{I2} + u_{I12}N_{It} + u_{F2I2}N_{Ft} + u_{M1I2}N_{Mt+1} + u_{M2I2}N_{Mt} \qquad (12.17)$$
$$+u_{I1I2}N_{It+1} + u_{I2Y}Y_{t+1} + u_{I2e}e_{t+1}) = \lambda P_{It}\delta_I \; .$$

Solving Eq. 12.14 for Y_t provides

$$u_{YY}Y_t = \lambda - u_Y - u_{F1Y}N_{Ft} - u_{F2Y}N_{Ft-1} - u_{M1Y}N_{Mt} - u_{M2Y}N_{Mt-1}$$
$$-u_{I1Y}N_{It} - u_{I2Y}N_{It-1t} - u_{Ye}e_t \; . \tag{12.18}$$

That is,

$$Y_t = (1/u_{YY})(\lambda - u_Y - u_{F1Y}N_{Ft} - u_{F2Y}N_{Ft-1} - u_{M1Y}N_{Mt}$$
$$-u_{M2Y}N_{Mt-1} - u_{I1Y}N_{It} - u_{I2Y}N_{It-1} - u_{Ye}e_t) \; . \tag{12.19}$$

Updating Eq. 12.19 by one period provides

$$Y_{t+1} = (1/u_{YY})(\lambda - u_Y - u_{F1Y}N_{Ft+1} - u_{F2Y}N_{Ft} - u_{M1Y}N_{Mt+1}$$
$$-u_{M2Y}N_{Mt} - u_{I1Y}N_{It+1} - u_{I2Y}N_{It} - u_{Ye}e_{t+1}) \; . \tag{12.20}$$

Further, substituting Eq. 12.19 and Eq. 12.20 into Eq. 12.15

$$u_{F1} + u_{FF1}N_{Ft} + u_{F12}N_{Ft-1} + u_{F1M1}N_{Mt} + u_{F1M2}N_{Mt-1} + u_{F1I1}N_{It}$$
$$+u_{F1I2}N_{It-1} + u_{F1Y}((1/u_{YY})(\lambda - u_Y - u_{F1Y}N_{Ft} - u_{F2Y}N_{Ft-1} - u_{M1Y}N_{Mt}$$
$$-u_{M2Y}N_{Mt-1} - u_{I1Y}N_{It} - u_{I2Y}N_{It-1} - u_{Ye}e_t)) + u_{F1e}e_t + \beta(u_{F2} + u_{FF2}N_{Ft}$$
$$+u_{F12}N_{Ft+1} + u_{F2M1}N_{Mt+1} + u_{F2M2}N_{Mt} + u_{F2I1}N_{It+1} + u_{F2I2}N_{It}$$
$$+u_{F2Y}((1/u_{YY})(\lambda - u_Y - u_{F1Y}N_{Ft+1} - u_{F2Y}N_{Ft} - u_{M1Y}N_{Mt+1}$$
$$-u_{M2Y}N_{Mt} - u_{I1Y}N_{It+1} - u_{I2Y}N_{It} - u_{Ye}e_{t+1})) + u_{F2e}e_{t+1}) = \lambda P_{Ft}\delta_F \; .$$

Simplifying provides

$$u_{F1} + \beta u_{F2} + ((\lambda - u_Y)/u_{YY})(u_{F1Y} + \beta u_{F2Y}) + \beta(u_{F12} - (u_{F2Y}/u_{YY})u_{F1Y})N_{Ft+1}$$
$$+(u_{FF1} - (u^2_{F1Y}/u_{YY}) + \beta(u_{FF2} - (u^2_{F2Y}/u_{YY})))N_{Ft} + (u_{F12} - (u_{F1Y}/u_{YY})u_{F2Y})N_{Ft-1}$$
$$+\beta(u_{F2M1} - (u_{F2Y}/u_{YY})u_{M1Y})N_{Mt+1} + (u_{F1M1} - (u_{F1Y}/u_{YY})u_{M1Y}$$
$$+\beta(u_{F2M2} - (u_{F2Y}/u_{YY})u_{M2Y}))N_{Mt} + (u_{F1M2} - (u_{F1Y}/u_{YY})u_{M2Y})N_{Mt-1}$$
$$+\beta(u_{F2I1} - (u_{F2Y}/u_{YY})u_{I1Y})N_{It+1} + (u_{F1I1} - (u_{F1Y}/u_{YY})u_{I1Y}$$
$$+\beta(u_{F2I2} - (u_{F2Y}/u_{YY})u_{I2Y}))N_{It} + (u_{F1I2} - (u_{F1Y}/u_{YY})u_{I2Y})N_{It-1}$$
$$+(u_{F1e} - (u_{F1Y}/u_{YY})u_{Ye})e_t - \beta(u_{F2e} + (u_{F2Y}/u_{YY})u_{Ye})e_{t+1} = \lambda P_{Ft}\delta_F \; .$$

Solving for N_{Ft} provides

$$(1/u_{YY})(u_{FF1}u_{YY} - u^2_{F1Y} + \beta(u_{FF2}u_{YY} - u^2_{F2Y}))N_{Ft} =$$
$$-u_{F1} - \beta u_{F2} - ((\lambda - u_Y)/u_{YY})(u_{F1Y} + \beta u_{F2Y})$$
$$-\beta(u_{F12} - (u_{F2Y}/u_{YY})u_{F1Y})N_{Ft+1}$$
$$-(u_{F12} - (u_{F1Y}/u_{YY})u_{F2Y})N_{Ft-1}$$
$$-\beta(u_{F2M1} - (u_{F2Y}/u_{YY})u_{M1Y})N_{Mt+1}$$
$$-(u_{F1M1} - (u_{F1Y}/u_{YY})u_{M1Y} + \beta(u_{F2M2} - (u_{F2Y}/u_{YY})u_{M2Y}))N_{Mt}$$
$$-(u_{F1M2} - (u_{F1Y}/u_{YY})u_{M2Y})N_{Mt-1}$$
$$-\beta(u_{F2I1} - (u_{F2Y}/u_{YY})u_{I1Y})N_{It+1}$$
$$-(u_{F1I1} - (u_{F1Y}/u_{YY})u_{I1Y} + \beta(u_{F2I2} - (u_{F2Y}/u_{YY})u_{I2Y}))N_{It}$$
$$-(u_{F1I2} - (u_{F1Y}/u_{YY})u_{I2Y})N_{It-1}$$
$$+\lambda P_{Ft}\delta_F - (u_{F1e} - (u_{F1Y}/u_{YY})u_{Ye})e_t + \beta(u_{F2e} + (u_{F2Y}/u_{YY})u_{Ye})e_{t+1} \ .$$

Hence, the demand for fixed-line telephone is

$$N_{Ft} = \theta_0 + \theta_1 N_{Ft+1} + \theta_2 N_{Ft-1} + \theta_3 N_{Mt+1} + \theta_4 N_{Mt} + \theta_5 N_{Mt-1}$$
$$+\theta_6 N_{It+1} + \theta_7 N_{It} + \theta_8 N_{It-1} + \theta_9 P_{Ft} + \theta_{10} e_t + \theta_{11} e_{t+1}$$

where

$$\theta_0 = (-u_{YY}u_{F1} - \beta u_{YY}u_{F2} - (\lambda - u_Y)(u_{F1Y} + \beta u_{F2Y}))$$
$$/(u_{FF1}u_{YY} - u^2_{F1Y} + \beta(u_{FF2}u_{YY} - u^2_{F2Y})) \ ,$$

$$\theta_1 = -\beta(u_{YY}u_{F12} - u_{F2Y}u_{F1Y})/(u_{FF1}u_{YY} - u^2_{F1Y} + \beta(u_{FF2}u_{YY} - u^2_{F2Y})) = \beta\theta_2 \ ,$$

$$\theta_2 = -(u_{YY}u_{F12} - u_{F1Y}u_{F2Y})/(u_{FF1}u_{YY} - u^2_{F1Y} + \beta(u_{FF2}u_{YY} - u^2_{F2Y})) \ ,$$

$$\theta_3 = -\beta(u_{YY}u_{F2M1} - u_{F2Y}u_{M1Y})/(u_{FF1}u_{YY} - u^2_{F1Y} + \beta(u_{FF2}u_{YY} - u^2_{F2Y})) \ ,$$

$$\theta_4 = -(u_{YY}u_{F1M1} - u_{F1Y}u_{M1Y} + \beta(u_{YY}u_{F2M2} - u_{F2Y}u_{M2Y}))$$
$$/(u_{FF1}u_{YY} - u^2_{F1Y} + \beta(u_{FF2}u_{YY} - u^2_{F2Y})) \ ,$$

$$\theta_5 = -(u_{YY}u_{F1M2} - u_{F1Y}u_{M2Y})/(u_{FF1}u_{YY} - u^2_{F1Y} + \beta(u_{FF2}u_{YY} - u^2_{F2Y})) \ ,$$

$$\theta_6 = -\beta(u_{YY}u_{F2I1} - u_{F2Y}u_{I1Y})/(u_{FF1}u_{YY} - u^2_{F1Y} + \beta(u_{FF2}u_{YY} - u^2_{F2Y})) \ ,$$

$$\theta_7 = -(u_{YY}u_{F1I1} - u_{F1Y}u_{I1Y} + \beta(u_{YY}u_{F2I2} - u_{F2Y}u_{I2Y}))$$
$$/(u_{FF1}u_{YY} - u^2_{F1Y} + \beta(u_{FF2}u_{YY} - u^2_{F2Y})),$$

$$\theta_8 = -(u_{YY}u_{F1I2} - u_{F1Y}u_{I2Y})/(u_{FF1}u_{YY} - u^2_{F1Y} + \beta(u_{FF2}u_{YY} - u^2_{F2Y})),$$

$$\theta_9 = u_{YY}\lambda\delta_F /(u_{FF1}u_{YY} - u^2_{F1Y} + \beta(u_{FF2}u_{YY} - u^2_{F2Y})) < 0,$$

$$\theta_{10} = -(u_{YY}u_{F1e} - u_{F1Y}u_{Ye})/(u_{FF1}u_{YY} - u^2_{F1Y} + \beta(u_{FF2}u_{YY} - u^2_{F2Y}))$$

and

$$\theta_{11} = \beta(u_{YY}u_{F2e} + u_{F2Y}u_{Ye})/(u_{FF1}u_{YY} - u^2_{F1Y} + \beta(u_{FF2}u_{YY} - u^2_{F2Y})).$$

The demand for fixed-line telephony, mobile telephony- and Internet are

$$N_{it} = \theta_{i0} + \theta_{ii1}N_{it-1} + \theta_{ii2}N_{it+1} + \sum_j \theta_{ij1}N_{jt-1}$$
$$+ \sum_j \theta_{ij2}N_{jt} + \sum_j \theta_{ij3}N_{jt+1} + \theta_{iP}P_{it} + \theta_{i1}e_t + \theta_{i2}e_{t+1},$$

(12.21)

where $i \ne j$ and $i, j = \{F, M, I\}$ and

$$\theta_{i0} = -(u_{YY}u_{i1} + \beta u_{YY}u_{i2} + (\lambda - u_Y)(u_{i1Y} + \beta u_{i2Y}))$$
$$/(u_{ii1}u_{YY} - u^2_{i1Y} + \beta(u_{ii2}u_{YY} - u^2_{i2Y})),$$

$$\theta_{ii1} = -(u_{YY}u_{i12} - u_{i1Y}u_{i2Y})/(u_{ii1}u_{YY} - u^2_{i1Y} + \beta(u_{ii2}u_{YY} - u^2_{i2Y})) = \theta_{ii},$$

$$\theta_{ii2} = -\beta(u_{YY}u_{i12} - u_{i2Y}u_{i1Y})/(u_{ii1}u_{YY} - u^2_{i1Y} + \beta(u_{ii2}u_{YY} - u^2_{i2Y})) = \beta\theta_{ii},$$

$$\theta_{ij1} = -(u_{YY}u_{i1j2} - u_{i1Y}u_{j2Y})/(u_{ii1}u_{YY} - u^2_{i1Y} + \beta(u_{ii2}u_{YY} - u^2_{i2Y})),$$

$$\theta_{ij2} = -(u_{YY}u_{i1j1} - u_{i1Y}u_{j1Y} + \beta(u_{YY}u_{i2j2} - u_{i2Y}u_{j2Y}))$$
$$/(u_{ii1}u_{YY} - u^2_{i1Y} + \beta(u_{ii2}u_{YY} - u^2_{i2Y})),$$

$$\theta_{ij3} = -\beta(u_{YY}u_{i2j1} - u_{i2Y}u_{j1Y})/(u_{ii1}u_{YY} - u^2_{i1Y} + \beta(u_{ii2}u_{YY} - u^2_{i2Y})),$$

$$\theta_{iP} = u_{YY}\lambda\delta_i / (u_{ii1}u_{YY} - u^2_{i1Y} + \beta(u_{ii2}u_{YY} - u^2_{i2Y})) < 0,$$

$$\theta_{i1} = -(u_{YY}u_{i1e} - u_{i1Y}u_{Ye}) / (u_{ii1}u_{YY} - u^2_{i1Y} + \beta(u_{ii2}u_{YY} - u^2_{i2Y}))$$

and

$$\theta_{i2} = \beta(u_{YY}u_{i2e} + u_{i2Y}u_{Ye}) / (u_{ii1}u_{YY} - u^2_{i1Y} + \beta(u_{ii2}u_{YY} - u^2_{i2Y})).$$

The consumer's utility maximizing demand for service i is defined in terms of the size of network i. Eq. 12.21 represents a system of linear difference demand equations. Note the requirement that $\theta_{iP} < 0$, and that the current size of network i is a function of size in both the immediate past and future periods. Hence, an anticipated fall in next-period price for network i yields an increase in current subscription if $\theta_{ii} > 0$. Moreover, $\theta_{ij3} > 0$ implies the anticipated fall in the price of network i induces a current period increase in subscription for network j. A permanent price decrease implies a larger increase in current subscription than a temporary price decrease since the permanent price decrease combines a fall in current and all future prices. The implication, should $\theta_{ij3} = 0$ is that next period price change is not systematically anticipated. Observed cross-network effects may be the result of a change in relative prices or an autonomous increase in the size of the originating service network. Relative price change and market growth influences may be opposing or mutually reinforcing, depending on whether the services are complements or substitutes. For the relatively mature fixed-line telephone network, the substitution effect may dominate the market growth effect.

The demand system specifies constant marginal utility of wealth in Eq. 12.21, where the representative consumer has perfect foresight. This raises two issues for estimation, viz, the need to allow heterogeneity in the marginal utility of wealth across individuals and to allow for unanticipated wealth changes. Following Becker et al. (1994), Eq. 12.21 is augmented by individual-specific dummy variables. Time-specific dummy variables capture the fixed-element of unanticipated changes in wealth, which in turn induce changes in the marginal utility of wealth. In addition, real per capita income is added to capture deviations around these fixed two-way effects associated with individual-specific changes in the marginal utility of wealth across time. Accordingly, Eq. 12.21 becomes

$$N_{it} = \sum_{t=1}^{T-1}\theta_{ii}d_t + \theta_{ii1}N_{it-1} + \theta_{ii2}N_{it+1} + \sum_j\theta_{ij1}N_{jt-1}$$
$$+ \sum_j\theta_{ij2}N_{jt} + \sum_j\theta_{ij3}N_{jt+1} + \theta_{iP}P_{it} + \gamma Y_t + \theta_{i1}e_t + \theta_{i2}e_{t+1}.$$

(12.21a)

Data and Variables

Biannual price data are collected for 30 OECD Member Countries for 1996, 1998, and 2000 from the OECD Communications Outlook (1997, 1999, 2001, 2003). Annual quantity (network size) data for 1996-2002 are obtained from International Telecommunication Union World Telecommunications Indicators Database (2003).[4] Fixed-line price data are the fixed component of the OECD's basket of residential telephone charges. Mobile telephone price is the fixed component of the OECD's basket of consumer mobile telephone charges. Internet price is the OECD's Internet access basket for 20 hours using discounted PSTN rates. Price data are denominated in US dollars (US$) according to OECD purchasing power parities. Prices are deflated by the US consumer price index to allow price comparison across time. The fixed-line telephone quantity variable is the number of main telephone lines in operation. The quantity variable for mobile telephone is the number of mobile telephone subscribers and Internet quantity is the number of Internet users. All quantity variables are per 100 persons. The resulting estimating data is comprised of 79 observations corresponding to the years 1996, 1998, 2000 and 2002.[5] Descriptive statistics are presented in Table 12.1.

Table 12.1. Descriptive Statistics for OECD Telecommunications Network

Variable	Mean	Std. Dev.	Coef. of Var.	Minimum	Maximum
N_F	52.06	14.34	27.55	9.28	75.48
N_M	32.87	22.34	67.96	1.08	77.00
N_I	17.24	14.38	83.41	0.19	59.79
P_F	165.48	60.02	36.27	34.29	314.44
P_M	305.90	198.58	64.92	2.10	940.89
P_I	49.77	20.43	41.05	18.09	98.85
Y	20,505	6,648	32.42	5,670	41,948

Note. Network size and price data for 2002 not included in calculations. Price and income are denominated in US$ PPP. Network sizes are per 100 persons.

As indicated, fixed-line telephone is the service most widely adopted with penetration averaging 52.06 mainlines per 100 persons across the OECD. How-

---

[4] An underlying assumption in using these data, following Becker et al. (1994), is that per capita telecommunications consumption reflects behavior of a representative consumer.

[5] An unbalanced panel data set is comprised of the following. Data are collected for the years 1996-2002 for: Australia, Austria, Belgium, Canada, Denmark, Finland, France, Germany, Iceland, Ireland, Italy, Japan, Mexico, Netherlands, New Zealand, Norway, Portugal, Spain, Sweden, Switzerland, Turkey, United Kingdom and the United States. These data are augmented by adding observations corresponding to the years 1998-2002 for: Czech Republic, Greece, Republic of Korea and Luxembourg. Further observations for the years 2000-2002 are added for Hungary. Note that the lack of network size data for 2003 prevents the use of 2002 price data in estimation.

ever, there is substantial variation with fixed-line penetration ranging from 9.28
mainlines per 100 persons (Mexico) to 75.48 mainlines per 100 persons (Luxembourg). Mobile telephony is the next most popular service with average penetration of 32.87 subscribers per 100 persons. Mexico and Austria have the lowest and
highest penetration rates, respectively. Internet adoption is substantially lower,
though more variable than both fixed-line and mobile telephony, recording an average 17.24 subscribers per 100 persons and a coefficient of variation of 83.41.[6]
Mean price comparison between services in Table 12.1 indicates that the price of
the most expensive service (mobile telephony) is typically 6.2 times greater than
the least expensive service (Internet access). The coefficient of variation indicates
that mobile telephone prices exhibit greater variation than fixed-line telephone or
Internet access. Casual inspection of scatter plots suggests fixed-line subscription
price exhibits greater variability in 1996 than for 1998 or 2000. Finally, Table 12.1
indicates consideration income variation, with the highest income per capita (Luxembourg) 7.4 times larger than the lowest income per capita (Turkey).

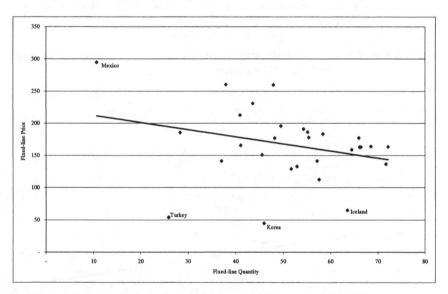

Fig. 12.1. OECD Fixed-line Telephone Price and Network Size. Note. Observations are
group means based on the years 1996, 1998 and 2000.

---

[6] Note that the coefficient of variation statistics follow the order of service maturity with
the oldest service (fixed-line telephone) exhibiting the least variation while the newest
services (Internet access) exhibits the most variation.

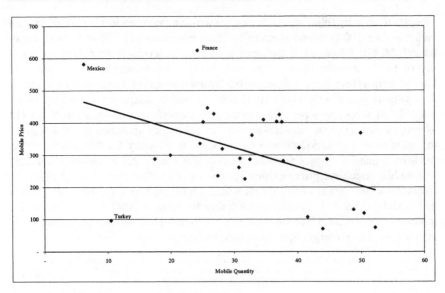

Fig. 12.2. OECD Mobile Telephone Price and Network Size. Note. Observations are group means based on the years 1996, 1998 and 2000.

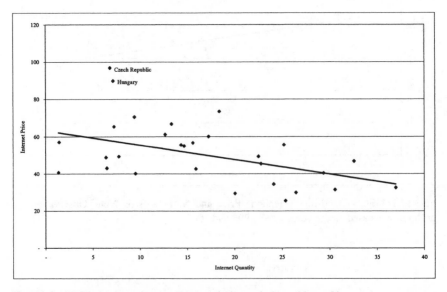

Fig. 12.3. OECD Internet Access Price and Network Size. Note. Observations are group means for the years 1996, 1998 and 2000.

Cross-country price variability is illustrated in group means scatter plots by service in Fig. 12.1 through Fig. 12.3. The scatter plots reveal a negative relationship between current price and network size, albeit with some apparently outlying observations. Fig. 12.1 confirms Mexico is the most expensive and least developed in terms of fixed-line penetration. Turkey, Korea and Iceland exhibit progressively greater penetration at similar average prices. Inspection of individual observations indicates Iceland and Turkey both experience real fixed-line price increases over the sampled time frame while Korea experienced a modest decline. Fig. 12.2 again identifies Mexico and Turkey with observed price-quantity combinations substantially away from the indicated trend-line. Nominal price data for France indicates a substantial price decline from US$ 911 in 1996 to US$ 197 in 2000. Finally, Fig. 12.3 shows the Czech Republic and Hungary among the most expensive and least developed OECD nations in terms of Internet access.

Despite the high degree of cross-country variability, partial correlation statistics presented in Table 12.2 indicate that the expected negative relationship between price and quantity holds. Moreover, these data indicate that service adoption reacts most to own-price changes, with mobile telephone adoption responding most to changes in its own price. Income exhibits the expected positive association with network size, particularly with fixed-line telephone penetration. These data also indicate a close relationship between current and previous period service adoption. Note that the correlation between current and previous period fixed-line telephone adoption (0.9973) is the strongest of the three services. The high correlation together with the relatively low coefficient of variation, indicated in Table 12.1, suggests a relatively mature service in comparison to mobile telephony or Internet access.

Table 12.2. Partial Correlation

	Subscription (t)		
	Fixed-line telephony	Mobile telephony	Internet
Price			
Fixed-line telephony	-0.24	-0.05	-0.05
Mobile telephony	-0.30	-0.54	-0.48
Internet	-0.30	-0.48	-0.53
Income	0.84	0.39	0.48
Subscription (t-1)			
Fixed-line telephony	1.00	0.47	0.57
Mobile telephony	0.44	0.97	0.73
Internet	0.59	0.82	0.97

Finally, given anticipated interdependency between service specific network effects, it is prudent to examine growth rates between the size of the fixed-line telephone, mobile telephone and Internet networks. Inspection of scatter plots indicates a large degree of cross-country variation. However, it is apparent that the growth rates of both the mobile telephone and Internet networks, averaging 55% p.a. and 62% p.a., respectively is substantially higher than the average fixed-line telephone network of 3% p.a.

Estimation and Results

The demand function specified in Eq. 12.21 is a perfect foresight model that holds the marginal utility of wealth constant for each individual, but varies across individuals. Thus, in the present context, the intercepts in the cross-country model capture, in part, country-specific variation in the marginal utility of wealth. The specification is relaxed further by allowing time-specific effects to capture unanticipated growth in wealth. Deviations in country- and time-specific means are captured by adding an argument for per capita income to the demand function, which is associated with changes in marginal wealth across countries and through time.[7] The resulting augmentation (Eq. 12.21a) is a two-way effects model. In addition, given there is a possibility for simultaneity between network effects Eq. 12.21a is specified as a standard form vector autoregressive (VAR) model.

The presence of the unobserved components means that two-stage instrumental variables estimation is required. Past network size is an endogenous variable because of the dependence of network size on the unobserved components. However, caution is required when implementing instrumental variables since, as Nelson and Startz (1990) warn, the use of lagged values as instruments when estimating stochastic Euler equations can lead to bias. Thus, instruments for network size are restricted to future access and use prices. Particular care is taken to ensure instruments are good predictors for network size. The resulting equations are estimated in Limdep 8.0 using the unbalanced panel data set described above.[8]

The final estimating equations, presented in Table 12.3 through Table 12.5, are maximum likelihood estimates of the two-way (country and time) effects using Limdep's random coefficient effects estimator. Random coefficient estimation for the country and time effects are necessary to accommodate the high degree of variation in these data, combined with the presence of unobserved components and possible non-linearity between the network size arguments results. Attempts to correct for possible autocorrelation result in smaller standard errors in the corrected model than the initial model.[9] The forward-looking specification failed due to co-linearity between the future and past network size arguments, largely due to the small time dimension. The problem is common with slowly-changing variables. Therefore, the extent to which consumers anticipate future network size remains an open question.

Interpretation of the estimated coefficients depends on the restrictions imposed on the standard form VAR system. If there is no contemporaneous feedback between the services, then the estimated coefficients can be read as representing the magnitude of the marginal effects. If contemporaneous feedback is assumed, then measuring the magnitude of the structural parameters requires the researcher to impose a recursive structure on the estimating equations. In effect, the identifica-

---

[7] This is based on the assumption that deviations in real per capita income around country- and time-specific means follow a random walk. For further details, see Becker et al. (1994).

[8] The estimators in Limdep 8.0 do not require group sizes to be equal (Greene 2002).

[9] Becker et al. (1994) report a similar outcome.

tion process means that one of the services is deemed to be completely exogenous and not affected by contemporaneous effects from the other two services. At the other extreme, the demand for one of the other two services is completely endogenous. Note, however, that the signs of the structural parameters are not affected by these assumptions.

Table 12.3. Fixed-line Telephone Network

Variable	Coefficient	Standard Error
P_{Ft}	-0.0016	0.0000
N_{Ft-1}	0.9624	0.0005
N_{Mt-1}	-0.0336	0.0005
N_{It-1}	-0.0218	0.0006
Y_t	0.0001	0.0000
Mean		
$\bar{\mu}$	-0.3671	0.0206
$\bar{T}_{1998}$	2.0575	0.0084
$\bar{T}_{2000}$	4.1204	0.0080
σ	0.0425	0.0024
Observations	79	
Log-likelihood	-131.0224	

Inspection of Table 12.3 through Table 12.5 indicates that price arguments in all models are correctly signed and significant. Per capita income is also correctly signed and significant. Importantly, own-network effects are all positive, significant and offer the most explanation for variation in network size. Cross network effects in Table 12.3 are both negative, suggesting the substitution effect dominates the possible network effects arising from the presence of increasing mobile and Internet subscribers. For mobile telephone demand (Table 12.4) cross-network effects are negative from fixed-line telephone and positive from the Internet. For the Internet (Table 12.5) both cross network effects are negative. In addition to the price and network arguments, Table 12.3 through Table 12.5 present mean estimates for the cross-country ($\bar{\mu}$) and the time effects ($\bar{T}_{1998}$ and $\bar{T}_{2000}$). These mean estimates reflect the high degree of heterogeneity across country and also appear to indicate wealth effects have changed substantially across time.

Table 12.4. Mobile Telephone Network

Variable	Coefficient	Standard Error
P_{Mt}	−0.0050	0.0015
N_{Ft-1}	−0.1204	0.0364
N_{Mt-1}	0.8626	0.0346
N_{It-1}	0.0507	0.0448
Y_t	0.0004	0.0000
Mean		
$\bar{\mu}$	−4.9569	1.3417
$\bar{T}_{1998}$	7.5490	0.6436
$\bar{T}_{2000}$	37.2528	0.7118
σ	3.2936	0.1811
Observations	79	
Log-likelihood	−257.7795	

Table 12.5. Internet Network

Variable	Coefficient	Standard Error
P_{It}	−0.0289	0.0005
N_{Ft-1}	−0.0039	0.0013
N_{Mt-1}	−0.0657	0.0011
N_{It-1}	0.9651	0.0014
Y_t	0.0002	0.0000
Mean		
$\bar{\mu}$	−4.4428	0.0544
$\bar{T}_{1998}$	5.0960	0.0206
$\bar{T}_{2000}$	17.5309	0.0216
σ	0.1053	0.0052
Observations	79	
Log-likelihood	−221.6762	

Identified price and income effects, reported in Table 12.6, assume that the fixed-line network is exogenous. Mobile telephony is affected by contemporaneous effects from fixed-line, but not from Internet service. Internet is assumed to be contemporaneously determined by both fixed-line and mobile telephone services. The results in Table 12.6 show that there is little difference between the estimated coefficients and the identified parameters.[10]

Table 12.6. Identified Price and Income Effects

	Fixed-line	Mobile	Internet
P_{it}	-0.0015	-0.0050	0.0006
Y_t	0.0001	0.0006	0.0013

Calculated at the sample means, fixed-line telephone, mobile telephone and Internet price elasticities are –0.0082, –0.0467 and –0.0855, respectively. While all price elasticities are inelastic, it is apparent that order of magnitude follows the order of service introduction with the fixed-line telephone elasticity the smallest and the Internet elasticity the largest. This reflects the standard pattern of the lowest elasticity for the most mature product and the highest elasticity (particularly at low penetration levels) for the least mature product. Given the unit of measurement is per 100 persons, the mobile telephone elasticity suggest a percent fall in the annual mobile subscription price in 2002 yields an extra 15,831,607 subscribers across the OECD.[11] Income elasticity is 0.0514 for fixed-line telephony, 0.2847 for mobile telephony and 0.3042 for the Internet.

Further insight is available by inspecting impulse response functions. Fig. 12.4 presents the Pesaran and Shin (1998) generalized impulse response functions for Internet demand.[12] The shape of the impulse response functions is similar across the three services, with the only substantial difference being the scale across service. Fig. 12.4 shows that an unexpected increase in the size of the Internet has the largest impact on the demand for Internet services. In effect, the figure suggests that every new Internet subscriber induces a further three new subscribers in the following year. This is not the usual network effect; and must be some form of demonstration or word-of-mouth effect. Moreover, the slow decay in the impulse function indicates that a once-off increase in the number of Internet subscribers will induce new subscribers to join the Internet in subsequent years. A similar initial effect is induced by adding new mobile telephone subscribers. That is, another mobile telephone subscriber induces a further three new Internet subscribers next year, followed by a further five new Internet subscribers over the subsequent three years. Thus, from a policy perspective, it doesn't matter which service is promoted

---

[10] Refer to the Appendix for further detail on the identification procedure.

[11] There were 58.47 mobile telephone subscribers per 100 persons in the OECD in 2002.

[12] The generalized impulse response function is not dependent on the ordering imposed on the system.

as both networks will still induce Internet growth. Note, however, that stimulating the size of the fixed-line network results in a substantially smaller impact on the growth of the Internet. Hence, short-lived policies aimed at stimulating the size of the Internet should focus on factors that affect the size of the Internet directly or indirectly via mobile telephony. Similar conclusions are drawn for fixed-line and mobile telephony, viz., stimulation of any part of the telecommunications network results in overall growth. However, stimulating fixed-line telephone offers the least return, so policy levers should be oriented towards mobile telephone and Internet service. Growth in these services will tend to induce growth in fixed-line demand.

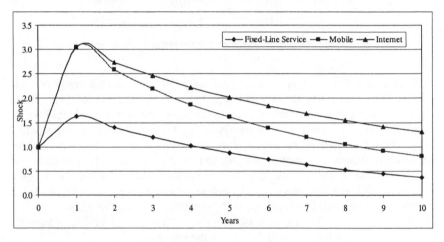

Fig. 12.4. Generalized Impulse Response Functions for Internet Demand

Conclusion

This paper measures the network effect for multiple telecommunications services using a model that is consistent with intertemporal utility maximization for the representative consumer. The estimated results show that the network effect is large for all three services. In addition, there is interaction between services, implying that the network effect should include all telecommunications services. In relation to specific services, the results suggest that mobile telephone and Internet service are most responsive to telecommunications policy changes that impact on price and network size. However, the lack of long country-specific time series prevents testing for the extent to which consumers anticipate network growth when making subscription decisions. The short time series also prevents tests for parameter stability and for more sophisticated approaches to control for heterogeneity across consumers. Nevertheless, the approach demonstrated here has proven fruitful and provides an indication of how estimation of the model could proceed should sufficient micro-data be made publicly available.

Appendix

The VAR system of equations Eq. 12.21a are under-identified due to an inability to estimate coefficients θ_{ij2}. The strategy adopted follows Cecchetti and Rich (2001) to establish a link between the structural and reduced-form representations of Eq. 12.21a. Beginning with Eq. 12.21a,

$$N_{it} = \sum_{t=1}^{T-1} \theta_{it} d_t + \theta_{ii} N_{it-1} + \beta \theta_{ii} N_{it+1} + \sum_j \theta_{ij1} N_{jt-1}$$
$$+ \sum_j \theta_{ij2} N_{jt} + \sum_j \theta_{ij3} N_{jt+1} + \theta_{iP} P_{it} + \gamma Y_t + \varepsilon_{it} , \tag{12.A1}$$

where $\varepsilon_{it} = \theta_{i1} e_t + \theta_{i2} e_{t+1}$. In matrix notation, this VAR is defined

$$\mathbf{B}_0 \mathbf{X}_t = \mathbf{\Phi} + \mathbf{B}_1 \mathbf{X}_{t+1} + \mathbf{B}_2 \mathbf{X}_{t-1} + \mathbf{\Psi} \mathbf{Z}_t + \mathbf{\varepsilon}_t , \tag{12.A2}$$

where $\mathbf{X}'_t = \begin{bmatrix} N_{Ft} & N_{Mt} & N_{It} \end{bmatrix}$, $\mathbf{Z}'_t = \begin{bmatrix} P_{Ft} & P_{Mt} & P_{It} & Y_t \end{bmatrix}$, $\mathbf{I}$ is a 3×3 identity matrix, $\mathbf{\Phi}$ is a 3×1 vector of intercepts, $\mathbf{B}_0$, $\mathbf{B}_1$, $\mathbf{B}_2$ are 3×3 matrices of structural coefficients and $\mathbf{\Psi}$ is a 3×4 matrix of coefficients corresponding to the exogenous variables in $\mathbf{Z}_t$. Eq. 12.A2 is represented as

$$\mathbf{B}(L)\mathbf{X}_t = \mathbf{\Phi} + \mathbf{\Psi} \mathbf{Z}_t + \mathbf{\varepsilon}_t , \tag{12.A3}$$

where $\mathbf{B}(L)$ is a 3×3 matrix of polynomial lags and L denotes the lag operator, i.e., $\mathbf{B}(L) = \mathbf{B}_0 - \mathbf{B}_1 L^{-1} - \mathbf{B}_2 L$. The infinite moving average representation of Eq. 12.A3 is

$$\mathbf{X}_t = \mathbf{A}(L)\mathbf{\varepsilon}_t + \tilde{\mathbf{A}}(L)\mathbf{Z}_t . \tag{12.A4}$$

It is necessary to estimate Eq. 12.A2 in reduced form

$$\mathbf{X}_t = \tilde{\mathbf{\Phi}} + \mathbf{D}_1 \mathbf{X}_{t+1} + \mathbf{D}_2 \mathbf{X}_{t-1} + \tilde{\mathbf{\Psi}} \mathbf{Z}_t + \mathbf{\mu}_t , \tag{12.A5}$$

where $\mathbf{\mu}_t = \mathbf{B}_0^{-1} \mathbf{\varepsilon}_t$. The infinite moving average representation of Eq. 12.A3 is

$$\mathbf{X}_t = \mathbf{C}(L)\mathbf{\mu}_t + \tilde{\mathbf{C}}(L)\mathbf{Z}_t . \tag{12.A6}$$

By expanding the polynomials in the moving average representation of both the structural and standard form VAR, Eq. 12.A4 and Eq. 12.A6, Cecchetti and Rich (2001) show that the first coefficient matrix in the structural VMA Eq. 12.A5 is equal to the inverse of the coefficient matrix in the structural VAR Eq. 12.A2. That is,

$$\mathbf{A}_0 = \mathbf{B}_0^{-1}. \tag{12.A7}$$

Eq. 12.A7 establishes the link between the variance-covariance matrices of both the structural and reduced form VAR representations. That is,

$$E\left[\mu_t \mu_t'\right] = \Sigma = \mathbf{B}_0^{-1} \Omega \mathbf{B}_0^{-1\prime}, \tag{12.A8}$$

where

$$\Sigma = \begin{bmatrix} \hat{\sigma}_{FF} & \hat{\sigma}_{FM} & \hat{\sigma}_{FI} \\ \hat{\sigma}_{MF} & \hat{\sigma}_{MM} & \hat{\sigma}_{MI} \\ \hat{\sigma}_{IF} & \hat{\sigma}_{IM} & \hat{\sigma}_{II} \end{bmatrix}.$$

For the representative consumer, the unobserved effects in ε_{it} are correlated across equations, but are orthogonal across individuals. Hence, the expected co-variance of the structural shocks is assumed zero. Hence the variance-covariance matrix of the unobserved components is defined as

$$\Omega = \begin{bmatrix} \sigma_{FF} & 0 & 0 \\ 0 & \sigma_{MM} & 0 \\ 0 & 0 & \sigma_{II} \end{bmatrix}. \tag{12.A9}$$

Assuming Eq. 12.A9 is diagonal it provides three of the six restrictions necessary to identify the structural VAR. To identify the structural model it is necessary to set three parameters in B_0 equal to zero. Since the zero restrictions imply a contemporaneous ordering, the zero restrictions is rotated so that

$$\mathbf{B}_0 = \begin{bmatrix} 1 & -\theta_{FM} & -\theta_{FI} \\ -\theta_{MF} & 1 & -\theta_{MI} \\ -\theta_{IF} & -\theta_{IM} & 1 \end{bmatrix}$$

requires a choice between four possible orderings, viz.,

$$\mathbf{B}_0 = \begin{bmatrix} 1 & 0 & 0 \\ -\theta_{MF} & 1 & 0 \\ -\theta_{IF} & -\theta_{IM} & 1 \end{bmatrix}, \qquad \mathbf{B}_0 = \begin{bmatrix} 1 & -\theta_{FM} & -\theta_{FI} \\ 0 & 1 & -\theta_{MI} \\ 0 & 0 & 1 \end{bmatrix},$$

$$\mathbf{B}_0 = \begin{bmatrix} 1 & -\theta_{FM} & -\theta_{FI} \\ 0 & 1 & 0 \\ -\theta_{IF} & 0 & 1 \end{bmatrix} \quad \text{and} \quad \mathbf{B}_0 = \begin{bmatrix} 1 & -\theta_{FM} & -\theta_{FI} \\ 0 & 1 & 0 \\ 0 & -\theta_{IM} & 1 \end{bmatrix}.$$

Choosing the first option $\mathbf{B}_0 = \begin{bmatrix} 1 & 0 & 0 \\ -\theta_{MF} & 1 & 0 \\ -\theta_{IF} & -\theta_{IM} & 1 \end{bmatrix}$, $\mathbf{B}_0^{-1} = \dfrac{1}{|\mathbf{B}_0|} Adj\mathbf{B}_0$, where

$Adj\mathbf{B}_0 = \hat{\mathbf{C}}'$, $\hat{C}_{ij} = -(1^{i+j})|M_{ij}|$ and M_{ij} are the minors of $\mathbf{B}_0$. The minors are:

$$M_{11} = \begin{bmatrix} 1 & 0 \\ -\theta_{IM} & 1 \end{bmatrix}, M_{12} = \begin{bmatrix} -\theta_{MF} & 0 \\ -\theta_{IF} & 1 \end{bmatrix}, M_{13} = \begin{bmatrix} -\theta_{MF} & 1 \\ -\theta_{IF} & -\theta_{IM} \end{bmatrix},$$

$$M_{21} = \begin{bmatrix} 0 & 0 \\ -\theta_{IM} & 1 \end{bmatrix}, M_{22} = \begin{bmatrix} 1 & 0 \\ -\theta_{IF} & 1 \end{bmatrix}, M_{23} = \begin{bmatrix} 1 & 0 \\ -\theta_{IF} & -\theta_{IM} \end{bmatrix},$$

and

$$M_{31} = \begin{bmatrix} 1 & 0 \\ 1 & 0 \end{bmatrix}, M_{32} = \begin{bmatrix} 1 & 0 \\ -\theta_{MF} & 0 \end{bmatrix}, M_{33} = \begin{bmatrix} 1 & 0 \\ -\theta_{MF} & 1 \end{bmatrix}.$$

The corresponding determinants of the minors are:

$$|M_{11}| = 1, |M_{12}| = -\theta_{MF}, |M_{13}| = \theta_{MF}\theta_{IM} + \theta_{IF}, |M_{21}| = 0,$$
$$|M_{22}| = 1, |M_{23}| = -\theta_{IM}, \text{ and } |M_{31}| = 0, |M_{32}| = 0, |M_{33}| = 1.$$

The corresponding cofactor matrix is:

$$\hat{\mathbf{C}} = \begin{bmatrix} 1 & \theta_{MF} & \theta_{MF}\theta_{IM} + \theta_{IF} \\ 0 & 1 & \theta_{IM} \\ 0 & 0 & 1 \end{bmatrix}.$$

Therefore,

$$Adj\mathbf{B}_0 = \hat{\mathbf{C}}' = \begin{bmatrix} 1 & 0 & 0 \\ \theta_{MF} & 1 & 0 \\ \theta_{MF}\theta_{IM} + \theta_{IF} & \theta_{IM} & 1 \end{bmatrix}$$

and $\left| \mathbf{B}_0 \right| = 1$. Therefore, the inverse is

$$
\mathbf{B}_0^{-1} = \begin{bmatrix} 1 & 0 & 0 \\ \theta_{MF} & 1 & 0 \\ \theta_{MF}\theta_{IM} + \theta_{IF} & \theta_{IM} & 1 \end{bmatrix}
$$

and Eq. 12.A8 becomes

$$
\begin{bmatrix} \hat{\sigma}_{FF} & \hat{\sigma}_{FM} & \hat{\sigma}_{FI} \\ \hat{\sigma}_{MF} & \hat{\sigma}_{MM} & \hat{\sigma}_{MI} \\ \hat{\sigma}_{IF} & \hat{\sigma}_{IM} & \hat{\sigma}_{II} \end{bmatrix}
$$

$$
= \begin{bmatrix} 1 & 0 & 0 \\ \theta_{MF} & 1 & 0 \\ \theta_{MF}\theta_{IM} + \theta_{IF} & \theta_{IM} & 1 \end{bmatrix} \begin{bmatrix} \sigma_{FF} & 0 & 0 \\ 0 & \sigma_{MM} & 0 \\ 0 & 0 & \sigma_{II} \end{bmatrix} \begin{bmatrix} 1 & \theta_{MF} & \theta_{MF}\theta_{IM} + \theta_{IF} \\ 0 & 1 & \theta_{IM} \\ 0 & 0 & 1 \end{bmatrix}
$$

$$
= \begin{bmatrix} \sigma_{FF} & 0 & 0 \\ \theta_{MF}\sigma_{FF} & \sigma_{MM} & 0 \\ (\theta_{MF}\theta_{IM} + \theta_{IF})\sigma_{FF} & \theta_{IM}\sigma_{MM} & \sigma_{II} \end{bmatrix} \begin{bmatrix} 1 & \theta_{MF} & \theta_{MF}\theta_{IM} + \theta_{IF} \\ 0 & 1 & \theta_{IM} \\ 0 & 0 & 1 \end{bmatrix}
$$

$$
= \begin{bmatrix} \sigma_{FF} & \theta_{MF}\sigma_{FF} & (\theta_{MF}\theta_{IM} + \theta_{IF})\sigma_{FF} \\ \theta_{MF}\sigma_{FF} & \theta_{MF}^2\sigma_{FF} + \sigma_{MM} & \theta_{MF}(\theta_{MF}\theta_{IM} + \theta_{IF})\sigma_{FF} + \theta_{IM}\sigma_{MM} \\ (\theta_{MF}\theta_{IM} + \theta_{IF})\sigma_{FF} & \theta_{MF}(\theta_{MF}\theta_{IM} + \theta_{IF})\sigma_{FF} + \theta_{IM}\sigma_{MM} & (\theta_{MF}\theta_{IM} + \theta_{IF})^2\sigma_{FF} + \theta_{IM}^2\sigma_{MM} + \sigma_{II} \end{bmatrix}.
$$

$$(12.A10)$$

Hence,

$$
\hat{\sigma}_{FF} = \sigma_{FF}, \tag{12.A11}
$$

$$
\hat{\sigma}_{FM} = \theta_{MF}\sigma_{FF}, \tag{12.A12}
$$

$$
\hat{\sigma}_{FI} = (\theta_{MF}\theta_{IM} + \theta_{IF})\sigma_{FF}, \tag{12.A13}
$$

$$
\hat{\sigma}_{MM} = \theta_{MF}^2\sigma_{FF} + \sigma_{MM}, \tag{12.A14}
$$

$$\hat{\sigma}_{MI} = \theta_{MF}(\theta_{MF}\theta_{IM} + \theta_{IF})\sigma_{FF} + \theta_{IM}\sigma_{MM} \quad (12.A15)$$

and

$$\hat{\sigma}_{II} = (\theta_{MF}\theta_{IM} + \theta_{IF})^2 \sigma_{FF} + \theta_{IM}^2 \sigma_{MM} + \sigma_{II} . \quad (12.A16)$$

To solve for the unknowns in Eq. 12.A11 to Eq. 12.A16

$$\hat{\sigma}_{FF} = \sigma_{FF}, \quad (12.A11a)$$

$$\begin{aligned}\theta_{MF} &= \hat{\sigma}_{FM} / \sigma_{FF} \\ &= \hat{\sigma}_{FM} / \hat{\sigma}_{FF} ,\end{aligned} \quad (12.A12a)$$

$$\begin{aligned}\hat{\sigma}_{FI} &= ((\hat{\sigma}_{FM} / \hat{\sigma}_{FF})\theta_{IM} + \theta_{IF})\hat{\sigma}_{FF} \\ \hat{\sigma}_{FI} / \hat{\sigma}_{FF} &= (\hat{\sigma}_{FM} / \hat{\sigma}_{FF})\theta_{IM} + \theta_{IF} \\ \theta_{IF} &= (\hat{\sigma}_{FI} / \hat{\sigma}_{FF}) - (\hat{\sigma}_{FM} / \hat{\sigma}_{FF})\theta_{IM} ,\end{aligned} \quad (12.A13a)$$

$$\begin{aligned}\hat{\sigma}_{MM} &= (\hat{\sigma}_{FM} / \hat{\sigma}_{FF})^2 \hat{\sigma}_{FF} + \sigma_{MM} \\ \sigma_{MM} &= \hat{\sigma}_{MM} - (\hat{\sigma}_{FM} / \hat{\sigma}_{FF})^2 \hat{\sigma}_{FF} ,\end{aligned} \quad (12.A14a)$$

$$\begin{aligned}\hat{\sigma}_{MI} &= (\hat{\sigma}_{FM} / \hat{\sigma}_{FF})((\hat{\sigma}_{FM} / \hat{\sigma}_{FF})\theta_{IM} + (\hat{\sigma}_{FI} / \hat{\sigma}_{FF}) - (\hat{\sigma}_{FM} / \hat{\sigma}_{FF})\theta_{IM})\hat{\sigma}_{FF} \\ &\quad + \theta_{IM}\hat{\sigma}_{MM} - (\hat{\sigma}_{FM} / \hat{\sigma}_{FF})^2 \hat{\sigma}_{FF} \\ \hat{\sigma}_{MI} &= (\hat{\sigma}_{FM} / \hat{\sigma}_{FF})^2 \theta_{IM}\hat{\sigma}_{FF} - (\hat{\sigma}_{FM} / \hat{\sigma}_{FF})^2 \theta_{IM}\hat{\sigma}_{FF} \\ &\quad + \theta_{IM}\hat{\sigma}_{MM} - (\hat{\sigma}_{FM} / \hat{\sigma}_{FF})^2 \hat{\sigma}_{FF} \\ \hat{\sigma}_{MI} &= \theta_{IM}\hat{\sigma}_{MM} - (\hat{\sigma}_{FM} / \hat{\sigma}_{FF})^2 \hat{\sigma}_{FF} \\ \theta_{IM}\hat{\sigma}_{MM} &= (\hat{\sigma}_{FM} / \hat{\sigma}_{FF})^2 \hat{\sigma}_{FF} + \hat{\sigma}_{MI} \\ \theta_{IM} &= (\hat{\sigma}_{FM} / \hat{\sigma}_{FF})^2 (\hat{\sigma}_{FF} / \hat{\sigma}_{MM} + (\hat{\sigma}_{MI} / \hat{\sigma}_{MM})\end{aligned} \quad (12.A15a)$$

and

$$\hat{\sigma}_{II} = ((\hat{\sigma}_{FM} / \hat{\sigma}_{FF})((\hat{\sigma}_{FM} / \hat{\sigma}_{FF})^2 (\hat{\sigma}_{FF} / \hat{\sigma}_{MM}) + (\hat{\sigma}_{MI} / \hat{\sigma}_{MM}))$$
$$+ (\hat{\sigma}_{FI} / \hat{\sigma}_{FF}) - (\hat{\sigma}_{FM} / \hat{\sigma}_{FF})\theta_{IM})^2 \hat{\sigma}_{FF} + ((\hat{\sigma}_{FM} / \hat{\sigma}_{FF})^2 (\hat{\sigma}_{FF} / \hat{\sigma}_{MM})$$
$$+ (\hat{\sigma}_{MI} / \hat{\sigma}_{MM}))^2 (\hat{\sigma}_{MM} - (\hat{\sigma}_{FM} / \hat{\sigma}_{FF})^2 \hat{\sigma}_{FF}) + \sigma_{II}$$

$$(12.A16a)$$

$$\sigma_{II} = \hat{\sigma}_{II} - (((\hat{\sigma}_{FM} / \hat{\sigma}_{FF})((\hat{\sigma}_{FM} / \hat{\sigma}_{FF})^2 (\hat{\sigma}_{FF} / \hat{\sigma}_{MM}) + (\hat{\sigma}_{MI} / \hat{\sigma}_{MM}))$$
$$+ (\hat{\sigma}_{FI} / \hat{\sigma}_{FF}) - (\hat{\sigma}_{FM} / \hat{\sigma}_{FF})\theta_{IM})^2 \hat{\sigma}_{FF} + ((\hat{\sigma}_{FM} / \hat{\sigma}_{FF})^2 (\hat{\sigma}_{FF} / \hat{\sigma}_{MM})$$
$$+ (\hat{\sigma}_{MI} / \hat{\sigma}_{MM}))^2 (\hat{\sigma}_{MM} - (\hat{\sigma}_{FM} / \hat{\sigma}_{FF})^2 \hat{\sigma}_{FF})) .$$

Identifying the Exogenous Structural Parameters

The standard form coefficients for the exogenous variable coefficient matrix is

$$\Psi = \begin{bmatrix} \gamma_{FF} & 0 & 0 & \gamma_{FY} \\ 0 & \gamma_{MM} & 0 & \gamma_{MY} \\ 0 & 0 & \gamma_{II} & \gamma_{IY} \end{bmatrix}$$

and is represented by the product $\Psi = B_0^{-1}\hat{\Psi}$, or

$$\begin{bmatrix} \gamma_{FF} & 0 & 0 & \gamma_{FY} \\ 0 & \gamma_{MM} & 0 & \gamma_{MY} \\ 0 & 0 & \gamma_{II} & \gamma_{IY} \end{bmatrix}$$

$$= \begin{bmatrix} 1 & 0 & 0 \\ \theta_{MF} & 1 & 0 \\ \theta_{MF}\theta_{IM} + \theta_{IF} & \theta_{IM} & 1 \end{bmatrix} \begin{bmatrix} \hat{\gamma}_{FF} & 0 & 0 & \hat{\gamma}_{FY} \\ 0 & \hat{\gamma}_{MM} & 0 & \hat{\gamma}_{MY} \\ 0 & 0 & \hat{\gamma}_{II} & \hat{\gamma}_{IY} \end{bmatrix}$$

$$= \begin{bmatrix} \hat{\gamma}_{FF} & 0 & 0 & \hat{\gamma}_{FY} \\ \theta_{MF}\hat{\gamma}_{FF} & \hat{\gamma}_{MM} & 0 & \theta_{MF}\hat{\gamma}_{FY} + \hat{\gamma}_{MY} \\ (\theta_{MF}\theta_{IM} + \theta_{IF})\hat{\gamma}_{FF} & \theta_{IM}\hat{\gamma}_{MM} & \hat{\gamma}_{II} & (\theta_{MF}\theta_{IM} + \theta_{IF})\hat{\gamma}_{FY} + \theta_{IM}\hat{\gamma}_{MY} + \hat{\gamma}_{IY} \end{bmatrix} .$$

Therefore,

$$\gamma_{FF} = \hat{\gamma}_{FF} , \qquad\qquad (12.A17)$$

$$\gamma_{FY} = \hat{\gamma}_{FY}, \tag{12.A18}$$

$$\gamma_{MM} = \hat{\gamma}_{MM}, \tag{12.A19}$$

$$\gamma_{MY} = \theta_{MF}\hat{\gamma}_{FY} + \hat{\gamma}_{MY}, \tag{12.A20}$$

$$\gamma_{II} = \hat{\gamma}_{II} \tag{12.A21}$$

and

$$\gamma_{IY} = (\theta_{MF}\theta_{IM} + \theta_{IF})\hat{\gamma}_{FY} + \theta_{IM}\hat{\gamma}_{MY} + \hat{\gamma}_{IY}. \tag{12.A22}$$

References

Artle R, Averous C (1973) The telephone system as a public good: Static and dynamic aspects. Bell Journal of Economics and Management Science 4: 89–100

Becker GS, Grossman M, Murphy KM (1994) An empirical analysis of cigarette addiction. American Economic Review 84: 396–417

Cecchetti SG, Rich RW (2001) Structural estimates of the US sacrifice ratio. Journal of Business and Economics Statistics 19: 416–27

Economides N, Himmelberg C (1995) Critical mass and network size with application to the US FAX market. Discussion Paper EC-95-11, Stern School of Business, New York University

Greene WH (2002) LIMDEP Version 8.0 Reference Guide. Econometric Software Incorporated, New York

International Telecommunication Union (2003) World telecommunications indicators database. International Telecommunication Union, Geneva

Littlechild SC (1975) Two-part tariffs and consumption externalities. Bell Journal of Economics and Management Science 5: 661–70

Madden G, Coble-Neal G (2002) Internet forecasting and the economics of networks. In Loomis D, Taylor D (eds) Forecasting the Internet: Understanding the data communications revolution. Kluwer Academic Publishers, Boston, pp 105–30

Madden G, Coble-Neal G (2004) Economic determinants of global mobile telephony growth. Information Economics and Policy 16, forthcoming

Madden G, Coble-Neal G, Dalzell B (2004) A dynamic model of mobile telephony subscription incorporating a network effect. Telecommunications Policy 28 (2): 135–44

Nelson CR, Startz R (1990) The distribution of the instrumental variables estimator and its t-ratio when the instrument is a poor one. Journal of Business 61: S125–40

OECD (1997, 1999, 2001, 2003) Communication outlook. OECD, Paris

Perl LJ (1983) Residential demand for telephone service 1983. Prepared for Central Service Organization of the Bell Operating Companies, Inc., White Plains NY

Pesaran MH, Shin Y (1998) Generalized impulse response analysis in linear multivariate models. Economics Letters 58: 17–29

Rohlfs J (1974) A theory of interdependent demand for a communications service. Bell Journal of Economics and Management Science 5: 16–37

Taylor LD, Kridel DJ (1990) Residential demand for access to the telephone network. In de Fontenay A, Shugard MH, Sibley DS (eds) Telecommunications demand modeling. North-Holland, Amsterdam

13 Open Access Rules and Equilibrium Broadband Deployment

Glenn A. Woroch

Introduction

Investment in advanced communications infrastructure promises such tantalizing payoffs as accelerated economic growth and enhanced national competitiveness. Relatively little disagreement arises in policy debates that such benefits are possible—usually only their size and distribution across the population are at issue. Bitter disputes, however, have occurred over which path will lead the communications sector to deploy these technologies most expeditiously and equitably.

While competition is now widely accepted as essential to effective broadband policy, policy makers have adopted a wide array of alternative rules to promote advanced network investment. Arguably, the most contentious of these policies forces incumbent network owners to share their facilities and equipment to enable rival broadband service providers. Opponents protest that such sharing destroys incentives to undertake the expense and risk of deploying new technologies. Such policies, they contend, will likely delay the rollout of innovative services, and possibly forestall deployment in some markets altogether. Proponents of 'open access', in contrast, emphasize how sharing creates competition in retail service without the waste of duplicate investment.[1] In its strong form, this view foresees future facilities-based competition which results in net increase in advanced network investment.

The impact of pro-competition policy on broadband deployment has generated a considerable body of empirical evidence. This literature predates the opening of many advanced service markets to competition.[2] Several early econometric investigations conclude that liberalized regulation of incumbent telephone companies had the effect of stimulating their investment in advanced technology.[3] The empirical evidence on this question is far less conclusive once those communications markets were open to competition, no doubt partly a result that the studies were

---

[1] Open access refers to policies that require facilities-based providers to share their networks with downstream service rivals. In the US, the term has been used to describe nondiscriminatory interconnection between a cable TV system, and affiliated and unaffiliated Internet service providers (ISPs).

[2] Woroch (1998) provides a review of earlier empirical studies of the competition-investment relationship in telecommunications and other deregulated industries.

[3] In particular, Greenstein et al. (1995) and Kridel et al. (1996) find that price cap regulation (or its variants) is associated with increased investment by local telephone incumbents in new digital technology, including optical fiber.

conducted on behalf of interested parties. For instance, Willig et al. (2002) conclude that aggregate incumbent local exchange carrier (ILEC) investment is inversely related to unbundled network element prices. Haring et al. (2002) take issue with this conclusion. In comparison, Crandall et al. (2002) offer some impressionistic evidence that leads them to conclude that unbundling reduces investment by ILECs. Gabel and Huang (2003) and Floyd and Gabel (2003) estimate simultaneous equations models of ILEC deployment of packet switching technology and of the presence of competitive local carriers as a function of regulatory treatment and market conditions. They find, contrary to the earlier analysis, that traditional rate of return regulation is associated with a greater propensity to deploy advanced technology.

A longer-run rationale for open access is the possibility that it will facilitate infrastructure competition by providing entrants a 'stepping stone' whereby they are able to market some kind of service while they are building their own network. By getting to market more quickly, entrants may be more potent competitors when infrastructure competition materializes. Examining an early period of competition, Woroch (2000) finds evidence that facilities-based entry triggers a virtuous cycle whereby incumbent carriers respond by deploying urban fiber rings as both compete for high-speed business access customers. Addressing a similar question, Crandall et al. (2002) estimate a relatively high elasticity of substitution between leased and purchased local loops, and conclude that low lease rates significantly discourage facilities-based entry.

In the end, no consensus emerges from the empirical literature examining the impact of regulation-mandated competition and the extent of firm-specific and industry-wide investment. Based on superficial modeling, the empirical studies are incapable of capturing the complex relationship between the pro-competitive policy and incumbents' and entrants' investment incentives. Equilibrium investment behavior—especially when it involves innovation—does not obey a simple direct relation to standard measures of competition.[4] Furthermore, broadband policy can be highly idiosyncratic, generating unique and unexpected consequences for broadband investment.

The principal contribution of this paper is to provide a formal model that links open access policies with equilibrium deployment of advanced networks which, in turn, generate hypotheses to guide empirical tests and inform broadband policy initiatives. The timing of deployment of broadband services is modeled as a 'technology race' among competing network owners and service providers. In equilibrium, firms decide if and when to deploy broadband technology. Depending on identifiable demand and cost conditions, any contestant may 'win' this race. The model with two firms depicts the vigorous contests taking place in most countries between incumbent local telephone companies and cable TV operators that are deploying digital subscriber line (DSL) and cable modem (CM) technology, respectively.

---

[4] See Boone (2000).

Properties of the equilibrium technology race are used to characterize the impact of various open access policies on the outcome of broadband race. Regulatory policy affects the outcome of this race by altering participants' investment incentives and, in turn, the pace of deployment and the winning technology. The analysis begins by examining the indirect role of non-broadband regulation on broadband deployment incentives. Rate regulation of traditional services, e.g., voice telephony and broadcast video, are irrelevant if it is uncoupled from the broadband deployment decision. Legacy regulation could nevertheless further constrain profits due to cost sharing rules and the like. In that case, non-broadband regulation tends to delay broadband deployment by both carriers even if they are treated symmetrically. Next, a simple policy of sharing a monopoly incumbent's advanced network and sharing it with a single pure reseller of broadband services is considered. In that case the outcome is unambiguous: a policy of open access delays deployment, and might even preclude it. On balance whether social welfare is reduced depends on the extent to which consumers benefit from lower broadband service prices.

Turning to a setting with two facilities-based contenders, a policy that mandates the leader share its broadband network with its rival while that latter builds its own facilities is examined. Again, both deployment dates are postponed relative to the outcome without sharing. This version of the open access is indicative to asymmetric treatment of local telephone companies and cable TV operators. Mandating that both incumbents must make their broadband facilities available to its rival may reverse this tendency and accelerate deployment. One last open access policy allows a follower to continue to lease the leader's network even after it has built its own network. Allowed to both lease and build broadband facilities, the follower's incentive to upgrade its own network is diminished, delaying the date of platform competition as well as delaying initial construction of broadband network. Similar impacts on deployment arise when regulators mandate pure resale, in which case resellers can lease broadband networks at regulated wholesale rates but with no opportunity to build their own facilities. When pure resellers are ruled out, and both carriers gain access to the other's broadband network, initial deployment is delayed even when profit regulation is applied symmetrically. In this case, however, it is possible that the order of deployment is reversed, in which case an inferior technology could be deployed before the superior one.

In the next section some distinguishing features of broadband technology are described. It is hoped that these features are captured by the technology race model, as well as several of the more prominent forms of broadband regulation. Then the technology race model of broadband deployment is constructed, while equilibrium and some comparative-static properties are briefly described. Much of the derivation is relegated to the Appendix. The subsequent sections are devoted to analyzing the variants of the open access rules. A final section concludes with remarks about a broader welfare issue of facility sharing policies.

Broadband Technology and Regulation

Broadband service is defined chiefly by its bidirectional data transmission speed.[5] Whereas bit rates exceeding 200 kbps were once considered broadband service, much faster thresholds are now required for that classification. In reality transmission speeds vary continuously even for the same technology deployed within a region. Minimum speeds to meet users' demands will increase further as content and applications become available that take full advantage of the greater bandwidth.

Many kinds of physical networks can deliver high-speed data. The most common technologies are DSL over the public switched telephone network and cable modem over cable TV systems.[6] The duopoly race that is the focus of much of the analysis is readily interpreted as competition between an incumbent local telephone company and a cable TV operator. Wireless networks—3G cellular, WiFi and two-way satellite—represent a smaller but rapidly growing share of the broadband access market. New wire-based networks are also being built in competition with embedded networks.[7] These broadband service providers sometimes undertake green field construction of fiber networks and at other times retrofit of existing cable TV systems.[8]

Broadband Technology and Cost Characteristics

Technologies for delivering broadband services continue to experience steady improvement, leading to increased speeds and falling costs. Progress of broadband technologies remain highly uncertain, however, as they depend on overcoming many technical challenges and on the shifting capabilities of embedded networks.

The investment needed to deploy broadband service is very lumpy. Although some expenditure varies with the scale of the broadband network, the vast majority of deployment costs are fixed and sunk[9]. The incremental cost of adding broad-

---

[5] Other key attributes include an always-on connection and unrestricted access to the Internet. Transmission rates may be further qualified by the constancy of the bit rate and the asymmetry in upload and download speeds.

[6] International Telecommunication Union (ITU, 2003) gives penetration of the leading broadband technologies among select developed countries.

[7] The electricity grid represents an existing wireline network that potentially can carry broadband services using power-line transmission technology.

[8] See US General Accounting Office (2004) for six case studies of such broadband deployments in the US.

[9] Deployment costs that vary with the size of customer base include customer premise equipment—modems, network cards and radio dishes—as well as network-side investments such as DSLAMs and cable head-end equipment. These costs will also fall over time along as, e.g., options for customer self-installation are perfected. Customer acquisition costs tend to vary with customer base as well, but they may not fall as quickly under competition with an expected increase in customer churn. They also differ across carriers

broadband capability to an incumbent's network tends to be much less than that for green field construction.[10] Given any specific geographic market, deployment cost of incumbent carriers also depends on the inherent characteristics of their respective broadband technologies as well as their network footprints. For instance, cable systems have a greater presence in residential neighborhoods and rural areas whereas the telephone network is relatively better represented in densely populated areas such as central business districts. DSL more often has an attenuation problem in residential areas, especially low-density suburbs and rural areas, where loop lengths are the longest. CM service requires an upgrade to hybrid fiber-coax network so as to enable two-way data transmission. Wireless broadband technologies typically necessitate smaller outlays for facilities and equipment than fixed networks though the cost of licensing radio spectrum can eliminate this advantage.

The model of a broadband race developed here captures the fixed, sunk nature of deployment costs, the steady decline in those costs, and their differences across contenders. Other aspects such as uncertainty over the cost of alternative technologies are suppressed in favor of simplicity. Depending on the scope of the geographic market, the model can accommodate cases where just one of the incumbent networks is viable, as well as the more interestingly case where both vie to serve the market.

Policy Promoting Broadband Competition

As broadband services only recently reached mass market appeal, a variety of regulatory treatments are found. Governments have taken a variety of policy approaches, ranging from forbearance to direct regulation of rates and investments. Even within the US, the 50 State commissions and the FCC exhibit disparity in their treatment of broadband services.[11] Some regulators require broadband tariffs—in some cases placing those services under price caps. Generally, ILEC provision of DSL service is the most common target of regulation, with cable modem and wireless broadband access receiving light regulation. Notably, implementation of the Telecommunications Act of 1996 (TA96) required ILECs to open their networks and unbundled network service elements (UNEs) for lease to competitive

---

as it is likely to be cheaper to migrate an existing non-broadband customer to broadband service than to either attract a first-time user or poach extant broadband customers from rivals.

[10] Alternatively, these embedded networks were optimized to provide services that differ in several respects from broadband data access: local and long-distance networks are designed for voice telephony, and cable's hybrid fiber-coax network for broadcast of multichannel video entertainment. In addition, green field deployment—such as the construction a next-generation optical fiber network—has much freedom to choose the serving territory, the network architecture and the transmission protocol without concern for compatibility with the legacy network.

[11] See Lee (2001) for a summary of State commission treatment of broadband services.

local exchange carriers (CLECs). One such unbundled element is the high-frequency digital portion of local loops. Such 'line splitting' enabled service-based providers to offer high-speed access along with other telephony and/or video services, possibly provided using other unbundled elements, and especially UNE platforms. The Telecom Act also identifies a potential exemption from the unbundling mandates for, among other services, high-speed Internet access. In the FCC's implementation of Section 706, they allow ILECs to avoid unbundling their high-speed facilities when they structurally separate their data affiliates from telephone operations, supplying competing carriers with parity service to their broadband affiliate.[12]

Typically, cable provision of high-speed Internet access is lightly regulated. Recently the FCC declared cable modem an 'information service', and hence beyond its regulatory reach, only to have the decision subsequently reversed by an appeals court. Whatever the outcome of Federal treatment, local municipalities have considerable power to regulate CM service deriving from their authority to award operating franchises. The principal concern here is how franchise boards have placed new conditions on their franchisees, and in several highly visible decisions, have forced cable franchisees to provide nondiscriminatory access to all ISPs as well as their affiliated ISPs. Neither cable franchise authorities nor State and Federal regulators have required operators to unbundle cable modem service and to offer high-speed data transmission to unaffiliated service providers. Nevertheless, as is shown below, regulation of cable video services can indirectly impact broadband deployment decisions.[13]

Opening of broadband facilities and unbundling high-speed services is the first of two essential elements of an open access policy. The other component is the pricing of the network services used by service-based providers. Pricing rules impact an incumbent's incentive to build a broadband network as well as the incentives of rival carriers to purchase its network services. These rates could be set unilaterally by the incumbent or privately negotiated between the carriers. It is assumed here that a regulatory authority sets wholesale rates although that process is not explicitly modeled. The analysis will be guided by two popular wholesale pricing methodologies: the efficient component pricing rule (ECPR) and long run incremental cost (LRIC) pricing. The FCC devised a version of LRIC, called Total Element LRIC to price network elements and imposed this methodology on states that did not develop their own approach. Generally speaking, ECPR ensures the facility owner is compensated for profits on sales lost to downstream competitors, whereas LRIC pricing makes no such guarantee. Nevertheless, the analysis as-

---

[12] SBC has, for instance, chosen to place all its broadband data services in a separate subsidiary, Advanced Solutions, Inc.

[13] While the TA96 eliminated (as of February 1999) Federal regulation of rates for cable TV services that had been enabled by the Cable Act of 1992, States and municipalities still retain some authority to regulate cable TV rates. Also, the FCC continues to impose 'must carry' obligations on cable operators whereby they retransmit qualifying local over-the-air programming.

sumes that facility owners will not voluntarily unbundle their broadband networks and lease them to service-based rivals.

Of course, in the short run, resale provides another means by which competing providers can exert discipline on incumbent carriers, at least in limited terms, based on price and service characteristics. Resellers, like all service-based providers, trade off the cost and risk associated with constructing a network against the wholesale rates they pay for using the incumbent's facilities, in addition to lack of control over network capabilities. The opportunity to provide broadband service without investing in facilities gives service-based providers a 'real option' that can be 'called' when realized demand or cost conditions are out of line with expectations.[14] Pricing rules that ignore this option value will skew incentives to undertake such large, sunk outlays, causing incumbents to curtail or delay the investment, or in the extreme, to forgo the expenditure entirely.[15]

A Technology Race Model of Broadband Deployment

Here the standard model of a technology race is modified to capture essential features of broadband deployment decisions.[16] The model is sufficiently flexible to accommodate key features of regulatory policies aimed at promoting broadband investment. Each firm has the strategic timing problem of deciding if and when to deploy a broadband network. If they should do so, they incur a one-time, fixed cost that varies by firm and the date of deployment: $c_i(t)$ is the nominal cost of broadband deployment for Firm i at date t.[17, 18] Due to continuous improvement in microelectronics, optics, radio technology, software and other enabling technol-

---

[14] Hausman (2000) and Pindyck (2004) show how unbundling creates a real option that derives from the sunk nature of investment and the uncertainty of future costs. They go on to demonstrate how TELRIC pricing under compensates an incumbent for its investment relative to competitive returns. This modeling approach does not draw conclusions about the equilibrium timing of deployment decisions.

[15] A related concern is that, compelled to open its facility, an owner will degrade access service to a competitor, or find another means to disadvantage rivals relative to its service. Open access rules seek to address such concerns via nondiscrimination provisions.

[16] See Katz and Shapiro (1987) and Fudenberg and Tirole (1985) for general formulations of a technology race between contestants. Here some of Fudenberg and Tirole's notations are adopted. For an application of the technology race model to a context similar to this one, see Riordan's (1992) model of video competition between telephone companies and cable operators and its regulation.

[17] In reality, construction of any network is a gradual process during which firms will incur adjustment costs. In that case, firms also must trade off the high cost of faster deployment costs more in present value (or sacrifice quality of the network) compared to that for slower deployment.

[18] Faulhaber and Hogendorn (2000) model staged broadband competition in which carriers decide how much capacity to build, as well as, the extent of territory served.

ogy, (nominal) deployment cost falls over time at a decreasing rate: $c_i'(t) < 0$ and $c_i''(t) > 0$. Deployment costs asymptotically approach a lower bound: $C_i > 0$. The bounds are low enough to allow at least one firm to find broadband deployment profitable. Costs are allowed to vary across firms reflecting differences in their embedded networks and the technology to upgrade to broadband. This variation is an important source of determining the equilibrium order of deployment. Finally, assume that deployment costs $c_i(0)$, are so high that no firm finds deployment profitable, even if it were the sole broadband service provider. Further, deployment costs are assumed to not depend on any firms' past occurrences of broadband investment—ruling out, among other phenomenon, the possibility of 'learning by building'.

Falling deployment costs pose broadband contenders with a tradeoff between lower costs from waiting against the risk of being preempted by a rival. Each firm's profit in a given period depends on the industry profile at the moment in terms of broadband supply. When more than one firm offers broadband service in a market, they share in the broadband revenue according to the intensity of retail competition.[19] The dependence of firms' operating profit on past deployment history is indicated by $\pi_i^h(t)$, where h captures the history of past deployment by all firms in the market. It is important to note that this expression includes operating profits from non-broadband operations. For instance, it would include traditional switched voice revenues for a local telephone company and video revenues for a cable TV operator. When there are two facilities-based firms, as in the case of a telephone and a cable company, let $h = n$ to indicate that neither one has deployed, $h = 1$ (2) that just Firm 1 (2) has deployed, and finally $h = d$ that there are dual networks deployed.

Generally, the $\pi_i^h(t)$ s are increasing over time, reflecting a steady increase in service demand—typically driven by growth in the number and quality of broadband applications and content (e.g., multimedia games, video conferencing and VoIP services), and also word-of-mouth communication and network effects stemming from user file sharing (e.g., digital photographs and videos). As written, dual-deployment profits π_i^d are invariant to the order in which firms choose to deploy, in which case no permanent advantages derive from being a leader or a follower. Since, furthermore, profits do not depend on the timing of the deployments, first-mover advantage gained through customer lock-in or brand equity is ruled out. The provision of broadband service adds to operating profit of the carrier who deploys the necessary facilities: $\pi_i^i(t) > \pi_i^n(t)$ and $\pi_i^d(t) > \pi_i^j(t)$. It is reasonable to assume that the baseline profits prior to deployment, $\pi_i^n(t)$, grows rather slowly since those revenues derive from mature service markets. In comparison, post-deployment profit, $\pi_i^i(t)$, should track the growth rate in aggregate demand for broadband services. A firm's operating profit is likely to fall when its

---

[19] In this way, the model differs from patent race models in which the winner takes all.

rival deploys broadband technology even if it has not entered that market yet: $\pi_i^j(t) < \pi_i^n(t)$ and $\pi_i^d(t) < \pi_i^i(t)$. Through marketing and bundling of narrowband and broadband services, a laggard suffers loss of customers to its broadband rival: $\pi_i^n(t) > \pi_i^j(t)$.

Duopoly Broadband Race Equilibrium

Contestants choose a timing strategy or rule specifying the build date given history to that time. With knowledge of deployment times, firms can compute their discounted profit. If there is no competitive threat, a firm chooses the best time to lead, denoted t_i^l. This occurs when the incremental operating profit from deployment equals a measure of the dynamic marginal cost of deployment:

$$L_i(t) = \pi_i^i(t) - \pi_i^n(t) = rc_i(t) - c_i'(t) .\tag{13.1}$$

Refer to $L_i(t)$ as the leader's incentive. It measures the increment in operating profits when the firm is the only one to deploy broadband services. The best time to follow, t_i^f, conditional on the other firm having deployed, obeys a similar rule:

$$F_i(t) = \pi_i^d(t) - \pi_i^j(t) = rc_i(t) - c_i'(t) .\tag{13.2}$$

Here $F_i(t)$ is the follower's incentive. The best time to follow is independent of the date of first deployment, signifying there are no spillovers from the leader to the follower.

The leader and follower dates do not determine the equilibrium dates because, while a firm might choose to deploy earlier whether it is a leader or follower, its rival may prefer not to assume the remaining position in the order of deployment. To sort this out, define a carrier's preemption time, t_i^0, as the time when Firm i is indifferent between leading at that date and, instead, allowing its rival to lead at that same date. With this notation it is stated that the main result on equilibrium in this two-player timing game, renumbering the firms if necessary: If $t_1^0 < t_2^0$ and $t_1^l < t_2^l$, then, at a sub-game perfect Nash equilibrium, Firm 1 deploys first at the earlier of the dates, t_2^0 and t_1^l, while Firm 2 deploys second at time t_2^f. In words, when Firm 1 is the more eager to deploy, it will deploy first in equilibrium, but not necessarily at the time that yields it maximal profits. Instead, it deploys an instant before its rival finds it preferable to be the leader rather than being relegated to the role of the follower. To wait beyond the rival's preemption date, the firm risks having the lead stolen. The assumption that contestants react quickly to deployment decisions is crucial to arriving at this equilibrium as Fudenberg and Tirole

(1987) emphasize.[20] Within a single market, the winner may take all in this race. Once a carrier has deployed a broadband network, a complete overbuild is not profitable as the providers enter into a price war. Households are unlikely to subscribe to more than one broadband Internet access service. Alternatively, dual subscription may not be rare among business customers. Furthermore, depending on the technology, the costs may not be great to build a network that merely 'passes' subscribers in an area, without incurring the additional investment necessary to connect users. Finally, differentiation in the broadband services is inevitable given the differences in technology.

Comparative-static Analysis

Clearly, both market demand and firm cost conditions affect deployment timing. Open access policy alters these conditions and, hence, the equilibrium timing of broadband deployment. To conduct these exercises, label, without loss of generality, Firm 1 as the leader when no open access is imposed, and Firm 2 the follower. Let the baseline market condition be such that both firms eventually deploy a broadband network in absence of open access. Then formally the effect of an open access rule expressed as a perturbation in the profit path of both firms during the monopoly period (when just one firm has built a network and offers broadband service) and the dual deployment period (when both firms have built networks):

$\pi_i^h(t) + \varepsilon \delta(t)$ where $\varepsilon > 0$ and $\delta(t) > 0$ for t in the relevant range and $h = i$ and d.

Next, examine the local impact of open access rules when ε is arbitrarily small.

Table 13.1. Comparative-static Analysis of Broadband Race Operating Profits

	π_i^i	π_i^n	π_i^d	π_i^j	π_j^j	π_j^n	π_j^d	π_j^i
t_i^l	−	+	0	0	0	0	0	0
t_j^0	0	0	+	−	−	0	+/−*	+
t_j^f	0	0	0	0	0	0	−	+

Note. * - according to whether $t_i^f < / > t_j^f$.

Simple comparative static exercises provide the impact of open access rules. Of particular interest is how the deployment pattern is affected by the levels of operating profits. In the Appendix some comparative static results are calculated. Those results are summarized above in Table 13.1. Effects of changes in deployment costs are somewhat less useful. By specifying a functional form for the de-

---

[20] Allowing firms to be imperfectly informed as to rivals' progress on its construction program, or informed with a lag, is more realistic, but greatly complicates the strategic interaction of the timing game.

ployment costs, comparative static effects are derived. The table entries are reasonably intuitive. For instance, lower construction costs—both in terms of absolute level and also the dynamic marginal cost—accelerate deployment dates.

Regulation of Non-broadband Service

Next, consider the impact that non-broadband regulatory policy can have on the equilibrium of the broadband race. Somewhat surprisingly, the analysis finds that incentives to make broadband investments are affected by the regulatory treatment of other services provided by the incumbent carriers. For instance, regulations governing cable operator's video sales and a telephone company's rates for voice services affect the equilibrium deployment of broadband services indirectly. To begin, assume that regulation has the effect of reducing the operating profits of the two incumbents by a fixed amount and that this amount is constant over time and independent of deployment history. So, for instance, Firm 1's operating profits are reduced by a constant amount Δ_1 regardless of which, if any, firms have deployed a broadband network; and similarly Δ_2 for Firm 2 where the profit impact of regulation may differ by carrier. Clearly, this regulation has no effect on the timing of deployment, whether or not the lead firm, Firm 1, engages in preemption.

First, the timing of the first deployment is determined by the incentive to lead: $L_i(t) = \pi_1^1(t) - \pi_1^n(t)$. If profit is reduced by Δ_1 under either scenario, i.e., Firm 1 leads or neither firm deploys, then there is no change in the time of first deployment, t_1^l. Second, a similar result holds when preemption occurs. To see this, note that preemption time t_2^0 depends on the operating profits π_2^2, π_2^1 and π_2^d. Totally differentiating the preemption date when these profits are reduced by an amount of time, cancel each other out, resulting in no change in any deployment date.[21]

Regulators are not likely to intervene in these markets in such a symmetric way, however. If broadband capabilities invite any regulatory intervention, overall firm profits will be more constrained after deployment than before. Broadband services are provided using at least some of the network assets built to provide narrowband services. Any regulation that allocates some portion of the cost of these facilities to broadband, in addition to investments that are directly attributable to broadband services, will make narrowband services appear more profitable. In that case, the constraint on profits will tighten after a firm has deployed broadband services.

To be precise, suppose regulation reduces profits by Δ' before a firm's broadband deployment and by Δ'' afterwards, where $\Delta'' > \Delta'$. Even when the two providers are treated symmetrically, so that their profits are reduced by the same

---

[21] Of course, profit regulation could be made conditional on the level of competition in the industry so as to either accelerate or impede deployment. The interest here is in policy that appears neutral toward broadband investment decisions.

amounts, this regulatory rule results in delay of deployment. It is easy to show that both t_1^l and t_2^0 increase, so that first deployment is delayed regardless of whether the leader engages in preemption. Even though broadband services are not directly regulated, by penalizing broadband deployment through the cost allocation rule applied to narrowband services, both firms are less eager to make the necessary investment. Direct regulation of retail and wholesale broadband services are not the only means to affect their deployment. Regulation of a service that was related to broadband services through an artificial cost allocation rule is also examined. Alternatively, regulation to a non-broadband service that is complementary in demand to broadband Internet access (or a close substitute such as ISDN) could have been applied to arrive at the same conclusion. This exercise suggests that, given existing regulatory institutions, non-regulation of broadband service may be a policy maker's chimera.[22]

Open Access Rules and the Equilibrium Broadband Race

Open access rules vary across several dimensions. Which facilities-based broadband providers must grant access, the first firm to build a broadband network, or any firm that eventually builds such a network? Who is assigned rights to access existing broadband facilities, other facilities-based carriers who have built (or potentially could build) a broadband network, or service-based providers who resell broadband services? When must access be granted? Open access could be required immediately upon completion of the broadband facilities, or there could be a delay. Additionally, mandates to share facilities could expire once infrastructure competition is realized. Finally, on what terms must facilities owners supply wholesale broadband services? A wide range of pricing methodologies lie between the ECPR and LRIC principles.

In this section alternative open access rules are specified and their impact on the broadband race is assessed. Comparing the outcome of the race with and without an open access rule reveals its impact on deployment pattern relative to the outcome without sharing. It is important to remember that this benchmark case is not necessarily welfare optimal. Since firms find themselves in a winner-take-most contest, preemptive tendencies may cause investment to occur too soon relative to the welfare optimum.[23]

---

[22] See, e.g., Oxman (1999).

[23] At bottom, the firms have no means to express the value they attach to deploying at specific times, except to actually make the expenditure. Suppose, for instance, the first firm derives its highest profit by deploying in year-1 which is much lower than the highest profit the second firm can generate, and which is realized only when that firm deploys in year-2. In absence of an auction of the right to deploy at the preferred time, the first firm may preempt the second firm if the first firm would earn much lower profit if it were to

Resale of Monopoly Broadband Service

The analysis begins by considering the case of a single infrastructure owner who is forced to share its network with a single service-based rival immediately upon completing the broadband upgrade. This situation arises in markets where the local telephone company does not encounter a cable TV system or a wireless network capable of upgrading to broadband service. In absence of an open access obligation, the incumbent network (Firm 1) deploys broadband services at its stand-alone profit-maximizing date t_1^I. Then suppose that it must lease the use of its network to a service-based firm who can then offer undifferentiated broadband services.[24] Provided lease rates do not preserve monopoly levels of incumbent profits, and assuming that the reseller enters at its first opportunity, deployment will reduce incumbent profits. Then, by inspection of Eq. 13.1, the decline in operating profits results in a delay in deployment date t_1^I.

It is entirely possible that the reduction in incumbent operating profits caused by open access will be so large that the incumbent will decline to upgrade to broadband altogether. This is likely to occur in markets that are marginally profitable on a long run basis to begin with, such as rural areas. Open access in this situation only decreases the likelihood these customers will be served by advanced services. It is important to note that any additional costs incurred as a result the opening by the facilities-based carrier is assumed away. In fact, the cost of deploying an operations support services to enable resale can be substantial and would further reduce incumbent incentives to deploy.

This conclusion reverses the otherwise robust result of Gilbert and Newbery (1982). In that paper, they find that an incumbent will adopt the innovation before an entrant and earlier than it would have but for the threat of entry. Their result derives from the fact that competition necessarily dissipates profits; the prospect of forgoing its monopoly rents spurs the incumbent to adopt earlier. In the current setting there is a competitor, but because it resells the incumbent's service, it cannot pre-empt the incumbent. The incumbent has complete control over if and when it faces retail competition—even if it does not control resale rates—because competition is possible if and only if the incumbent deploys a broadband network. The delay caused by open access would not be eliminated, but it may be reduced, were the incumbent to enjoy a monopoly over broadband services for some period of time before resale was required. The interim monopoly profits would then reward it for earlier deployment. Indeed, as the grace period is extended indefinitely,

---

wait until year-2. This would be inefficient if total welfare was roughly proportional to firm profits.

[24] Were the reseller to differentiate its broadband service some how, overall industry profits could increase with competition, and the incumbent could take a share of the incremental profits via higher lease rates. In fact the opportunity for product differentiation is minimal under pure resale, in which case retail competition necessarily reduces industry profit below the monopoly level.

the timing of deployment would converge to the monopoly case. The rationale for such an approach is exactly the same as the patent system which gives the patent holder a monopoly over the use or licensing of the technology.

Interim Facility Sharing

Consider now an open access rule that allows a follower to use the leader's broadband facility until that time when it builds its own facilities.[25] No pure reseller is allowed to enter to use the available broadband facility; only a 'committed entrant' that already owns a network and has the potential to upgrade to broadband has that right—though it may choose never to do so.[26] The interim period during which access to the facility is required ends when broadband platform competition is realized with a second deployment. This scenario, translated into technology race framework, raises profitability of the latecomer during the monopoly period: $\Delta \pi_i^j > 0$. Notice that the profits of the leader increase for both firms—even while Firm 1 is assumed never to follow. This out-of-equilibrium strategy will nevertheless alter deployment incentives. Next, consider the case when profits of the leader are unaffected by open access: $\Delta \pi_1^1 = \Delta \pi_2^2 = 0$. The interpretation of this additional assumption is that the facilities-based provider remains 'whole' as if prices are set according to ECPR, either by a regulator or the self-interested firm. Besides these changes, profits after second deployment are assumed to be unchanged since the follower can no longer lease first-mover's facilities: $\Delta \pi_1^d = \Delta \pi_2^d = 0$. In this case the impact of open access is quite simple to derive and intuitive to explain. Firm 2's preemption date, the timing of the first deployment, is delayed as a result of the increase in profits derived from resale: $\Delta t_2^0 > 0$. Furthermore, the date at which Firm 2 follows is also delayed: $\Delta t_2^f > 0$. Thus, the option to lease the leader's network slows down both deployments.[27] The follower is now more profitable prior to deploying its broadband facilities, and as a result, defers its de-

---

[25] Derivations supporting the comparative dynamic claims appear in the Appendix.

[26] Note that it may be less costly for one of the firms to lease the other's network, and this should be reflected by some differential cost of service-based broadband provision. An example would be the fact that the telephone network in most cases is designed around industry technical standards that do not vary from one region to another. Cable TV systems, in contrast, adhere to a variety of technical specifications which would make it more costly for a service-based competitor to make use of a rival's network. This situation is changing, e.g., Cable Laboratories has defined and promoted its DOCSIS (Data Over Cable Service Interface Specification) standard. Standardization could make it more attractive to lease access from a cable operator were they compelled to unbundle their network services.

[27] If, instead of preemption, Firm 1 would lead at its preferred time, then this policy would have no effect on the initial deployment. That would not, however, change the fact that the second deployment is delayed.

ployment date. The leader also delays its deployment—which occurs at the follower's preemption date—because the follower is less threatening.[28] Of course, the magnitude of the impact of open access rule depends on how profitable it will be for the follower to lease the leader's network, i.e., the size of $\Delta\pi_2^1$. This profit increment would shrink, e.g., if the follower was precluded from leasing the leader's broadband network for some fixed period, akin to the exclusivity period of a patent. If the period of exclusivity is short, then the results continue to hold qualitatively. Since the profit increment from following does not change as a result of an exclusivity period, Firm 2 will not alter the date at which it follows. However, forgone profit during the exclusivity period will reduce its overall profit from following (relative to open access without an exclusivity period), and so the preemption date will occur a bit earlier.[29]

Now suppose that Firm 1's profit is also increased to reflect the possibility that it, too, could earn higher profits should it instead follow the lead of Firm 2: $\Delta\pi_1^2 > 0$. In fact, assuming the order of deployment does not change, Firm 1 would never be in a position to realize these profits. But the prospect that Firm 2 could lead, and that an open access rule would make Firm 1 a slower follower, has the effect of advancing Firm 2's preemption date. If Firm 1, the actual leader, engages in preemption, then this effect taken by itself will speed up the initial deployment, $\Delta t_2^0 < 0$, just the opposite direction from the previous case. The net effect of this more symmetric open access rule is ambiguous, depending on several factors. The earlier conclusion that both deployment dates are delayed is preserved if the Firm 2 derives relatively more profits from reselling than Firm 1.

Further, suppose that during the interim period of sharing, the leader's profits are not unaffected but are reduced below their unregulated levels: $\Delta\pi_1^1 < 0$ and $\Delta\pi_2^2 < 0$. The interpretation of this possibility is that access prices shift some of the profits from the facilities-based leader to the service-based follower. This occurs, e.g., if regulators imposed some form of LRIC pricing. The effect of this form of open access, as might be expected, delays deployment as now the leader has reduced incentives to invest. To simplify analysis of this case, return to the asymmetric rule where Firm 1 alone is required to open up its network. The rule then redistributes profits from Firm 1 to Firm 2: $\Delta\pi_1^2 > 0$ and $\Delta\pi_1^2 < 0$. In that case, all three critical deployment dates are delayed: $\Delta t_2^0 > 0$, $\Delta t_2^1 > 0$ and

---

[28] The higher profits to Firm 2 during the monopoly period raises the value of following independent of the first deployment because all other profit levels are unchanged for Firm 2, and the only way it takes advantage of the reselling profits is by being a follower. The equality of Firm 2's profits of leading and following is restored when the first deployment is delayed because the analysis starts from a time earlier than Firm 2's monopoly deployment date.

[29] Of course, if the exclusivity period grows indefinitely long (beyond the duration of the unregulated monopoly period), then the open access rule becomes irrelevant.

$\Delta t_2^f > 0$. Consequently, once again the first and second deployments occur later than at the unregulated equilibrium as a result of the open access rule.

Pure Broadband Resale

Another open access rule would reserve use of an incumbent's network to a pure reseller, denying access to another facilities-based incumbent. In practical terms, this would say the cable and telephone companies cannot gain access to each other's unbundled network services before or after they upgrade to broadband capabilities, but an independent service-based provider could lease either broadband network. This condition promotes open access as a means to stimulate downstream service competition, and not specifically to encourage potential infrastructure competitors. In terms of the broadband race, this rule uncouples the follower's decision to deploy broadband from its option to use the leader's network to provide service beforehand. If the reseller simply earns a profit, and the leader remains whole, e.g., using ECPR, then the pattern of deployment will not depart from the unregulated outcome. This rule becomes more interesting when, possibly through LRIC access pricing, the leader's profits are reduced during the monopoly period: $\Delta \pi_1^1 < 0$. This effect of this condition is to slow down deployment. In all cases the follower date is unchanged because incremental profits are independent of how much the leader earns. However, the leader's monopoly date and its preemption date are delayed, so assuming the open access rule is applied to both firms, the initial deployment will be delayed. This is true whether the leader preempts the follower, or is able to deploy at its monopoly time.

This characterization of this open access policy may be too limited. It could be the case that by signing up many customers for a particular technology, a reseller could aid the facilities-based carrier in defending against later competition from an alternative technology. Arguably, ILECs may stem the loss of DSL customers once cable modem service becomes available in a market if broadband data CLECs have signed up many DSL subscribers in the meantime. To capture this feature profits need to be redistributed from the follower to the leader after the second deployment.

Symmetric and Asymmetric Facility Sharing

Next, we examine the effects of allowing the follower to continue to lease the leader's network after the follower builds its broadband facilities. A justification for such a rule rests on the need for additional profit to reach platform competition. The rule gives the follower greater freedom in making its buy-build decisions, thereby lowering its overall costs at the expense of the leader's profits. Below, a more symmetric version of this rule gives both firms rights to lease the other's facilities. When just the follow has this right, the open access rule increases the follower's profits during the dual-deployment period as well as during the leader's monopoly period: $\Delta \pi_2^1 > 0$ and $\Delta \pi_2^d > 0$. The analysis is simplified by assuming no reduction in the leader's profit in either period, as if the ECPR was applied. It can then be shown using the comparative-static results that the effect of these profit changes will unambiguously delay the follower's preemption date: $\Delta t_2^0 > 0$. Assuming that the leader will preempt the follower, this open access rule works to delay initial deployment. Also, if the increase in profits is roughly the same in both cases, then there is no effect on when the follower deploys since the net effect on the incentive to follow is unchanged: $\Delta t_2^f = 0$. Contrary to intent, the rule does not accelerate subsequent deployment.

This latter open access rule is highly asymmetric. If its intent is to improve incumbent networks' buy-build decisions, the sharing of the broadband facilities should be more symmetric. To examine this possibility, suppose that there is no open access during the monopoly period, but once the two broadband networks have been built, both firms can lease the other's broadband network.[30] Here the focus of the rule is directly on improving the buy-build decisions. The effects of this rule are characterized by assuming: $\Delta \pi_1^d > 0$ and $\Delta \pi_2^d > 0$. By increasing the profits from dual-deployment, the date at which either firm would follow the leader is advanced since the profit incentive to follow has increased. The effect on the follower's preemption date depends on the exact size of the profit changes. In fact, if the two carriers are treated symmetrically (in that their profits rise the same absolute amount), then there is no change in the preemption date. However, these same changes will unambiguously delay the preemption date for either firm: $\Delta t_i^0 > 0$. This will not be critical unless it had the effect of reversing the order of the leader and follower. If Firm 1's preemption date is delayed long enough, then in equilibrium it could become the follower, Firm 2 would assume the role of the leader, deploying at preemption date t_i^0. Here is one case when an open access rule could result in the apparently more efficient technology being deployed second, or potentially, not at all.

Conclusions

This study analyzed the impacts of alternative open access rules using an equilibrium model of the broadband deployment race. The rules altered deployment timing by two or more contestants, typically resulting in delay in either the first or second deployments, or both. Delays can be traced back to reduced incentives to invest in broadband facilities relative to service-based alternatives, or relative to no investment. It is also found that asymmetric treatment of carriers, and hence their corresponding technology, can have significant effects on the pattern of de-

---

[30] As before, assume that leasing broadband network services does not reduce the facility owner's profit as if ECPR pricing is used.

ployment even while intervention in this new service market can be quite subtle and indirect.

The analysis of the open access rules is limited to assessing their impact on timing of broadband deployment, and its implications for technology choice. A more complete analysis would evaluate the rules in terms of their impact on social welfare. As formulated, the technology race model lacks sufficient detail to conduct a full welfare analysis of the different open access rules.[31] It is reasonable to assume, however, that the more broadband carriers that serve the same market—whether they own facilities or resell incumbent services—the lower prices are for broadband service. Assuming narrowband rates are not affected, it is concluded that welfare rises with either form of competition. To the extent that open access rules tend to delay deployment of broadband service, consumer welfare is foregone. Alternatively, this outcome must be compared against the alternative where a monopolist deploys earlier, but also sets monopoly prices throughout.

Certainly the welfare costs associated with deployment delays should factor importantly in the debate over the form of open access rules applied to broadband infrastructure. Indeed, the cost of delays in deploying other new telecommunications technology is large.[32] Nevertheless, this calculus may miss other welfare costs stemming from reduced pace of innovation in broadband technologies. First, it is likely that each generation of broadband technologies builds upon the previous generation, learning from earlier mistakes. Deployment delays only retard the rate at which this knowledge accumulates. Furthermore, since new technology often rides on at least some portion of existing infrastructure and customer equipment (as in the case of DSL and cable modem technology), when investment is delayed and the current network is being amortized, so too may be the date when new technologies are deployed by upgrading or retrofitting existing infrastructure. Finally, the incentives for investing in R&D may be blunted depending on how open access rules alter rates of return on broadband investment. All of these considerations argue for an expanded analysis of the broadband race beyond modeling incentives to commercialize proven technology.

Acknowledgement

Comments on various drafts by Francois Bar, Johannes Bauer, Alain de Fontenay, Joe Farrell, Dennis Weller and Steve Wildman are greatly appreciated. Several of the propositions in this analysis are informally described in Woroch (2002).

---

[31] Owen and Rosston (2003) undertake a welfare analysis of an open access called 'net neutrality,' focusing on the implications for transaction costs.

[32] See Rohlfs et al. (1991), Baer (1995) and Hausman (1997).

Appendix

Suppose that the first and second deployments occur at t_1 and t_2, respectively. Then, if firm k ($=i$ or j) deploys first at t_1, and the remaining firm deploys at t_2, Firm i's discounted operating profit is given by:

$$\Pi_i^k(t_1,t_2) = \int_0^{t_1} \pi_i^n(t)e^{-rt}dt + \int_{t_1}^{t_2} \pi_i^k(t)e^{-rt}dt + \int_{t_2}^{\infty} \pi_i^d(t)e^{-rt}dt \qquad (13.\text{A}1)$$

where $j = i$ when Firm i leads, and $k = i$ when it follows. Notice how firms visit each of the three levels of operating profit, $\pi_i^n(t)$, $\pi_i^k(t)$ and $\pi_i^d(t)$ during the intervals $[0,t_1)$, $[t_1,t_2)$ and $[t_2,\infty)$, respectively. To arrive at Firm i's net payoff, simply deduct the present value of deployment cost from Eq. 13.A1:

$$\Pi_i^j(t_j,t_k) - c_i(t_i)\exp(-rt_i). \qquad (13.\text{A}2)$$

Suppose, for the moment, that Firm i was certain that Firm j would deploy at t_j. Then Firm i's best time to lead maximizes:

$$\Pi_i^i(t,t_j) - c_i(t)\exp(-rt) \qquad (13.\text{A}3)$$

over the range $t < t_j$. The solution, t_i^l, satisfies a simple first-order condition:

$$(\pi_i^i(t) - \pi_i^n(t))e^{-rt} = (rc_i(t) - c_i'(t))e^{-rt}. \qquad (13.\text{A}4)$$

The left-side of Eq. 13.A4 is the (discounted) incremental operating profit from deployment. It is graphed as an increasing curve in Figure 13.1. The right-side of Eq. 13.A4 is the (discounted) incremental savings in deployment cost from waiting, and equals the amortized cost less the marginal cost of deploying. It is represented by the falling curve in Figure 13.1. Call $L_i(t) = \pi_i^i(t) - \pi_i^n(t)$ the leader's incentive. From the figure, observe that t_i^l decreases as the curve shifts up. This makes sense since higher rewards for innovation should speed up their introduction. Reductions in incremental deployment cost have the same effect. The firm will choose never to lead if, in all periods t, $L_i(t) < rC_i$, the amortized minimum deployment cost. The optimal time for Firm i to follow (assuming Firm j will lead at a predetermined time) is defined analogously. The date t_i^f is found by replacing the leader's incentive with the follower's incentive $F_i(t) = \pi_i^d(t) - \pi_i^j(t)$. Finally the firm never follows if $F_i(t) < rC_i$ for all t. Notice that t_j is entirely absent from the marginal condition Eq. 13.A4. As a result, the rival's timing has no effect on either solution, t_i^f or t_i^l. Nevertheless, a firm's total profit is a function of t_j as can easily be seen from Eq. 13.A1. For this reason, a firm's fortunes crucially depend on when its rival deploys because that determines whether it will be a leader or a follower. Thus, firms battle for position in the order of deployment, but given that position, the timing of their deployment is not a strategic concern. This distinction becomes clearer when the strategic game is constructed. But before doing so, one last critical date is needed.

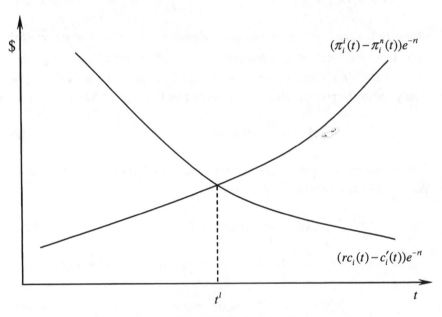

$$(\pi_i^i(t) - \pi_i^n(t))e^{-rt}$$

$$(rc_i(t) - c_i'(t))e^{-rt}$$

t^l

t

Fig. 13.1. Savings from Waiting and Marginal Deployment Costs

Given times when a firm wants to lead and to follow, there is some date when it is indifferent between the two roles. Offered the choice of being the leader by deployment at this preemption date t_i^0 or having its rival deploy at that time, payoffs are the same. The payoffs from each of these scenarios are expressed as follows:

$$\tilde{L}_i(t) = \Pi_i^i(t, t_j^f) - c_i(t)\exp(-rt),\qquad (13.A5)$$

$$\tilde{F}_i(t) = \Pi_i^j(t, t_i^f) - c_i(t_i^f)\exp(-rt_i^f).\qquad (13.A6)$$

Equation 13.A5 gives cumulative payoff to Firm i if it deploys at t while Firm j deploys later at t_j^f, and Eq. 13.A6 the payoff if Firm j deploys first at t while Firm i follows at t_i^f.

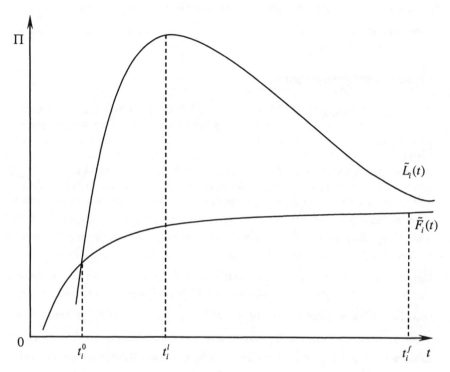

Fig. 13.2. Leader and Follower Incentives

As in Figure 13.2, $\tilde{L}_i(t)$ is single peaked at the preferred leader date, t_i^l. $\tilde{F}_i(t)$ increases over the region $[0, t_i^f)$ since a delayed first deployment postpones the date when the follower registers a profit decline. The preemption date equates the functions: $\tilde{L}_i(t) = \tilde{F}_i(t)$. Writing this equation and rearranging yields:

$$P_i(t)\exp(-rt)/r - M_i(t)\exp(-rt_i^f)/r - F_i(t)\exp(-rt)/r$$
$$= c_i(t)\exp(-rt) - c_i(t_i^f)\exp(-rt_i^f) \tag{13.A7}$$

where $P_i(t) = \pi_i^i(t) - \pi_i^j(t)$ is Firm i's preemption incentive, the difference between being a leader and a laggard; $M_i(t) = \pi_i^i(t) - \pi_i^d(t)$ is the imitation penalty imposed by the follower on the leader. This complicated expression simply balances the incremental revenue from taking the lead with the incremental cost, with all values properly discounted.

It is shown below that the preemption date t_i^0 is a decreasing function of $P_i(t)$ and $M_i(t)$ as well as $F_j(t)$ (through its dependence on t_j^f). Finally, t_i^0 is increasing (decreasing) in $F_i(t)$ depending on whether Firm i is a faster (slower) follower than j, i.e., $t_i^f < (>) t_j^f$.

Equilibrium Deployment

A strategy for each firm is a rule specifying whether or not to deploy in the current period given the history up to that time and given other firms' strategies. Within a period, firms move simultaneously. As usual players' strategies form a sub-game perfect Nash equilibrium when, in each and every sub-game, the continuation of players' strategies together form a Nash equilibrium. This equilibrium concept rules out deployment plans that would never be carried out but, if believed by rivals, could alter the equilibrium outcome. For instance, Firm i could not credibly claim to deploy in any period $t < t_i^0$ regardless of the history leading up to that time. Quite simply, it will lose money compared to never deploying. If Firm i's preferred deployment date is very early (t_i^l occurs long before either t_j^l or t_j^0, then it will deploy at that time provided leading is more profitable than following, i.e., $\tilde{L}_i(t_i^l) > \tilde{F}_i(t_i^l)$. Its plans might be upset if its rival were also to find leading at that date more profitable than following, i.e., $\tilde{L}_j(t_i^l) > \tilde{F}_j(t_i^l)$, or equivalently, $t_j^0 < t_i^l$) since then Firm j will prefer to lead some time before that date t_i^l. By the same reasoning, Firm j cannot credibly commit to delay its deployment beyond its preemption date, t_j^0, because it could do better by taking the lead. As a result, Firm i is induced to advance its deployment date up to t_j^0 provided that it still prefers to lead. To sum, the general conclusion states that, if $t_i^0 < t_j^0$ and $t_i^l < t_j^l$, then equilibrium has Firm i deploy first at $\min\{t_i^l, t_j^0\}$ and Firm j follows at t_j^f .[33] Of course, the identity of the leader is an open issue. It depends on how operating profits and deployment costs affect the firms' relative leader and preemption dates. These dates depend, in turn, on the specific context of the application.

Comparative Statics

Comparative-static exercises establish the relation between critical dates, levels of operating profit and deployment costs. To simplify derivations, assume operating profits are constant over time: $\pi_i^h(t) = \pi_i^h$ for all t. From Fig. 13.1 we observe that, even when operating profits vary over periods, t_i^l and t_i^f decrease in

---

[33] Proved in Katz and Shapiro (1987).

$L_i = \pi_i^i - \pi_i^n$ and $F_i = \pi_i^d - \pi_i^j$, respectively. This implies t_i^l is decreasing in π_i^i and increasing in π_i^n. Similarly, t_i^f is decreasing in π_i^d and increasing in π_i^j.

Pictures are less convincing for the effects on the preemption date. Therefore, totally differentiating $\tilde{L}_i(t) = \tilde{F}_i(t)$ around t_i^0, and applying the Envelope Theorem provides:

$$dt_i^0 / d\pi_i^i = (\exp(-rt_j^f) - \exp(-rt_i^0))/rD,$$ (13.A8)

$$dt_i^0 / d\pi_i^j = (\exp(-rt_i^0) - \exp(-rt_i^f))/rD,$$ (13.A9)

$$dt_i^0 / d\pi_i^d = (\exp(-rt_i^f) - \exp(-rt_j^f))/rD,$$ (13.A10)

$$dt_i^0 / dt_i^f = 0,$$ (13.A11)

$$dt_i^0 / dt_j^f = -(\pi_i^i - \pi_i^d)\exp(-rt_j^f)/D,$$ (13.A12)

where $D = -(\pi_i^i - \pi_i^j - (rc_i - c_i'))\exp(-rt_i^0)$. As long as $\tilde{L}_i(t)$ cuts $\tilde{F}_i(t)$ from above in the vicinity of t_i^0, as in Fig. 13.2, then $d\tilde{L}_i(t_i^0)/dt > d\tilde{F}_i(t_i^0)/dt$ which ensures that the bracketed term in the expression for D is negative, and hence, D itself is positive. From Eq. 13.A8 and Eq. 13.A9 it is concluded that $dt_i^0 / d\pi_i^i < 0$ and $dt_i^0 / d\pi_i^j > 0$, provided that the preemption date precedes the follower's date for both firms, i.e., $t_i^0 < t_i^f$ and $t_j^0 < t_j^f$. Examining Eq. 13.A10, finds that $dt_i^0 / d\pi_i^d$ is positive or negative depending on whether t_i^f is earlier or later than t_j^f, respectively. Therefore, an increase in dual-deployment profits will delay the preemption date if, in popular terminology, the firm is a 'fast second' compared to its rival. Finally, from Eq. 13.A12, $dt_i^0 / dt_j^f < 0$ since $\pi_i^i > \pi_i^d$. These results are collected in the table found in the main text.

To assess the effects of changes in entry costs, the function $c_i(\ )$ could be perturbed however, instead an exponential form $c_i(t) = a_i \exp(-b_i t) + C_i$ is specified, and changes in its parameter values are considered. Substituting this expression into Eq. 13.A4 yields a closed-form solution to the leader and follower dates:

$$t_i^l = (1/b_i)\log(a_i(b_i + r)/(L_i - rC_i)),$$ (13.A13)

$$t_i^f = (1/b_i)\log(a_i(b_i + r)/(F_i - rC_i)) , \qquad (13.\text{A}14)$$

Inspection reveals both t_i^l and t_i^f are increasing in entry cost parameters a_i and C_i. The effect of b_i is slightly less transparent. Differentiating Eq. 13.A13 gives:

$$dt_i^l / db_i = (t + 1/(b_i + r))b_i , \qquad (13.\text{A}15)$$

which is positive. The same holds for dt_i^f / db_i. The effects on t_i^f require some tedious calculations:

$$dt_i^f / da_i = (\exp(-b_i + r)t) - \exp(-(b_i + r)t))/D > 0 , \qquad (13.\text{A}16)$$

$$dt_i^f / db_i = (a_i t \exp(-b_i + r)t) - a_i t \exp(-(b_i + r)t))/D, \qquad (13.\text{A}17)$$

$$dt_i^f / dC_i = (\exp(-rt) - \exp(-rt))/D > 0, \qquad (13.\text{A}18)$$

$$dt_i^f / da_j = -M_i \exp(-rt)(dt_i^f / da_i)/D < 0, \qquad (13.\text{A}19)$$

$$dt_i^f / db_j = -M_i \exp(-rt)(dt_i^f / db_i)/D < 0, \qquad (13.\text{A}20)$$

$$dt_i^f / dC_j = -M_i \exp(-rt)(dt_i^f / dC_i)/D < 0, \qquad (13.\text{A}21)$$

where $D = -r(P_i - a_i \exp(-rt) - C_i) > 0$ by second-order conditions. An ambiguity is in Eq. 13.A17: $dt_i^f / db_i > (<)0$ according to $\log(t_i^l / t_i^f) < (>)(b_i + r)(t_i^l - t_i^f)$. These results are summarized in Table 13.2:

Table 13.2. Comparative Statics for Deployment Costs

	a_i	b_i	C_i	a_j	b_j	C_j
t_i^l	+	+	+	0	0	0
t_j^0	−	−	−	+	+/−*	−
t_j^f	0	0	0	+	+	+

Note. * - according to whether $\log(t_i^l / t_i^f) < (>)(b_i + r)(t_i^l - t_i^f)$.

References

Baer W (1995) Telecommunications infrastructure competition: The costs of delay. Telecommunications Policy 19(5): 351–63

Boone J (2000) Competitive pressure: The effects on investments in product and process
innovation. Rand Journal of Economics 31(3): 549–69

Crandall R, Sidak G, Singer H (2002) The empirical case against asymmetric regulation of
broadband Internet access. Berkeley Law and Technology Journal 17(1): 953–87

Faulhaber G, Hogendorn C (2000) The market structure of broadband telecommunications.
Journal of Industrial Economics 48(3): 305–29

Floyd E, Gabel D (2003) An econometric analysis of the factors that influence the deployment of advanced telecommunications services. Draft August 31

Fudenberg D, Tirole J (1985) Preemption and rent equalization in the adoption of new
technology. Review of Economic Studies 52(3): 383–401

Fudenberg D, Tirole J (1987) Understanding rent dissipation: On the use of game theory in
industrial organization. American Economic Review 77(2): 176–81

Gabel D, Huang G-L (2003) Promoting innovation: Impact of local competition and regulation on deployment of advanced telecommunications services for businesses. Draft

Gilbert RJ, Newbery DMG (1982) Preemptive patenting and the persistence of monopoly.
American Economic Review 72(3): 514–26

Greenstein S, McMaster S, Spiller P (1995) The effect of incentive regulation on infrastructure modernization: Local exchange companies' deployment of digital technology.
Journal of Economics and Management Strategy 4(2): 187–236

Haring J, Rettle M, Rohlfs J, Shooshan H (2002) UNE prices and telecommunications investment. Strategic Policy Research Inc. July 17

Hausman J (1997) Valuation and the effect of regulation on new services in telecommunications. Brookings Papers: Microeconomics pp 1–38

Hausman J (2000) Regulation by TSLRIC: Economic effects on investment and innovation.
In: Sidak JG, Engel C, Knieps G (eds) Competition and regulation in telecommunications. Kluwer Academic Publishers, Boston

International Telecommunication Union (2003) The birth of broadband. ITU, Geneva

Katz M, Shapiro C (1987) R&D rivalry with licensing or imitation. American Economic
Review 77(3): 402–20

Kridel D, Sappington D, Weisman D (1996) The effects of incentive regulation in the telecommunications industry: A survey. Journal of Regulatory Economics 9: 269–306

Lee CH (2001) State Regulatory Commission treatment of advanced services: Results of a
survey. National Regulatory Research Institute, March

Owen BM, Rosston GL (2003) Local broadband access: *Primum non nocere* or *primum
processi*? A property rights approach. Stanford University, SIEPR Discussion Paper
02-37, July

Oxman J (1999) The FCC and the unregulation of the Internet. FCC, Office of Plans and
Policy Working Paper No. 31, July

Pindyck RS (2004) Mandatory unbundling and irreversible investment in telecom networks. National Bureau of Economic Research, Working Paper No. 10287, Cambridge, February

Riordan M (1992) Regulation and preemptive technology adoption: Cable TV versus telephone companies. Rand Journal of Economics 23(3): 334–49

Rohlfs J, Jackson C, Kelley T (1991) Estimation of the loss to the US caused by the FCC's delay in licensing cellular telecommunications. National Economic Research Associates, Washington DC, November

Willig R, Lehr W, Bigelow J, Levinson S (2002) Stimulating investment and the Telecommunications Act of 1996. Draft, October 11

Woroch G (1998) Facilities competition and local network investment: Theory, evidence and policy implications. Industrial and Corporate Change 7(4): 601–14

Woroch G (2000) Competition's effect on investment in digital infrastructure. Draft, July

Woroch G (2002) Open access rules and the broadband race. The Law Review of Michigan State University–Detroit College of Law 3: 719–42

14 Spectrum Management and Mobile Telephone Service Markets

Johannes M. Bauer

Introduction

In 2002 global mobile telephony subscription surpassed fixed-line subscription.[1] Within this context, emerging mobile computing and data services are expected to enhance existing services with the blending of wireless, computing and digital information industry innovation. Visions of opportunity arising from new wireless platforms (e.g., meshed networks), intelligent devices (e.g., agile radio) and wireless applications abound (Lightman and Rojas 2002). Accordingly, industry experts predict mobile data communication to generate more than half wireless service market revenue by 2010 (Wireless Data 2000). Additionally, in many countries, including Western Europe, parts of Asia and many developing countries, mobile telephony networks are more widespread than fixed-line service. In these countries, mobile devices rather than personal or laptop computers are likely to be the dominant Internet access devices. Despite wireless platform bandwidth limitations, mobile Internet subscription is growing rapidly. However, substantial variations in mobile Internet activity exist internationally. For instance, 72.3% of mobile subscribers in Japan and 59.1% in Korea access the Internet via mobile telephone, whereas the corresponding use for the US and Western Europe is 7.9% and 6.4%, respectively (ITU 2002: 44). Network operators are currently upgrading network platforms to provide bandwidth necessary to support more advanced multi-media applications. Evolving second generation (2G) mobile service, or 2.5G service, that provide to 171.2 kbps are well suited to deliver mobile data applications, including basic Internet access. Some network operators are offering third generation service (3G)—designed to provide 384 kbps in fully mobile mode and to 2 Mbps in stationary mode—sufficient to allow video streaming and multimedia applications. Wireless local area networks, e.g., WiFi or Hiperlan, typically operate in unlicensed spectrum bands, provide to 54 Mbps and are deployed for local connectivity. Platforms such as Bluetooth provide very short range personal area networks. While such technology requires further testing and experimentation, and their compatibility and interoperability pose challenges, they promise to radically transform communication markets.

---

[1] See International Telecommunication Union, 'World Main Telephone Lines', available at http://www.itu.int/ITU-D/ict/statistics/at_glance/main02.pdf and 'World Cellular Subscribers', available at http://www.itu.int/ITU-D/ict/statistics/at_glance/cellular02.pdf (last visited July 29, 2003).

Wireless service and application markets have evolved in a less regulated and more entrepreneurial environment than wire-line telephony markets (Galambos and Abrahamson 2002). Mobile network operators are at most lightly regulated. Furthermore, most handset and network equipment manufacture, software development, applications service provision and mobile portal industry that contribute to the innovative dynamics in mobile service business markets, are effectively unregulated. Not surprisingly, most industry leaders and user groups consider government policy should retreat and release gales of 'creative destruction'. This position recognizes that legal and institutional environments shape industry, and potentially hinder development. In particular, for the wireless industry the framework governing spectrum access and use is such an issue. The proliferation of wireless applications has placed strain on traditional administrative licensing systems and revealed their shortcomings. Many countries introduced spectrum auctions to accelerate license assignment however, these limited reforms created problems. Clearly, more fundamental spectrum management changes must be considered. Debate has focused primarily on the design of frameworks to ensure efficient spectrum use. Solutions under consideration include spectrum privatization, establishing spectrum markets and allowing open access to spectrum. Also discussed is the establishment of managed spectrum commons. Rosston and Hazlett (2001), Kwerel and Williams (2002) and Faulhaber and Farber (2002) favor a privatization model. In contrast, Reed (2002) supports an open access model. Still another group, argue that not enough is known to rigidly set policy. To generate necessary information, this group suggests an experimentation period during which the alternative models are monitored (Noam 1998; Lessig 2001; Benkler 2002).

With the exception of Benkler (2002), research has not addressed the impact of spectrum management regimes on wireless industry evolution. This approach is reasonable when spectrum efficiency is the sole concern, i.e., when Arrow-Debreu conditions of complete information, complete present and future markets and no transaction costs hold. In this world, markets are efficient co-ordination mechanisms and competition assures economy-wide efficiency. However, sector efficiency improvements are not necessarily translated into economy-wide efficiency improvements when these conditions fail to hold (Lipsey and Lancaster 1956). Coase (1960) shows the presence of transaction cost voluntary exchange does not necessarily lead to efficient allocation of resources. Under plausible assumptions it matters how rights governing markets are assigned (North 1990; Aoki 2001).[2] Alternative spectrum management regimes have fundamentally different implications for the mobile telephone services industry. Policy should view spectrum as a resource market and so consider inter-industry effects. In the US, mobile voice telephony markets are exempt from many provisions applied to the fixed-line telephone industry. Spectrum policy has helped shape the organization of mobile te-

---

[2] This is evident from wire-line industry developments when competing paradigms coexist. In the US, telephony markets are structured on a common carrier model, while cable markets are based on a private contract carrier model. These arrangements resulted in different service quality, pricing, investment and third-party access performance.

lephony markets, i.e., granting exclusive rights have facilitated the development of an effectively closed market structure. In the emerging mobile service industry, spectrum policy is shaping industry evolution and communications process enabled by platforms. This chapter examines the implications of alternative spectrum policy options on innovation within the mobile telephone services industry. The following section discusses the characteristics of innovation in advanced mobile telephony service markets. The task of spectrum management and approaches to pursue them are then reviewed. Next, the nexus between spectrum policy, market structure and the effect on innovation trajectory are analyzed. The study concludes with a summary of findings and policy consequences, and recommendations for further research.

Mobile Telephone Service Market Innovation

Mobile voice markets require semi-conductor, equipment, acoustics and network operator inputs. Technical advance in these industries, especially component miniaturization and microchip processing power enhancement, has facilitated advanced cellular telephony development. Further, modern wireless networks are layered systems that combine physical network, logical and application layer functions. The physical layer encompasses mobile devices, antennas, base stations, backhaul networks and core network fast data backbones. Traffic that flows through the physical network is controlled by logical layers that configure service, provide user authorization and route traffic. In the US market, alternative configurations (standards) are used at physical and associated logical network layers.[3] The applications layer must be configured so as to work with the physical and logical layer. As many wireless services, e.g., m-banking, information services, are software driven this layer is a source of much potential innovation. Mobile telephony innovations are realized at related layers of the value chain. For example, process innovations affect network platforms, either from devices, network hardware or logical components, such as protocols. That is, increased radio technology spectral efficiency, network topology (such as cellular) that expands infrastructure capacity, digital signal representation and improved antenna design. Further, software defined radio technology that replace dedicated with general purpose radio devices create more flexibility in configuring applications has substantial innovation potential (Lehr et al. 2003).[4] Product innovations include mobile devices, e.g., personal digital assistants, wristwatch cellular telephones and improvements in the

---

[3] Time division multiple access (TDMA) and global systems for mobile communications (GSM) networks are based on time-division multiple access, Code division multiple access' (CDMA) networks use Qualcomm's code division multiple access technology, and integrated dispatch enhanced network (iDEN), Motorola's standard, is used only by Nextel.

[4] However, radio technology is built more efficiently, e.g., with lower energy consumption and longer battery life when devices are engineered for special purposes. Thus, there is a trade-off between flexibility in configuring services and battery life.

functionality of devices, e.g., color screens. Networks and terminals enable process and product innovation. Moreover, there exists vast potential for innovation in bundling and pricing services. For example, the mobile telephony industry introduced multi-part self-selection tariffs, prepaid service and off-peak promotions such as the bucket plans offered at zero incremental cost in the US.

Technology change and innovation originate from different segments of the advanced mobile telephony market value chain, e.g., by manufacturers, application providers and from beyond the market. The revealed innovation pattern depends on conditions ruling in sub-markets and complementary activity among value chain participants. Many configurations are conducive to innovation, and they result in different industry trajectory. For earlier generation technology, designed principally for closed or limited group use such as taxi dispatch, innovation originated from equipment manufacturers.[5] With cellular telephony, control has shifted to network operators that hold licenses and so control user gateways. In mobile voice markets, network capacity and coverage are important innovation and service attributes. Substantial sunk network infrastructure investment provides an incentive to develop closed business models in which network operators, within the terms of the license, control value chains. For example, Noam (2001) indicates that mobile devices are approved by network operators and so choice is limited. In contrast, wire-line terminal equipment is removed from fixed-line network operator control initiating innovative activity. Spectrum policy is among factors shaping the configuration of mobile innovation. Accordingly, it is not surprising that LMRS services are organized on a shared basis while the more concentrated cellular service is based on exclusive licenses to network operators. Further, despite the need for industry interaction the 2G mobile industry value chain remains reasonably simple (Maitland et al. 2002). However, next generation mobile service value chains are more complex with software developers, application service providers, mobile portals and content providers having important roles (see Fig 14.1). Interaction among mobile telephony market participants, conditions in complementary good markets and markets that deploy mobile service as an intermediate good ensure economy-wide diffusion of returns to innovation.

Sources of innovation risk include high sunk costs such as non-recoverable research and development (R&D) and marketing costs, and uncertain market conditions. To encourage firms to innovate require potential supra-normal compensation premiums. Premiums are obtained from intellectual property right systems and firm competitive strategy, i.e., first-mover advantage and the imposing of switching costs. Such strategy creates market power. Incentives to innovate are higher when such market power is temporary and open to challenge (Schumpeter 1942). However, the risk premium—and corresponding degree of market power— most favorable to innovation depends on innovation process characteristics. Namely, an innovation process with low sunk costs and little uncertainty has an optimal market power lower than that for a very uncertain innovation with high sunk costs. Additionally, within the mobile data communications industry innova-

---

[5] The FCC classified services as land mobile radio services (LMRS). They are typically not interconnected with the public switched telephone network or sold publicly.

tive activity is influenced by network effects and complementary market participant behavior. Consumption effects arise from new subscribers augmenting the calling opportunity of existing subscribers. Further, as equipment, applications and software must be compatible, technical effects exist. Pecuniary network effects occur as mobile industry firm profits depend on producer activity, e.g., equipment manufacturers. Finally, diffusion effects arise when innovation adoption is advantageous to potential users, e.g., by lowering search costs and complementary product price (Antonelli 1992: 7). Stakeholders often are able to internalize such effects through collaboration and ownership (Liebowitz and Margolis 2002).

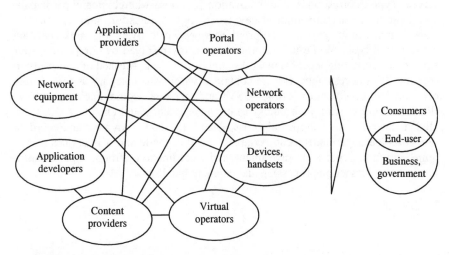

Fig. 14.1. Advanced Mobile Services Value Chain

When strong network effects are present decentralized markets are poorly coordinated. Additional quantity and quality signals, and new coordination mechanisms may improve the functioning of the system. Alliance formation among firms located in the different value chain segments can also improve operations. Vertical and horizontal coordination is through inter-firm committees and technical groups (Antonelli 1992: 14). Further, when technology is modular in design, firms like Palm are able to interface and provide voluntarily information to value chain participants to internalize complementary externalities (Farrell and Weiser 2002, Langlois 2002). When voluntary agreements are not successful, regulatory oversight is required to ensure adequate standardization, interoperability, interconnection and access conditions. Moreover, industry innovation paths are influenced by installed infrastructure. When the economic life of an installed infrastructure is reasonably long and technology idiosyncratic, as is the case with mobile network platforms, innovation is mostly incremental (localized). This tendency is reinforced by incumbents wanting to prevent the erosion of asset values (Antonelli 2001). The strategy is undermined when a major innovation originates outside an industry segment (such as WiFi), or when a fundamentally new technology provides a decisive economic advantage and justifies the replacement of an installed base. There is more room for discretionary innovation when the planned life of an installed base is short or sunk costs are low.

Figure 14.2 integrates these perspectives and illustrates that different coordination configurations and sunk cost are possible. Innovation conditions include: Type I characterized by low sunk and coordination costs, Type II has either high sunk or coordination costs, and Type III has both high sunk and coordination costs. Conditions for successful innovation under these scenarios, in particular the appropriation of any innovation risk premium, differ. Depending on existing appropriation conditions, certain innovation processes are more likely to be observed. Type I corresponds to an innovation commons whereby many participants contribute to the accumulation of new technology, applications and services. For Type II innovation, greater appropriation is needed to overcome sunk costs and coordination costs. While this situation is possibly resolved by competitive strategy, some supporting legal frameworks simplify the task. Meshed or nationally integrated WiFi access networks are contained within this category as they require considerable coordination effort. GSM and personal communications service (PCS) networks, with their idiosyncratic investment requirements are also included. Type III innovations only arise when joint risks from sunk and coordination costs are apparent. Recent examples include i-mode and 3G services that require substantial network investment and coordination efforts, e.g., technical integration, transaction processing and payment handling.

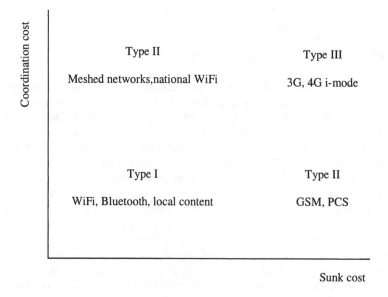

Fig. 14.2. Mobile Communication Innovation Processes

The link between innovation incentive and conditions to appropriate an adequate risk premium are illustrated in Fig. 14.3, which also shows that certain innovations only coexist when particular appropriation conditions prevail. For instance, Type III innovations do not arise under conditions favorable to Type I innovations, and vice versa. Spectrum policy directly and indirectly frames these conditions. Particular spectrum management regimes, e.g. exclusivity of rights and conditions of enforcing them, determine the ability of spectrum users to appropriate innovation premiums. Spectrum policy also affects upstream gateways to consumers, and sunk market entry costs. The interaction of these factors with demand influences the viability of competition and appropriation conditions.

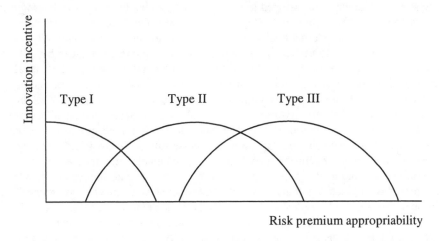

Fig. 14.3. Innovation Incentive and Risk Premium

Approaches to Spectrum Management

Electromagnetic spectrum is a pervasive resource that spans the 3 kHz to 300 GHz band. While this resource is not depleted through use, as radio signals overlay in an additive manner, interdependence exists among rival users. At the extreme, interference damages participant radio signals and makes them unrecognizable. Furthermore, frequency bands have different propagation characteristics and are not perfectly substitutable. Longer radio waves (lower frequency) travel great distances, through ground and sky, to bypass obstacles. Shorter radio waves (higher frequency) depend on line of sight and cannot easily penetrate or bypass objects. With no system of rights in place to ensure orderly use at the time of invention, spectrum bands became occupied in an ad hoc manner. This situation remained until commercial radio growth led to an explosion in spectrum use and an associated escalation in interference. In 1906, European governments attempted to coordinate international spectrum use by signing an international radio convention. Interestingly, US attempts to coordinate spectrum use voluntarily did not succeed. After several radio conferences, pressure from the military and broadcasters led Congress to treat spectrum as a national resource controlled by the government. This policy is codified in the Radio Act of 1927 and subsequently in Title III of the Communications Act of 1934. The latter Act established the FCC to execute spectrum management tasks. From these national and international initiatives an administrative spectrum management system emerged. Spectrum allocation is by use or use class. When conflicts between regions arise, e.g., in satellite communication, separate bands are allotted. Further, temporary spectrum use rights, not ownership, are assigned to users or user groups. Finally, as technology and use evolve, allocations and assignments are adjusted. The ITU coordinates spectrum allocation internationally. National governments reserve authority to manage spectrum domestically—generally within parameters set by the international community. Within this framework, national authorities draft legal and technical rules and regulations that govern services and equipment providing protection against interference and alleviating congestion. This system allows the pursuit of non-market objectives such as national security, safety and equal access in addition to efficient spectrum use. The basic design is rooted in simple 1920s radio technology that required high signal-to-noise ratios. To avoid interference, licensees are given exclusive privilege while channels are spaced apart at appropriate distances. Progress in information theory and microprocessor power led to radio engineering that does not require exclusive channel assignment. Through this approach, sophisticated communications devices and protocols avoid interference. For instance, agile radio devices sense their radio environment and select communications parameters (frequency, polarization and coding) to control interference. Unlike analog radio, digital mobile communications allow more advanced forms of error correction. Communications are accordingly less sensitive and tolerate higher interference levels. Moreover, recent technical development questions whether spectrum is indeed scarce—undermining a fundamental tenant of the exclusive licensing approach. Innovative network architecture, such as meshed networks, automatically reconfigures using any active device to relay information. Reed (2002) claims for such network topology, network capacity grows with users more than proportionally.

An explosion in license applications revealed weakness in administrative process, e.g., delay, inappropriate proposal selection and inefficient spectrum allocation. In response, national experiments with more flexible licensing systems occurred. Such responses include: administrative licensing, flexible licensing, ownership, spectrum commons and open access. The regimes differ in rights they bestow and in spectrum management (see Table 14.1). Administrative licensing, flexible licensing and ownership regimes, although structured differently, grant exclusive rights to licensees. A license grants temporary use.[6] Spectrum ownership establishes complete property rights, viz., owners have discretion to use, lease, sell

---

[6] As licenses are transferable, rights are more comprehensive.

or otherwise dispose of spectrum. This model is based on the notion that exclusive control of communications channels is necessary to avoid interference.[7] In contrast, spectrum commons grant use rights to defined groups, and establish exclusive rights regarding the external environment and shared use within the commons. Open access, established to avoid interference and reduce congestion, allows anybody meeting certain conditions spectrum use (see Table 14.1). Further, in traditional licensing frameworks alternative proposals reduce government involvement through decentralized coordination mechanisms. In radical ownership models, viz., commons and open access, government involvement is reduced to law enforcement and conflict resolution. Alternatively, government remains involved by making basic allocation decisions, including conflict resolution, and setting interference protection and congestion management parameters.

Table 14.1. Spectrum Management Regime

	Licensing		Ownership	Commons	Open access
	Administrative	Flexible			
Exclusiveness	Yes	Yes	Yes	Mixed	No
Allocation	Public	Public	Mixed	Mixed	Private
Assignment	Administrative	Auction	Mixed	Mixed	None
Adjustment	Government	Government	Market	User	User
Protection	Government	Government	Ownership	User	Protocol
Congestion	Government	Government	Protocol	Protocol	Protocol
Selection	Government	Government	Market	Government	Technology

Spectrum Management Regimes

In practice, spectrum management has resulted in underutilized licensed spectrum, with certain spectrum bands crowded. Spectrum administration is typically biased in favor of incumbents, so delaying innovation. Additionally, administrations lack

---

[7] In practice, many frequency bands allow limited shared use, e.g., of US spectrum to 300 GHz, only 6.9% is assigned to government or the private sector on a non-shared basis. This stake is higher in prime spectrum bands between 300 MHz and 3 GHZ, where 64% is allocated on a non-shared basis. However, primary licensees typically have strong protection against interference and do not have to protect secondary users.

enough information to properly judge applications, and license processing remains political, slow and inefficient (Melody 1980; Hazlett 2001). Recent wireless application lodgment increase has further slowed the process. In response, more flexible spectrum allocation approaches have been introduced, viz., traditional 'beauty contests' replaced by auctions (Bauer 2003). At their best, spectrum auctions allow market values to be determined by transparent processes not prone to manipulation. Further, auctions streamline assignment and create substantial government revenue.[8] However, auctions are not universally successful in their application, with the European universal mobile telecommunications system (UMTS) auctions exposing carriers to financial crisis, whilst the US C-Block PCS auction ended in legal dispute. Among the reasons for such failures are that theoretical auction principles do not translate readily to practical auction situations (Melody 2001; Cramton 2002; Klemperer 2002a, b). Namely, assumptions that future demand conditions are known, efficient capital markets are present, and license price is irrelevant for market evolution, are implausible (Bauer 2002). Moreover, auctions remain embedded in rigid administrative frameworks, and when too little spectrum is allocated auctions provide artificial scarcity premiums to government (Melody 2001). While auctions are based on conditions and expectations applicable at the time of auction, such rigidity is ameliorated when secondary spectrum markets are established (Valletti 2001). Clearly, it is desired that auctions generate spectrum fees that constitute true mobile telephony market entry costs, otherwise licensees tend to pursue a 'walled garden' strategy to recover costs. As such, poorly designed auctions increase industry consolidation and reduce gateways between application services, service providers and users.

Establishing private spectrum ownership implies government withdraws from the process (Herzel 1951; Coase 1959). Privatization requires an initial assignment of private property rights using some market-based mechanism, such as spectrum auctions. For existing licensees, Spiller and Cardilli (1999) suggest an auction of warrants to convert use privileges into ownership rights. In principle, private property models solve spectrum management problems—with market forces driving allocation, assignment and dynamic adjustment. Markets also address interference protection and congestion management through decentralized negotiated arrangements. Conceptual objections include the pervasiveness of externalities, noncompetitive nature of wireless markets and that large portions of spectrum are used by non-profit organizations (Melody 1980). To function properly, spectrum markets must be liquid. In practice, assembling spectrum bands in contiguous geographic areas are subject to high transaction costs. That is, spectrum owners may resist making unused spectrum available, especially when for use by competing service providers. Typically, owners facing competition release spectrum only when the gap between private opportunity cost and foregone profit is positive. Finally, when international frequency coordination is required and spectrum is allotted regionally, private markets may result in inequity.

---

[8] Since 1994, the US government earned US$ 41.8 billion in spectrum auctions. Auctions shortened the licensing process from 48 to 4 months. See FCC, 'Auctions Summary': http://wireless.fcc.gov/auctions/summary.html.

Spectrum commons and open access proposals accept advanced radio engineering premises, and are based on successful common property management regimes. That is, open access resources due to the separation of private and social costs (and benefits) are typically used in an inefficient manner (Ostrom 1990; Stevenson 1991). However, when such resources have clear common ownership and management rights they are utilized more effectively. Buck (2002) suggests principles to be employed in creating spectrum commons. The approach allows user driven spectrum organization, with open access among a group but not necessarily all spectrum users. Government, auctions or spectrum markets are employed to make initial allocations. Additionally, spectrum commons form part of a nested system of management—from local commons through to more aggregated markets. Further, as open access models do not require licensing, government is released from any spectrum management obligations, with decisions made by suppliers and users. However, less radical proposals dealing with interference and congestion make access contingent on technical or economic prerequisites, and so retain a role for government or self-regulatory body. Particular models are defined by how such issues are addressed. Such an approach is that currently unlicensed bands available in the US 100 MHz of spectrum in the 2.4 GHz range—industrial scientific and medical band—be designated unlicensed. In 1997, the FCC designated several MHz of spectrum in the 5 GHz band as unlicensed national information infrastructure (U-NII). In May 2003, the FCC proposed to allocate another 255 MHz in the 5.470-5.725 GHz band to the U-NII.[9] To ensure effective market operation, unlicensed band users must comply with etiquette rules and technical specifications, e.g., power limits and communication protocols to ameliorate any interference.

Finally, proponents of more radical approaches rely on self-regulation and trusting industry with the drafting of specifications. However, Benkler (2002) argues that with open access regimes, the equipment industry captures communication service value through equipment prices. Ideally this process increases equipment manufacturer R&D spending and assures appropriate technology solutions for coping with interference. While Noam (1998) advocates opening access to spectrum, market-based fees are proposed that reflect spectrum opportunity and congestion control costs. The approach allows establishment of spectrum futures and derivative markets, and hence users are able to enter into long-term contracts. As the proposal converts sunk license fees determined in an auction to variable payments determined in spot and future spectrum markets, potential distortions from auctions are avoided. As yet, no workable mechanism is available for determining and collecting spectrum fees. To sum, open access models emphasize spectrum as a public good, and thus any assignment of exclusive rights is inefficient and creates an anti-commons problem due to user exclusion (Buchanan and Yoon 2000).

---

[9] See FCC, 'FCC Proposes Additional Spectrum for Unlicensed Use,' press release, May 15, 2003.

Spectrum Management and Market Organization

Potentially high transaction costs are associated with untested regimes. Alternative management regimes imply different ownership and disposition right configurations that impact on mobile network market evolution. In particular, spectrum management directly affects the organization of network platforms, and indirectly affects opportunity available to vertically related industry. Further, spectrum management interacts with business strategy (e.g., to the extent that suppliers create modular technology), cost and demand conditions and public policy (e.g., third-party access rules and standardization) variables to determine industry performance.

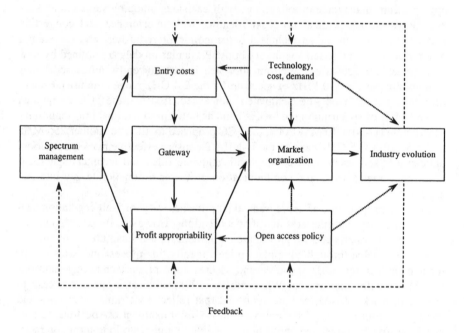

Fig. 14.4. Spectrum Management and Industry Evolution

Spectrum management regimes also impact on subscriber gateways. That is, for exclusively licensed services gateway numbers are fixed by government, e.g., in the US from 1983 to 1994 a duopoly cellular service market structure prevailed. At the introduction of PCS in 1994, per market licenses increased from three to seven. Alternatively, gateways are determined endogenously by market forces—within competition law limits—and depend on technology, cost and demand conditions within a market segment. When sunk costs are high relative to market demand, only a few service providers can operate profitably. Spectrum management also affects market entry costs, a determinant of horizontal and vertical industry structure. Higher sunk entry costs typically leads to more concentrated markets (Sutton 1991). Alternatively, when sunk costs are low more diverse market structures result. Historically, although administrative licensing systems incur substantial costs in processing applications, no fee is collected for spectrum use. With alternative and rival spectrum uses, it is desirable that spectrum prices reflect opportunity cost. Accordingly, many countries introduced spectrum fees that are either determined administratively or by an auction. Such auction fees raise concerns that exclusive rights may create distortions contributing to deviations between spectrum opportunity costs and sunk license prices (Noam 1998; Melody 2001; Bauer 2003). Any deviation is an additional entry barrier, and may reduce competitor numbers (Gruber 2001). Secondary spectrum markets can ameliorate any undesirable effect by reducing the sunk part of an access fee. However, spectrum owner incentives do not necessarily lead to efficient trades. Further, when use right futures markets exist it is optimal for mobile service providers to purchase access tokens in both real-time and futures markets so as to ensure prices. Effectively, this mechanism is an auction model combined with secondary and spot markets. That is, the only regime not imposing direct spectrum-related entry costs is pure open access.

Table 14.2. Spectrum Management and Industry Evolution Determinants

	Licensing		Ownership	Commons	Open Access
	Administrative	Flexible			
Entry cost	Indirect	Bid price	Market price	Access rules	Equipment
Gateways	Government	Government	Market	Self regulate	Market
Risk	High	High	High	Low/ high	Low

Note. Entry costs are infinite for unsuccessful applicants. The appropriability of a risk premium is influenced by factors such as an open access regime providing little protection via control of spectrum.

Licensing regimes, in addition to policy and strategy dimensions, influence mobile industry appropriation conditions—a precondition for risky investment. Alternatively, regimes with exclusive rights allow easier service provider appropriation of returns. For non-exclusive regimes, spectrum policy is less effective in supporting appropriation, e.g., setting minimal acceptable interference standards to protect service providers from intrusion that deteriorate service quality.[10] Similarly, congestion rules can emerge from self-regulation. From an investor perspective, non-exclusive regimes increase uncertainty as interference and congestion threaten business models. Namely, too few risky long-term projects are pursued.

---

[10] In principle, it is possible to negotiate, however, this entails prohibitive transaction cost unless supported by a collective rule.

260   Johannes M. Bauer

Open access critics argue that experiments with unlicensed spectrum result in low investment and local overuse (Hazlett 2001).

Spectrum Management and Innovation

Spectrum policy is related to innovation patterns and processes. When technology is exogenous to spectrum management, there is a correspondence between innovation forms, investment process and spectrum policy. However, when technology is endogenous the relation is recursive, viz., spectrum management shape the innovation path. Furthermore, spectrum management influences value chain layers and innovation locus relations. Specifically, assume technology is exogenous to spectrum management regimes. Factors influencing risk premium appropriation include the ability of suppliers to gain competitive advantage, e.g., based on generated user value, service functionality and service design. While regimes granting exclusive spectrum rights facilitate long-term planning they have undesirable effects. Regimes that secure temporary use rights via market mechanisms support appropriation (Noam 1998; Oniki 2003). Open access regimes are based on promises that the technology assures spectrum availability. As this proposition is uncertain investors and innovators demand higher returns. Clearly, under open access, Type II and Type III innovation is less likely given the higher required returns. In this sense, regimes allowing exclusive spectrum use are more conducive to Type II and Type III innovation. However, these frameworks are not conducive to Type I innovation and create an anti-commons problem. Spectrum management models allowing secure use rather than ownership are compatible with the Table 14.2 scenarios, as service providers acquire use rights only when the likelihood is above some threshold jeopardizing service quality. When technology is flexible, prevailing spectrum management shapes its path and a system of exclusive rights strengthens rights holder positions in the value chain. The approach applies to any value chain player, such as equipment manufacturers or mobile portals. Experience suggests this situation with investment and innovation organized by network operators in licensed bands and by equipment manufacturers in unlicensed bands. Further, license fees imply distortion in the allocation of R&D funds. Moreover, the entry cost aspect of license fees strengthen right holder incentives to limit vertically related market participation so as to enhance revenue and profit opportunity. Clearly, frameworks based solely on open access bias innovation toward projects compatible with the approach, and so lead to underinvestment in Type II and Type III infrastructure. Like exogenous technology scenarios, spectrum management that allows flexible use rights appears more capable of supporting a range of innovation.

Further, as open access models increase user gateways they must stimulate innovation in vertically related application markets (Bar et. al. 2000; Lemley and Lessig 2001). Securing spectrum facilitates network level risk premium appropriation, and is a prerequisite for network infrastructure investment. When vertically related activity requires highly specialized skills not available in the physical network layer, firms have an incentive to open their networks to grow and internalize network effects (Farrell and Weiser 2002). Likewise, when innovation conditions in vertically related industry entail economies of scale and scope there are advantages in having access to fewer and larger networks. Lastly, when coordination costs between application and network layers are high fewer network layer players are required for market development. Alternatively, the fewer are gateways the greater is concern that access conditions are structured to capture network rents that weaken innovation advantage from vertically integration.[11] When these conditions do not apply, innovation in the application and service layer benefits from more open network layers populated by gateway service providers. Empirical experience for alternative spectrum management regimes is provided by recent mobile communication innovations. For example, consider mobile Internet services— i-mode in Japan and Nate in Korea—as innovation systems based on cooperation between networks service providers, manufacturers and service providers (Fransman 2002). This business model is characterized by high sunk costs and coordination needs, and the combination of exclusive spectrum control, proprietary technology platform to organize business process and modular network interface to enable content creation facilitated growth. The proprietary network platforms facilitate the appropriation of innovation rents. It appears that standardization of modular interfaces is sufficient to establish required compatibility between manufacturers and network operators. Substantial potential for innovation is apparent for Type I innovation, i.e., for services with relatively low sunk (compared to 3G) and coordination costs based on system architectures that place intelligence at the network fringe. Accordingly, the most conducive innovation framework is a network layer with many gateways running on a transparent protocol. For many applications secure spectrum access is less important. This is particularly so for mobile data applications that employs advanced error correction. Such communications are fault tolerant and function satisfactorily under considerable interference. Exclusive licensing reduces gateways, although proponents of the applications such as Apple and Intel serve as license holders and make spectrum available. Applications in these bands, such as WiFi or Bluetooth, are complementary to solutions offered by licensed service providers. WiFi has a lower range and in the 5 GHz services is susceptible to obstacles. So far, WiFi service is not mobile and has security issues. Moreover, alternative unlicensed technology do not co-exist in a benign fashion and interference problems persist, e.g., Bluetooth and 802.11b. As use increases, interference protection and coordination may be a challenge. However, perhaps these issues will be addressed by innovative technology, including voluntary standardization, appropriate network topology such as standardized IP access points that control traffic when congestion is a problem.

---

[11] When this concern arises other policy measures may be adopted to remedy it. Sweden has imposed open access condition, while the Japanese government is considering it when there are only a few network platforms.

Conclusions

Although there is a strong push toward privatization, spectrum policy is not limited by this option. Alternative approaches define characteristic property and disposition rights that influence investment and innovation process within the wireless industry in complex but tractable ways. Market-based, commons and open access models are compatible with particular innovation processes. This analysis indicates that a framework of spectrum use markets fits broadly with innovation scenarios. Insights gained suggest that blending approaches provides a superior framework. Faulhaber and Farber (2002) suggest a model in which private ownership coexists with open access rights. Many proponents of free spectrum access argue that ubiquitous underlay rights are sufficient for emerging mobile technology. An advantage of this proposal is administrative simplicity. However, the proposal has potential problems as satellite radio, military radar and low power FM radio demonstrate. Namely, incumbent users have an incentive to keep underlay users to a minimum to prevent viable competition and protect exclusive use. Thus, strong conflict of interest exists between primary and underlay users. An alternative is to create regimes for designated bands however this requires a metric for regime comparison. In principle, spectrum markets solve this issue, although the approach remains incompatible with open access.[12] The rationality of spectrum planning is increased as comparative information on the relative performance of different regimes is generated. The advanced mobile service industry value chain is more complex than that for present generation mobile services. Spectrum policy is only a component shaping the innovation processes in these markets. Other factors include policy governing access to a communications gateway or business strategy that interacts with spectrum policy. The configuration of spectrum management and these factors is crucial for mobile industry evolution. The complexity of the value chain makes open access regimes neither necessary nor sufficient for vibrant upstream industry. Open access policy is neutralized by the creation of exclusive rights rooted in the real estate on which access points are located and proprietary technology. A system of exclusive rights in spectrum is conducive to innovation. As a best regime does not exist, policy makers may be well advised to allow competition between institutional regimes for spectrum management.

Acknowledgements

This study benefited from discussions with Steven S. Wildman, Carol Ting and Kevin Huang. National Science Foundation grant CCR-0205281 research support is acknowledged.

---

[12] Markets require clearly defined tradable rights. Such rights have a dual role—while they are necessary to cope with scarcity, their introduction creates scarcity. Further, when price is zero owners may not make use available.

References

Aoki M (2001) Toward a comparative institutional analysis. MIT Press, Cambridge

Antonelli C (1992) The economic theory of information networks. In: Antonelli C (ed) The economics of information networks. Elsevier Science, Amsterdam, pp 5–27

Antonelli C (2001) The microeconomics of technological systems. Oxford University Press, Oxford

Bar F, Cohen S, Cowhey P, DeLong B, Kleeman M, Zysman J (2000) Access and innovation policy for third-generation Internet. Telecommunications Policy 24: 489–518

Bauer JM (2002) A comparative analysis of spectrum management regimes. Presented at 30th Research Conference on Communication, Information and Internet Policy, Alexandria, September 28-30

Bauer JM (2003) Impact of license fees on the prices for mobile service. Telecommunications Policy 27: 417–34

Benkler Y (2002) Some economics of wireless communications. Presented at 30th Research Conference on Communication, Information and Internet Policy, Alexandria, September 28-30

Buchanan JM, Yoon YJ (2000) Symmetric tragedies: Commons and anticommons. Journal of Law and Economics 43: 1–13

Buck S (2002) Replacing spectrum auctions with a spectrum commons. Stanford Technology Law Review, 2, http://stlr.stanford.edu/STLR/Articles/02_STLR_2

Coase RH (1959) The Federal Communications Commission. Journal of Law and Economics 2: 1–40

Coase RH (1960) The problem of social cost. Journal of Law and Economics 3: 1–44

Cramton P, Kwerel E, Williams J (1998) Efficient reallocation of spectrum incumbents. Journal of Law and Economics 41: 647–75

Cramton P (2002) Spectrum auctions. In: Cave ME, Majumdar SK, Vogelsang I (eds) Handbook of telecommunications economics vol 1. North Holland, Amsterdam, pp 606–39

Farrell J, Weiser PJ (2002) Modularity, vertical integration, and open access policies: Towards a convergence of antitrust and regulation in the Internet age. Competition Policy Center, Paper CPC02-035, at http://repositories.cdlib.org/iber/cpc/CPC02-035

Faulhaber GR, Farber DJ (2002) Spectrum management: Property rights, markets and the commons. Presented at 30th Research Conference on Communication, Information and Internet Policy, Alexandria, September 28-30

Fransman M (2002) Telecommunications in the Internet age: From boom to bust to...? Oxford University Press, Oxford

Galambos L, Abrahamson EJ (2002) Anytime, anywhere: Entrepreneurship and the creation of a wireless world. Cambridge University Press, Cambridge

Gruber H (2001) Spectrum limits and competition in mobile markets: The role of license fees. Telecommunications Policy 25: 59–70

Hazlett TW (2001) The wireless craze, the unlimited bandwidth myth, the spectrum auction faux pas, and the punchline to Ronald Coase's 'Big Joke': An essay on airwave allocation policy. Harvard Journal of Law and Technology 14: 335–567

Herzel L (1951) Public interest and the market in color television regulation. University of Chicago Law Review 9: 802–16

ITU (2002) Internet for a mobile generation. ITU, Geneva

Klemperer P (2002a) How (not) to run auctions: The European 3G telecom auctions. European Economic Review 46: 829–45

Klemperer P (2002b) What really matters in auction design. Journal of Economic Perspectives 16: 169–89

Kwerel E, Williams J (2002) A proposal for a rapid transition to market allocation of spectrum. OPP Working Paper Series 38, FCC, Washington

Langlois RN (2002) Modularity in technology and organization. Journal of Economic Behavior and Organization 49: 19–37

Lehr W, Merino Artalejo MF, Eisner Gillett S (2003) Software radio: Implications for wireless services, industry structure, and public policy. Communications & Strategies 49: 15–42

Lemley M, Lessig L (2001) The end of end-to-end: Preserving the architecture of the Internet in the broadband era. UCLA Law Review 48: 925–72

Lessig L (2001) The future of ideas: The fate of the commons in a connected world. Random House, New York

Liebowitz SJ, Margolis SE (2002) Network effects. In: Cave ME, Majumdar SK, Vogelsang I (eds) Handbook of telecommunications economics vol 1. North Holland, Amsterdam, pp 75–96

Lightman A, Rojas W (2002) Brave new unwired world: The digital big bang and the infinite Internet. Wiley, New York

Lipsey RG, Lancaster K (1956) The general theory of second best. Review of Economic Studies 24: 11–32

Maitland CF, Bauer JM, Westerveld R (2002) The European market for mobile data: Evolving value chains and industry structure. Telecommunications Policy 26: 485–505

Melody WH (1980) Radio spectrum allocation: Role of the market. American Economic Review 70: 393–7

Melody WH (2001) Spectrum auctions and efficient resource allocation: Learning from the 3G experience in Europe. Info 3: 5–10

Noam EM (1998) Spectrum auctions: Yesterday's heresy, today's orthodoxy, tomorrow's anachronism: Taking the next step to open spectrum access. Journal of Law and Economics 56: 765–90

Noam EM (2001) Access issues for wireless content. Presented at 29th Research Conference on Communication, Information and Internet Policy, Alexandria, October 27-29

North DC (1990) Institutions, institutional change and economic performance. Cambridge University Press, Cambridge

Oniki H (2003) Modified lease auction and relocation: Proposal of a new system for efficient allocation of radio-spectrum resources. Revised version of paper presented at 14th Biennial Conference of the International Telecommunications Society, Seoul, Korea, August 18-21

Ostrom E (1990) Governing the commons. Cambridge University Press, Cambridge

Reed DP (2002) How wireless networks scale: The illusion of spectrum scarcity. Presented at International Symposium on Advanced Radio Technology, Boulder, March 4

Rosston GL, Hazlett TW (2001) Comments of 37 concerned economists. Filed before the FCC, in the matter of promoting efficient use of spectrum through elimination of barriers to the development of secondary markets, WT Docket No. 00-230, February 7

Schumpeter JA (1942) Capitalism, socialism, and democracy. Harper, New York

Spiller PT, Cardilli C (1999) Towards a property rights approach to communications spectrum. Yale Journal on Regulation 16: 53–83

Stevenson GG (1991) Common property economics: A general theory and land use applications. Cambridge University Press, Cambridge

Sutton J (1991) Sunk cost and market structure: Price competition, advertising, and the evolution of concentration. MIT Press, Cambridge

Valletti TM (2001) Spectrum trading. Telecommunications Policy 25: 655–70

Wireless Data (2000) Wireless data: The world in your hand. JP Morgan and Arthur Andersen, London

15 Rational Explanations of ICT Investment

Russel Cooper and Gary Madden

Introduction

After the 2000 NASDAQ decline, and subsequent downturn in global information and communications technology (ICT) markets, the telecommunications industry remains an important source of productivity growth and technology diffusion (OECD 2002, 2003). For example, packaged software sales, whose estimated value in 2001 is US$ 196 billion, are still expanding at 16% p.a. in OECD Member Country markets. Further, global mobile telephony subscription increased from 11 million subscribers at 1990 to 945 million subscribers at end-2001. Finally, Internet subscription reached a penetration of 8.2 users per 100 persons globally at end-2001, i.e., equaling 1999 mobile telephony market penetration (ITU 2002b). Reasons proffered for the NASDAQ decline, and the bursting of the dot.com bubble more generally, include unrealistic market growth expectations, firm activity not aligned with core competency and the accumulation of unsustainable debt. However, such rationalizations themselves require explanation. That is, whose expectations are at issue, those of firms making infrastructure investment decisions or those of shareholders making portfolio investment choices? With asymmetric information these expectations may differ, and so should policy response. Additionally, for both the dot.com and shareholder groups, unrealistic growth expectations may arise from premature timing in an unfamiliar environment. While it is the nature of uncertainty that expectations are mostly wrong, such expectations are still rational should they prove correct on average, i.e., averaged over the circumstances that might occur. Based on this criterion, over a time interval and in an environment subject to technology change it is difficult to confirm that expectations are irrational. Specifically, it is uninformative to characterize management decisions as poor ex post. Further, a claim that firm activity is not aligned with core competency is at variance with the view that, in a modern economy, it is prudent to operate a portfolio of interests. Finally, conventional wisdom in the form of advice to management is often fashion driven, viz., debt is only a problem when it is unsustainable, and assuming that foreseen unsustainable events are avoided, this explanation does not add anything to the unrealistic expectations variant.

Having failed to accept demand shifts, lower prices and new market conditions as adequate explanations for the telecommunications crisis, analysts have proposed the notion of irrational exuberance (OECD 2003). This term, introduced by US Federal Reserve Board Chairman Alan Greenspan in a speech on December 5, 1996, initially intended to portray shareholder behavior on the stock market (Shiller 2000). Subsequently, according to Shiller, Greenspan retreated from this view in favor of a more optimistic 'new era' position. Nevertheless, the term as applied to the stock market has developed an aura of its own. Given the link between stock market behavior and the ICT sector with the birth and subsequent bust of the dot.com bubble, it is perhaps understandable that irrational exuberance is now being offered as a reason for the telecommunications crisis. However, this idea is potentially misleading since it characterizes as illogical what may be complex but logical reactions to uncertainty. Acquisition excess, the payment of license fees that are too high viewed ex post, and unrealistic business models have all been proffered as examples of irrational exuberance (OECD 2003). However, such firm behaviors, even when ex post misguided, need not be irrational. Focusing solely on reasoning previously attributed to shareholder psychology offers no direct insight into the infrastructure investment decisions of incumbent firms when technology changes. However, because such situations are of ongoing relevance, e.g., with many incumbent firms having the potential to enter into joint ventures with application sector firms, so the need to understand incumbent firm infrastructure investment decisions remains. The approach taken here is to endow firms with the standard of rationality on which to base their decisions. The observed acquisition of low price foreign assets from firms in financial distress provides some support for this agenda, as does the augmentation of technological competence through cross-border strategic alliance, merger and acquisition (OECD 2003).[1]

This chapter constructs a model of investment behavior by a typical ICT firm that infers a complex pattern of investments through time. As investment decisions are made in an uncertain environment, occasionally 'bad news' dominates. To rationally explain the recent ICT sector investment experience, components of investment are decomposed in a manner analogous to that for the growth accounting perspective. However, here ICT investment components are identified with aspects of logical decision making. In particular, this rational accounting approach identifies factors that influence optimal investment decision making as: (i) profitable production; (ii) optimal portfolio choice; (iii) strategic merger and acquisition; (iv) shareholder satiation; and (v) futures preparation. Strands of the received investment literature are associated with these factors; however it is the combination of influences that leads to an investment outcome. Further, the model is capable of identifying technology parameters, preference parameters, business environment indicators and potential government policy levers. These parameters are allocated in systematic fashion to components of the rational growth accounting representation of investment. Finally, interaction among factors driving the components may lead to complex and possibly undesirable outcomes that can nevertheless be characterized as the result of rational decision making. Contrasted with the irrational explanation, this has important consequences for the identification of appropriate ameliorating policy.

---

[1] OECD (2003) argues that country-by-country investment analysis needs to be supplemented with company-by-company analysis so as to better capture industry dynamics.

Recent OECD ICT Investment

Table 15.1 and Table 15.2 show OECD Member Country annual telecommunications investment from 1980 through 1990 and 1991 through 2001, respectively. These data indicate substantial variation in investment levels through time and cross country. Casual inspection reveals most OECD Member Country annual investment increased substantially through the period. However, the rates of investment differed widely by Member Country, with higher growth rates typically reported for countries with smaller networks at the beginning of the sample period.[2]

Table 15.1. OECD Annual Telecommunications Investment (US$ billions), 1980–1990

	1980	1981	1982	1983	1984	1985	1986	1987	1988	1989	1990
Australia	1.1	1.3	1.3	1.4	1.2	1.1	1.3	1.7	1.8	2.0	2.3
Austria	0.5	0.5	0.5	0.5	0.6	0.6	0.7	0.9	1.0	1.0	1.4
Belgium	0.6	0.5	0.4	0.4	0.4	0.4	0.5	0.5	0.5	0.6	0.8
Canada	2.2	2.4	2.1	2.3	2.1	2.1	2.1	2.4	3.2	3.5	3.7
Czech Rep.											
Denmark	0.3	0.3	0.2	0.2	0.2	0.2	0.4	0.5	0.6	0.5	0.5
Finland	0.3	0.3	0.2	0.3	0.2	0.3	0.4	0.5	0.6	0.6	0.8
France	5.6	4.2	3.8	3.5	3.2	4.0	4.5	5.1	4.9	4.8	4.8
Germany	5.5	5.0	4.9	4.7	4.9	5.3	7.4	9.2	9.7	9.4	11.9
Greece	0.2	0.3	0.3	0.3	0.3	0.2	0.2	0.2	0.2	0.3	0.4
Hungary	0.1	0.1	0.1	0.1	0.1	0.1	0.1	0.2	0.2	0.2	0.2
Iceland	0	0	0	0	0	0	0	0	0	0	0
Ireland	0.3	0.4	0.3	0.2	0.2	0.1	0.2	0.2	0.2	0.2	0.3
Italy	2.8	2.4	2.7	2.9	2.8	2.8	3.7	4.6	5.9	7.5	8.3
Japan	7.7	8.1	7.2	7.4	7.5	6.9	9.9	13.0	14.9	15.7	15.7
Korea		1.5	1.4	1.3	1.3	1.7	1.4	1.8	3.0	3.0	
Luxembourg	0	0	0	0	0	0	0	0	0	0	0.1
Mexico	0.4	0.6	0.4	0.3	0.5	0.5	0.4	0.5	0.6	0.8	1.4
Netherlands	0.7	0.5	0.4	0.4	0.4	0.5	0.6	0.8	1.0	1.4	1.5
New Zealand		0.1	0.1	0.1	0.1	0.1	0.2	0.3	0.4	0.5	0.4
Norway	0.5	0.4	0.4	0.4	0.4	0.4	0.5	0.6	0.7	0.5	0.4
Poland									0.1	0.1	0.2
Portugal	0.1	0.2	0.2	0.2	0.2	0.2	0.2	0.3	0.4	0.5	0.7
Slovak Rep.											
Spain	1.6	1.4	1.4	1.1	1.1	1.1	1.5	2.1	3.0	4.8	7.1
Sweden	0.4	0.7	0.7	0.6	0.6	0.6	1.1	1.3	1.3	1.3	1.1
Switzerland	0.7	0.6	0.7	0.7	0.7	0.7	1.2	1.5	1.6	1.5	2.2
Turkey	0.2	0.2	0.2	0.3	0.3	0.5	0.7	0.9	0.6	0.4	0.8
UK	2.9	3.0	2.7	2.3	2.5	2.4	3.1	3.9	5.2	5.1	4.9
US	19.8	20.7	19.5	17.3	17.0	22.3	22.4	22.1	21.1	20.2	20.6
OECD ex USA	34.8	33.6	32.9	32.3	31.6	32.5	42.7	52.6	60.6	66.5	74.8
OECD Total	54.6	54.3	52.4	49.6	48.6	54.8	65.1	74.7	81.7	86.7	95.4

Source: ITU (2001, 2002a, b)

---

[2] Countries with lower penetration rates are expanding their networks, whereas those countries with established networks are modernizing. As the sector becomes digitized these strategies are converging, with network expansion undertaken by employing digital equipment.

Table 15.2. OECD Annual Telecommunications Investment (US$ billions), 1991-2001

	1991	1992	1993	1994	1995	1996	1997	1998	1999	2000	2001
Australia	2.3	2.2	1.8	1.8	2.8	3.7	3.7	2.4	3.4	3.7	4.7
Austria	1.6	1.5	1.3	1.5	1.6	0.9	1.2	1.7	1.8	0.8	1.6
Belgium	0.8	0.8	1.3	1.0	1.5	1.1	1.5	0.9	0.7	1.0	0.6
Canada	3.4	3.5	2.8	2.8	2.6	3.0	4.2	4.3	4.0	4.9	5.0
Czech Rep.	0.1	0.1	0.3	0.4	0.8	1.1	1.4	1.2	0.8	1.2	1.2
Denmark	0.4	0.4	0.4	0.4	0.5	0.7	0.9	1.1	1.0	1.1	1.3
Finland	0.6	0.5	0.4	0.6	0.8	0.8	0.8	0.7	0.9	0.9	
France	6.2	5.8	6.3	5.5	6.3	5.4	6.6	6.9	6.7	7.2	6.4
Germany	14.9	18.2	15.9	12.7	10.8	11.7	8.9	8.8	9.9	5.9	6.9
Greece	0.5	0.7	0.8	0.6	0.7	0.7	0.8	0.7	0.7	1.2	1.2
Hungary	0.4	0.5	0.5	0.8	0.7	0.6	0.5	0.5	0.5	0.5	0.8
Iceland	0	0	0	0	0	0	0	0.1	0.1	0.1	0
Ireland	0.3	0.3	0.2	0.3	0.3	0.4	0.5	0.5	0.6	0.4	
Italy	10.3	8.9	6.7	5.6	4.7	5.9	6.7	5.5	4.3	8.6	7.3
Japan	17.6	19.7	23.7	26.4	35.1	37.9	32.8	28.3	29.8	32.7	24.6
Korea	3.2	3.1	3.2	3.6	4.4	5.8	8.1	4.5	7.0	7.8	6.6
Luxembourg	0.1	0.1	0.1	0.1	0.1	0.2	0.1	0.1	0.1	0.1	0.1
Mexico	1.9	3.1	2.7	2.5	1.5	1.8	1.9	3.2	4.0	5.1	5.7
Netherlands	1.6	1.6	1.5	1.5	1.7	1.6	1.6	2.2	2.7	3.1	2.6
New Zealand	0.4	0.3	0.2	0.3	0.4	0.5	0.4	0.3	0.3	0.3	0.4
Norway	0.5	0.6	0.3	0.5	0.8	0.7	0.8	1.5	1.8	2.1	
Poland	0.3	0.5	0.7	0.6	0.9	1.2	1.3	1.2	1.4	1.4	
Portugal	0.9	1.0	0.6	0.7	1.2	1.1	1.3	1.5	1.7	2.0	1.4
Slovak Rep.	0	0.1	0.1	0.1	0.2	0.3	0.3	0.3	0.1	0.1	0.1
Spain	6.0	4.1	3.0	2.9	3.1	2.8	2.4	2.0	5.7	6.7	4.9
Sweden	1.2	0.9	0.7	0.8	1.1	1.0	0.8	0.9	1.3	2.5	1.5
Switzerland	2.2	2.2	1.6	1.6	1.8	1.8	1.6	1.3	2.0	2.2	4.0
Turkey	0.9	0.9	1.2	0.6	0.4	0.4	0.5	0.6	0.6	0.6	0.4
UK	4.3	4.9	4.0	4.9	7.2	9.5	12.5	12.5	14.6	16.3	13.8
US	20.9	21.7	23.3	22.8	23.6	22.4	23.2	24.2	26.3	28.8	29.6
OECD ex USA	82.9	86.4	82.4	81.0	94.1	102.8	104.3	95.7	108.8	120.4	103.0
OECD Total	103.8	108.1	105.8	103.8	117.7	125.2	127.5	119.9	135.1	149.2	132.6

Source: ITU (2001, 2002a, b)

For example, Mexico increased its average per annum (p.a.) investment seven-fold (from US$ 443m in 1980-1986 to US$ 3,304m in 1995-2001), whereas the US increased investment by approximately 28% (from US$ 19,815m in 1980-1986 to US$ 25,440m in 1995-2001). Particular events explain certain investment spikes, e.g., Germany extended its infrastructure nationally post-reunification, and Spain upgraded and expanded its network leading to the Olympic Games. Further, the former centrally planned economies of the Czech Republic, Poland and the Slovak Republic report massive increases in domestic investment since the late-1980s.

To aid comparison, smoothed five year percentage growth rates are constructed from overlapping seven year average annual investment levels centered on 1983, 1988, 1993 and 1998 (see Table 15.3 and Table 15.4). The geometric mean of the five-year growth rates is presented as a ranking indicator for Member Country ICT investment growth and GDP in US$ billions for 2000 is provided as a comparative economy size indicator.

Table 15.3. OECD Telecommunications Investment and GDP (US$ billions)

	Average annual investment (7 year centered)				GDP
	1983	1988	1993	1998	2000
Australia	1.3	1.8	2.4	3.5	390.1
Austria	0.6	1.0	1.4	1.4	189.0
Belgium	0.5	0.6	1.0	1.0	226.6
Canada	2.2	2.9	3.1	4.0	687.9
Czech Rep.	0	0.1	0.5	1.1	50.8
Denmark	0.3	0.5	0.5	0.9	162.3
Finland	0.3	0.6	0.7	0.8	121.5
France	4.1	4.9	5.8	6.5	1,294.2
Germany	5.4	9.7	13.7	9.0	1,873.0
Greece	0.3	0.3	0.6	0.9	112.6
Hungary	0.1	0.2	0.5	0.6	45.6
Iceland	0	0	0	0	8.5
Ireland	0.2	0.2	0.3	0.4	93.9
Italy	2.9	6.1	7.2	6.1	1,074.0
Japan	7.8	13.4	25.2	31.6	4,841.6
Korea	1.5	2.2	3.8	6.3	457.2
Luxembourg	0	0	0.1	0.1	18.9
Mexico	0.4	0.9	2.1	3.3	574.5
Netherlands	0.5	1.1	1.6	2.2	364.8
New Zealand	0.1	0.3	0.4	0.4	49.9
Norway	0.4	0.5	0.5	1.3	161.8
Poland	0	0.2	0.6	1.2	157.7
Portugal	0.2	0.4	0.9	1.5	105.1
Slovak Rep.	0	0	0.1	0.2	19.1
Spain	1.3	3.7	4.1	4.0	558.6
Sweden	0.7	1.1	1.0	1.3	227.3
Switzerland	0.8	1.6	1.9	2.1	239.8
Turkey	0.3	0.7	0.8	0.5	199.9
UK	2.7	4.1	5.7	12.3	1,414.6
US	19.9	21.4	22.2	25.4	9,837.4
OECD ex USA	34.3	58.9	86.3	104.1	15,720.8
OECD Total	54.2	80.3	108.5	129.6	25,558.2

Source: Constructed from Table 15.1 and Table 15.2; GDP 2000 from ITU (2001)

These data indicate substantial within-country volatility and cross-country differences not evident from aggregate OECD series. The relative smoothness of the
aggregate investment series is substantially influenced by US data magnitudes.
Abstracting from the US, a more substantial trend investment growth for the remainder of the OECD is apparent. However, this trend growth is an aggregation of
substantially different Member Country experiences, with individual country volatility not synchronized cross country. For most Member Countries, and to a lesser
extent in the OECD aggregate, investment volatility is more evident through the
1990s. However, this experience is not universal, and in countries where apparent
increased volatility has occurred there is nevertheless, with few exceptions, a con-

tinued trend increase in investment. Given this substantial cross country variation the challenge remains to find a coherent and comprehensive explanation for such behavior. The model presented here provides a potential explanation that does not concede the postulate of rationality. To do so, the investment motive is disaggregated into factors that are allowed to either be reinforcing or self-canceling. This feature of the model provides the potential to explain complex observed relationships cross country and through time.

Table 15.4. OECD Telecommunications 5-year Investment Growth Rates (%)

	1988/1983	1993/1988	1998/1993	Average
Australia	42	36	43	40
Austria	83	34	-2	34
Belgium	32	75	-2	32
Canada	33	7	29	22
Czech Republic	0	418	139	132
Denmark	64	8	90	50
Finland	97	17	28	44
France	19	18	13	17
Germany	80	42	-35	19
Greece	11	117	36	48
Hungary	144	155	16	93
Iceland	29	89	67	60
Ireland	-13	36	54	22
Italy	113	17	-15	29
Japan	71	88	26	59
Korea	52	70	68	63
Luxembourg	168	118	8	85
Mexico	95	144	56	95
Netherlands	105	48	42	63
New Zealand	196	8	2	48
Norway	16	3	141	42
Poland	0	268	98	94
Portugal	154	97	64	102
Slovak Republic	0	248	47	72
Spain	178	13	-4	44
Sweden	70	-15	34	25
Switzerland	105	22	9	40
Turkey	100	14	-33	15
UK	53	37	117	66
US	8	4	15	9
OECD ex US	72	46	21	45
OECD Total	48	35	19	34

Source: Constructed from Table 15.1 and Table 15.2

Optimal Investment Strategy

This chapter views telecommunications firms as providers of services necessary for effective 'old economy' operation. As such, when frequent and sometimes radical ICT sector innovation occurs it has productivity implications beyond the sector. In particular, within this context, strategic alliance is viewed as a vehicle to create and reinforce mutually beneficial links between ICT sector firms and innovative firms operating in the 'new economy'. Clearly, the outcome of radical innovation, whether in the form of new economy products or in improved old economy production techniques, is more problematic or uncertain than that for more mature business activity—because difficult to evaluate risks that are hard to insure are produced (Rosenberg 1996). Also because of its newness, radical innovation faces higher adoption costs and makes the achievement of required critical mass more difficult than for traditional investment in established markets.

Early rational (neoclassical) dynamic analysis of firm investment behavior considered situations where firms instantaneously adjust their capital stock in a costless manner. Jorgenson (1963) shows, in this environment, that even with a dynamic forward-looking criterion investment activity is reduced to a static decision, whereby the firm invests until its marginal product equals the user cost of capital.[3] A more recent literature introduced adjustment costs or frictions to explain slow adjustment (Eisner and Strotz 1963; Lucas 1967; Gould 1968). Despite this literature not paying explicit attention to uncertainty it served as the theoretical forerunner of investment specification in macroeconomic models. An alternative finance literature views investment decision making as an optimal portfolio allocation problem (Merton 1971). Abel (1983) integrated aspects of the Merton (1971) approach into the adjustment cost investment literature, emphasizing the link between the marginal optimal value function and q theory. Finally, the modern financial options pricing literature is introduced into real investment theory by McDonald and Siegal (1985, 1986), and popularized by Dixit and Pindyck (1994) as real options theory. These developments not only extend the generality of the rational approach, as applied to investment decision making models, but also represent a change in emphasis, viz., alternative explanations of the dot.com bubble have some validity. That is, a rational ICT firm takes explicit account of the opportunity cost of capital in profit maximizing situations, gives consideration to its financial portfolio position, and is aware that it must allow for costly reversibility and so consider the future implications of current decisions. Further, by developing a model with antecedents in growth, economic development and optimal saving literatures, and the more recent new growth resurgence—with an emphasis on the implications of drastic innovation in general purpose technology—the context facing an ICT firm making such investments is extended to allow for shareholder-owner pressure to save and recognizes the need to anticipate the consequences of

---

[3] The corresponding Marshall Criterion or investment trigger is to invest when the project has positive net worth. The criterion is not necessarily optimal when waiting is possible (Dixit 1992: 110-1).

technology innovation (Ramsey 1928; Solow 1956; Swan 1956; Romer 1986; Barro and Sala-i-Martin 1995; Bresnahan and Trajtenberg 1995; Helpman 1998).

A rational investment strategy is characterized here as the outcome of a representative firm's dynamic optimization program. The representative telecommunications firm is modeled as utilizing ICT capital to produce old economy services. When an innovation occurs, it raises a technology index s and allows the firm to operate more productively. The firm has the option to borrow or issue equity to finance network expansion. Such expansion, e.g., can involve new economy activity such as the acquisition of shares in firms operating in the software applications sector. This investment is inherently more risky. The representative firm's objective is to maximize the present value of the stream of utility flows to consumer-shareholders. Representative consumer preferences are defined by the instantaneous utility function $U(c)$

$$\max E_0 \int_0^\infty e^{-\rho t} U(c) dt ,$$ (15.1)

where c is real per capita consumption expenditure.

The representative firm makes both an optimal consumption c and portfolio choice x by solving Eq. 15.1 subject to the capital stock transition equation,

$$dk = (F(sk) - rx - c)dt + (dz/z)x - \omega ds ,$$ (15.2)

applications sector valuation equation,

$$dz/z = \mu dt + \sigma d\xi ,$$ (15.3)

technology transition equation,

$$ds = \vartheta F(sk)d\varphi, \quad 0 < \vartheta < 1$$ (15.4)

and initial conditions $k(0) = k_0$ and $s(0) = s_0$. Uncertainty is represented by stochastic processes. The applications sector return is influenced by $d\xi$, a Weiner process satisfying $E(d\xi) = 0$ and $E(d\xi)^2 = dt$. Technology shocks are governed by a Poisson process $d\varphi$ satisfying:

$$d\varphi = \begin{cases} 0 & \text{with probability } 1 - \lambda dt \\ 1 & \text{with probability } \lambda dt \end{cases}$$

where $0 < \lambda < 1$, so that $E(d\varphi) = \lambda dt$ and $E(ds) = \lambda \vartheta F(sk)dt$.

The control variable c is interpreted as either the consumption of the representative firm owner or a dividend paid to shareholders. Control variable x indicates the level of firm portfolio investment in applications sector equity. Borrowing is repaid at rate r, or equivalently diverting x internally generated revenue at opportunity cost r to acquire rights to an uncertain revenue stream with per unit valuation dz/z.[4] Also, attention is restricted to interior solutions $c > 0$ and $0 < x < k$. $F(sk)$ is interpreted as a normalized restricted profit function that is quasi-fixed in predetermined state variables k and s as it is conditioned on k and s, where the combined variable sk measures the effective capital stock. Given the restricted profit function interpretation, $F(sk)$ depends on output and variable input prices.[5] Further, assume $F_k > 0$ and consequently $F_s > 0$. Finally, to allow investigation of technology for which $F(sk)$ is non-linear, assume F is concave in k, viz., $F_{kk} < 0$.[6]

The innovation process, Eq. 15.4, relates innovation shock magnitudes to production activity, measured by $F(sk)$, and is consistent with Arrow's (1962) view that technical advance results from learning by doing.[7] Stochastic technology shocks ds have both temporary and permanent effects on productive capacity. An innovation $ds > 0$ delivers an immediate but temporary technology adoption cost, viz., a reduction in funds available for immediate investment, with scale ω, $0 < \omega < 1$ occurs. A larger ω value indicates more drastic innovation.[8] The technology shock also generates a permanent positive effect on productivity as the firm adapts to the new technology—this effect is captured by the argument s in the restricted profit function.[9]

---

[4] Assume firms attempt to exploit all profit opportunity. The absence of additional arbitrage opportunity implies that the external debt repayment rate r is equated to the internal opportunity cost of the firm.

[5] This dependence is ignored for notational simplicity. Additionally, for expositional simplicity ignore depreciation.

[6] This is equivalent to an assumption of decreasing returns to scale in the production technology. The model can be generalized to allow for increasing returns through external or network effects. While an important feature of the ICT industry, this does not have any effect on the issues discussed here.

[7] Arrow (1962) assumed the productivity of a firm increases with cumulative industry investment. He further argued that increasing returns arise from new knowledge discovered as investment and production occurs. This feature of the model is generalized, e.g., by replacing $F(sk)$ in Eq. 15.4 by a function of human capital, so as to allow for technology change through human capital development.

[8] A generalized model would have the size of ω modeled as dependent on the type of R&D activity. With the probability of an innovation occurring λ and its scale ϑ endogenized.

[9] The technology shock assumption is rationalized by supposing, like Nelson and Wright (1992) that technology, or know-how on how to do things is embedded in the firm's organizational structures, e.g., and is it difficult and costly to transfer to different contexts.

Next, consider alternative opportunity cost of capital scenarios facing the firm. In Case 1, the firm is treated as operating in a competitive environment with r given. In Case 2 the firm has market power and r depends on firm size. In general equilibrium, r must depend on predetermined k and s values. The scenarios are distinct, viz., in the market power case the firm recognizes that its size, represented by k, influences r. In that case, write $r = R(k,s)$ and assume the firm recognizes this in making decisions. In the competitive case, represent this relationship by $r = \overline{R}(\overline{k},s)$, i.e., r is not recognized by the competitive firm in making decisions, as influenced by its own capital stock, k. This holds, even though in general equilibrium r depends on the size of the capital stock of the representative firm $\overline{k}$. In equilibrium, $k = \overline{k}$ but R and $\overline{R}$ differ. To make the opportunity cost r endogenous, first note that r is the old economy rate of return on capital for which no instantaneous arbitrage opportunity exists. Also r is risk free except for occasional technology shocks. Next, recognizing the role k plays in the restricted profit function and in influencing technology shock magnitudes, absence of instantaneous arbitrage is equivalent to maximizing instantaneous expected profits with ICT capital costing r per unit when allowance is duly made for technology shocks. On this reasoning r is made endogenous in general equilibrium via the first-order condition:

$$\Pi(s) = \max_k \left\langle F(sk) - \tfrac{1}{dt}\omega E(ds) - rk \right\rangle, \tag{15.5}$$

where $r = \overline{R}(\overline{k},s)$ in the competitive case and $r = R(k,s)$ in the market power case. In the competitive case, and given that $\tfrac{1}{dt}E(ds) = \lambda\vartheta F(sk)$, the first-order condition is:

$$r = (1 - \omega\lambda\vartheta)F_k(sk), \tag{15.6}$$

while for the market power case it is $r = (1 - \omega\lambda\vartheta)F_k(sk) - R_k(k,s)k$. By construction, $1 - \omega\lambda\vartheta > 0$.[10] In the competitive case, the nonlinear technology $F(sk)$ with $F_{kk} < 0$, imply $\Pi(s) = (1 - \omega\lambda\vartheta)[F(sk) - F_k(sk)k] > 0$, and there is economic rent available for further investment. In the market power case, the first-order condition is integrated to provide:

$$r = (1 - \omega\lambda\vartheta)F(sk)/k, \tag{15.7}$$

---

[10] The parameters $\omega, \lambda, \vartheta$ lie in the $(0,1)$ interval.

implying $\Pi(s) = 0$ and the economic rent is appropriated.

Merger and acquisition activity by telecommunications firms can be analyzed within the model by adopting a real options approach. Considering the acquisition of another ICT firm's assets implies the existence of a buy price r_B to compare with the firm's internal opportunity cost r.[11] Also, to divest capital implies a sell price r_S exists that may be compared to r.[12]

Consider next the implications of the relationships between secondary market buy and sell prices, and the opportunity cost r. $r_B \geq r_S$ allows $r_S = 0$ when investment is irreversible, and costly reversibility and expandability in general, viz., when either $r < r_S$, $r_S \leq r \leq r_B$ or $r_B < r$.[13] Specifically, with $r > r_B$ the firm is able to buy capital more cheaply than its internal opportunity cost. Then the rational firm exercises a call option, and in doing so causes r_B to rise.[14] Alternatively, with $r < r_B$ the firm does not buy capital externally, since the market price exceeds the shadow price. When capital is not fully sunk it can be resold at price $r_S > 0$. In particular, when $r < r_S$ the firm can sell capital at a price greater than its shadow price, and so rationally exercises a put option. However, with $r > r_S$ the firm does not sell capital. Finally, when $r_S \leq r \leq r_B$, there is no incentive for a firm with opportunity cost r to either buy or sell ICT capital. This means the firm does not exercise any options. However, this behavior does not imply zero investment, viz., real options explain the reason for only a component of ICT investment.

To see this, let:

---

[11] For simplicity, to analyze call or put options by comparing their trigger values with the opportunity cost of capital, it is convenient to define triggers as rental rates. It is also convenient to refer to rentals as prices, e.g., the equivalent rental that a firm pays when it borrows the use of capital is called the buying price, and the equivalent rental it receives when it loans its capital is the selling price. This approach avoids introducing additional variables to distinguish prices from rental rates. The approach is comparable with that of Abel et al. (1996), where a distinction is made between the rental rate and price, but a one-to-one correspondence is assumed, in their two-period discrete time model.

[12] In practice, there may be several buying and selling prices due to variations in quality and risk regimes cross countries. Assume here that, normalizing for quality and risk, there is a unique buying and selling price in a common currency unit at any given time. Without loss of generality assume $r_B > r_S$.

[13] In the model, changes in r_B and r_S are (mostly) generated endogenously through the elimination of arbitrage. This outcome is important in generating 'exuberant' results from rational decision making. The effect arises from interaction among alternative investment motivations when faced with asymmetric price responses to disequilibria.

[14] It is also worth noting that the increase in k in the hands of an active firm, which by assumption is subject to decreasing returns to scale, would also typically lead r to fall. This effect is explicitly recognized by the firm in its own decision making if it has market power, but it is present in any event, albeit to a different extent, in the competitive case in general equilibrium.

$$\hat{r} = \begin{cases} r_B, & r > r_B \\ r, & r_S \leq r \leq r_B \\ r_S. & r < r_S \end{cases} \tag{15.8}$$

Next, consider optimal consumption and portfolio choice behavior within the model. In the results below, Eq. 15.11 associates the firm's ability to generate ICT investment to returns from portfolio choice, while Eq. 15.13 shows an ability to generate funds for ICT expansion from retained earnings by optimal consumption choice. For present purposes, it is sufficient to provide optimal consumption and portfolio choices by enforcing their first-order conditions without presenting explicit closed-form solutions.[15] Combining Eq. 15.2 through Eq. 15.5, and using Eq. 15.8, generates ICT investment:[16]

$$dk = dk_{OE} + dk_{NE} + dk_{RO} + dk_{RE} + dk_{TS} \tag{15.9}$$

where:

$$dk_{OE} = \Pi(s)dt \tag{15.10}$$

$$dk_{NE} = (\mu - r)xdt + \sigma x d\xi \tag{15.11}$$

$$dk_{RO} = [r - \hat{r}]kdt \tag{15.12}$$

---

[15] This approach avoids the necessity of working with particular functional forms for technology and preferences, and preserves the generality of conclusions. In the representative consumer-firm context, stochastic inter-temporal optimization problems require selection of optimal feedback controls to determine consumption c and debt x so as to maximize Eq. 15.1 subject to Eq. 15.2, Eq. 15.4 and initial conditions $k(0) = k_0$ and $s(0) = s_0$. With $J(k,s)$ denoting the optimal value function, Bellman's principle of optimality dictates the evolution of J and determines the first-order conditions for c and x, viz.,

$U_c(c) = J_k \Rightarrow c = U_c^{-1}(J_k)$ and $x = -(\frac{\mu - r}{\sigma^2})\frac{J_k}{J_{kk}}$. To simplify discussion of the

dk_{NE} component, define the Arrow-Pratt coefficient of relative risk aversion $\wp \equiv -kJ_{kk}/J_k$ and rewrite Eq. 15.11 as:

$$dk_{NE} = (\frac{\mu - r}{\sigma})^2 \frac{k}{\wp}dt + (\frac{\mu - r}{\sigma})\frac{k}{\wp}d\xi \tag{F.15.11}$$

[16] To demonstrate the equivalence of Eq. 15.2 with Eq. 15.9, note Eq. 15.10 through Eq. 15.14 gives Eq. 15.9 as:
$dk = \Pi(r,s)dt + (\mu - r)xdt + \sigma x d\xi + (r - \hat{r})kdt + (\hat{r}k - c)dt + \omega\vartheta F(sk)(\lambda dt - d\varphi)$. Canceling terms, using Eq. 15.5 and rearrangement gives:
$(F(sk) - \frac{1}{dt}\omega E(ds) - rk)dt - rxdt + (\mu dt + \sigma d\xi)x + rkdt - cdt + \omega\vartheta F(sk)(\lambda dt - d\varphi)$.
Further, using Eq.15.3 and Eq. 15.4 reduces this to Eq.15.2.

$$dk_{RE} = (\hat{r}k - c)dt \tag{15.13}$$

$$dk_{TS} = \omega \vartheta F(sk)(\lambda dt - d\varphi). \tag{15.14}$$

Implications of Rational Accounting

Summarizing the components of Eq. 15.10 through Eq. 15.14 for rational firm ICT investment:

(a) dk_{OE} is implicit profit from old economy activity, where profit is defined by Eq. 15.5 and ICT capital is valued at its opportunity cost. This is the Jorgenson component of rational ICT investment. In the market power case $dk_{OE} = 0$, but $dk_{OE} > 0$ in the competitive case, as payments at marginal factor cost do not exhaust the product due to technology non-linearity.

(b) dk_{NE} is explicit new economy investment, or the Merton or traditional finance theory component of rational ICT investment. In Eq. F.15.11, the expected rate of return $\mu - r$ is assumed positive; otherwise no financial investment occurs in the more risky new economy. However at any instant, as a result of this investment, the firm's capacity to invest in real ICT is reduced when $d\xi$ is sufficiently negative. This ICT investment component is the most likely source of erratic aggregate ICT investment as it is the main component subject to continuous Weiner-type shocks. This suggests investment volatility from this source is not accurately depicted as irrational exuberance; rather it is an outcome from rational decision making that results in volatile returns. The degree of acceptable volatility is controlled by the firm's optimal choices. Both drift and volatility are inversely related to the risk aversion coefficient $\wp$. Whether noise dominates drift depends on the volatility of the underlying new economy stochastic returns relative to the sector's risk premium. In particular Eq. F.15.11 demonstrates that drift dominates volatility when $(\mu - r)/\sigma > 1$. The parameters μ and σ appear exclusively in this component, and are potentially amenable to government policy.

(c) dk_{RO} is the real options component of the explanation of rational ICT investment, and $dk_{RO} = 0$ when $r = \hat{r}$. If $r > \hat{r}$ then $\hat{r} = r_B$ by Eq. 15.8 and the firm buys existing available capital. While relevant to a firm as it drives merger and acquisition activity, with national data much of this variation is netted out. What should remain in the national ICT investment data, however, is the effect of merger and acquisition across country borders. When $r < \hat{r}$ then $\hat{r} = r_S$ and the firm sells its capital. At the national level this involves foreign investment in the domestic economy.

(d) dk_{RE} indicates retained earnings are used for growth when the return to capital $\hat{r}k$ exceeds the endogenously determined dividend payout c. This is the traditional growth theory or Ramsey component. When retained earnings are positive the firm allocates this to growth. Note $\hat{r}k$, where $\hat{r}$ is defined by Eq. 15.8, is the appropriate valuation for economic earnings. For example, when the firm's internal opportunity cost exceeds $\hat{r}$ it buys further capital via dk_{RO} at a price such that $r_B = \hat{r}$. For $r_S < r < r_B$ the internal opportunity cost is an appropriate economic valuation metric.

(e) dk_{TS} is an allowance for expected technology shocks, and reflects an adjustment cost approach to investment analysis. However, this allowance is set in a Bresnahan and Trajtenberg (1995) drastic innovation to general purpose technology context, and results in a steady upward drift of capital stock to counter occasional losses when an innovation occurs. The term has an expectation of zero, with realizations mostly small and positive, and is interpreted as building capacity to withstand technology change shocks.

With investment components classified by alternative rationality-based approaches it is potentially useful to explore interaction among the components. The components are not completely independent in that they form part of an integrated optimization objective. However, to the extent that individual components are influenced by different variables there is scope for variation that is not completely offset. Variables explaining some components are buy and sell prices for existing capital r_B and r_S, and the expected return on new applications μ. The complexity of aggregate ICT investment depends indirectly on relationships among explanatory variables. Such relationships themselves depend on the economic environment, market characteristics, and the sophistication of firms and shareholders. To highlight these issues, consider the stylized representation of influences on the capital buy price r_B in a competitive market setting:

$$dr_B = r_B(\theta_B^- \min\{0, r - r_B\} + \theta_B^+ \max\{0, r - r_B\})(1 - d\varphi) + dr_B(t)d\varphi \qquad (15.15)$$

where $0 \le \theta_B^- < \theta_B^+ \le 1$. The term in square brackets in Eq. 15.15 applies when there is no technology change. However, whenever $r > r_B$ the firm profits by exercising a call option. In view of profits forgone when this option is exercised capital acquisition occurs swiftly. Conversely, when $r < r_B$ there is no incentive for the firm to exercise a call option. As a result r_B drifts downward toward r as firms adjust their ICT capital bid price in line with the ICT shadow price. This probably slow downward drift is represented by the speed of adjustment parameter θ_B^- in Eq. 15.15. The parameter θ_B^+ is assumed greater in magnitude than the downward adjustment parameter θ_B^-. Additionally, when technology change occurs at t, the contemporaneous response of r_B depends on whether the change enhances or reduces the productivity of extant capital. Consequently, Eq. 15.15 does not explain the change in r_B at t when technology change occurs, but simply records an event specific change at t.

Based on similar considerations the sell price is modeled as:

$$dr_S = r_S\,(\theta_S^+ \max\{0, r - r_S\} + \theta_S^- \min\{0, r - r_S\})(1 - d\varphi) + dr_S(t)d\varphi \qquad (15.16)$$

where $0 \le \theta_S^+ < \theta_S^- \le 1$. Since it is profitable to exercise a put option when $r < r_S$, it is reasonable to expect the speed of adjustment θ_S^- is large. Alternatively, when $r > r_S$ there is no incentive to sell. Provided investment is partially recoverable, viz., $r_S > 0$, gradual convergence of r and r_S is expected as sellers adjust their prices to the shadow price of capital. Slower upward adjustment for r_S is represented by a lower θ_S^+ value. When technology change occurs, these adjustment processes are undermined by changes that take into account the nature of change, e.g., a shock may adversely affect extant capital values.

To close the model, consider the likely relationship between expected applications sector returns and the ICT infrastructure bid price. Cogent modeling of this relationship must consider old economy ICT incumbent firms entering into joint ventures with new economy entrants seeking to profit from applications that utilize ICT infrastructure. Since an applications success is related to ICT network support, when ICT capital bid prices are rising for incumbents engaged in joint ventures, μ will rise. Furthermore in upswings, μ rises faster than r_B as new applications increase potential profits, while in a downswing μ falls more slowly than r_B and is associated with a lessening in the introduction of new applications, rather than reduced applications sector activity. Additionally, modeling the change in μ requires that only changes for which μ remains above r are observed.[17] At times when new technology is introduced, indicated by t, the determination of $d\mu(t)$ is not explained as it depends on the nature of technology change. These considerations suggest:

$$d\mu = \max\{d\hat{\mu},\, dr - (\mu - r)\}(1 - d\varphi) + d\mu(t)d\varphi \qquad (15.17)$$

where

---

[17] With $\mu > r$ there is an incentive to undertake risky portfolio investment. Enforcing this through time, $\mu + d\mu > r + dr \Rightarrow d\mu > dr - (\mu - r)$.

$$d\hat{\mu} = \mu(\theta_\mu^- \min\{0, dr_B\} + \theta_\mu^+ \max\{0, dr_B\}) \tag{15.18}$$

with $\theta_\mu^+ > 1 > \theta_\mu^-$.

For this specification, capital bid price rises can result in μ increasing—as it indicates firms are attaching greater value to their ICT capital. This rise in μ can be misplaced if market participants are unable to distinguish firms' motivations for investment. For example, exercising a real option can mean the representative firm is aligning its opportunity cost of capital with the capital buy price, and not that unexploited profit opportunity exists in the applications sector. Further, with the firm's capital stock expanding through acquisitions, r falls simultaneously as the acquisitions experience diminishing marginal productivity. Consequently the $\mu - r$ gap widens more rapidly, increasing the incentive for further portfolio investment. Because of this reinforcing interaction, a rational bubble may be created by underlying real options and portfolio investment activity.

Conclusion

Several explanations are proffered for OECD Member Country ICT investment behavior. OECD ICT investment series show substantial commonality with strong underlying growth, but considerable variation in timing and volatility. As complete explanations of this behavior are not available, there is a tendency to associate ICT market decline and subsequent downturn with irrational exuberance. Conversely, developments in the real investment literature support alternative rational reasons for differences in OECD Member Country ICT investment behavior. This chapter suggests a stylized model that decomposes investment motivation into factors associated with cogent theoretical underpinnings, viz., old economy or Jorgenson investment that is influenced by the user cost of capital; new economy or Merton investment rationally explainable by finance theory—though it may be driven by exuberant growth expectations following an innovation; merger and acquisition activity explained by the real options approach; growth through earnings retention and explained rationally through shareholder-orientated optimal saving; and investment intended to allow the firm to withstand adoption costs that follow an innovation. The components are endogenous in the model, and depend on different technology, preference, environment and policy oriented parameters. Thus, the model provides an opportunity to estimate the separate contribution of components and search for policy parameter settings.

The rational accounting approach allows the exploration of whether perceived irrational response is due to inherent incompatibility between investment components. Perceived incompatibility may arise because of the nature of optimization under uncertainty. For example, the new economy component depends on expected applications sector returns, while the real options component depends on the relationship between an incumbent firms' opportunity cost of capital and the buy and sell prices for existing capital. Observing a firm exercising a call option can lead shareholders to believe the expected return should be revised upward, at least for applications compatible with the supporting infrastructure of the acquired firm. However, call options can be exercised because the acquired firm's capital buy price is lower than the acquiring firm's opportunity cost, not because expected returns are undervalued. Of course, new economy application sector expected returns and old economy buy price should be linked. However, this link cannot be precise in the context of technological uncertainty. Furthermore, as a call option is exercised, the increased capital stock of the acquiring firm should reduce its shadow price of capital through diminished marginal productivity. This outcome widens the risk premium and suggests it is optimal for the firm to engage in additional portfolio investment. Upward revision of the expected return in such circumstances exaggerates the risk premium and can lead to overinvestment. These exaggerated cyclical effects should be regarded as due to asymmetric price responses rather than irrational exuberance. Whether cyclical effects are substantially based on rational responses to asymmetric price effects remains to be explored empirically. The possibility of rational bubbles arising from the interaction of component rational responses in the presence of asymmetric price adjustment suggests different implications for policy. In this circumstance irrational exuberance is an irrational explanation.

References

Abel AB (1983) Optimal investment under uncertainty. American Economic Review 73: 228–33

Abel AB, Dixit AK, Eberly JC, Pindyck RS (1996) Options, the value of capital, and investment. Quarterly Journal of Economics 111: 753–77

Arrow KJ (1962) The economic implications of learning by doing. Review of Economic Studies 29: 155–73

Barro R, Sala-i-Martin X (1995) Economic growth. McGraw-Hill, New York

Bresnahan TF, Trajtenberg M (1995) General purpose technologies: Engines of growth? Journal of Econometrics 65(1): 83–108

Dixit AK (1992) Investment and hysteresis. Journal of Economic Perspectives 6(1): 107–32

Dixit AK, Pindyck RS (1994) Investment under uncertainty. Princeton University Press, Princeton

Eisner R, Strotz R (1963) Determinants of business investment. In: Impacts of monetary policy. Prentice-Hall, Englewood Cliffs

Gould JP (1968) Adjustment costs in the theory of investment of the firm. Review of Economic Studies 35(1): 47–55

Helpman E (ed) (1998) General purpose technologies and economic growth. MIT Press, Cambridge

ITU (2001) World telecommunication indicators '01. CD-ROM, ITU, Geneva

ITU (2002a) World telecommunication development report. ITU, Geneva

ITU (2002b) Internet for a mobile generation. ITU, Geneva

Jorgenson DW (1963) Capital theory and investment behavior. American Economic Review 53(2): 247–59

Lucas RE (1967) Adjustment costs and the theory of supply. Journal of Political Economy 75(4): 321–34

McDonald RL, Siegel DR (1985) Investment and the valuation of firms where there is an option to shut down. International Economic Review 26(2): 331–49

McDonald RL, Siegel DR (1986) The value of waiting to invest. Quarterly Journal of Economics 101(4): 707–27

Merton RC (1971) Optimum consumption and portfolio rules in a continuous-time model. Journal of Economic Theory 3: 373–413

Nelson RR, Wright G (1992) The rise and fall of American technological leadership: The postwar era in historical perspective. Journal of Economic Literature 30: 1931–64

OECD (2002) OECD information technology outlook. OECD, Paris

OECD (2003) OECD communications outlook. OECD, Paris

Ramsey F (1928) A mathematical theory of saving. Economic Journal 38: 343-59

Romer P (1986) Increasing returns and long run growth. Journal of Political Economy 94(5): 1002–37

Rosenberg N (1996) Uncertainty and technological change. In: Landau R, Taylor T, Wright G (eds) The mosaic of economic growth. Stanford University Press, Stanford

Shiller R (2000) Irrational exuberance. Princeton University Press, Princeton

Solow R (1956) A contribution to the theory of economic growth. Quarterly Journal of Economics 70: 65–94

Swan T (1956) Economic growth and capital accumulation. Economic Record 32: 334–61

Part V: Development Imperative

16 North African Information Networks

Andrea L. Kavanaugh

Introduction

Coleman et al. (1957), Dutton et al. (1987) and Rogers (1995) establish that social networks are important for understanding innovation diffusion through social systems. Social network interaction exposes individuals to technology. Additionally, new technology is commonly adopted when acquaintances reinforce mass media messages (Katz and Lazarsfeld 1955). That is, adopters rely on personal networks to gain new technology insights. In this manner, social networks assist in building network technology critical mass (Markus 1987; Rogers 1995; Valente 1995; Kavanaugh and Patterson 2001). Namely, networked communication technology is more likely employed when it is of demonstrative value. Such demonstrations are reinforced by the social nature of computing activity, i.e., network users are linked by personal relationship and common interest (Wellman et al. 1996). Adopters involved in social networks, e.g., through work and community, increasingly integrate computing into regular activity. Formal education is also associated with computer adoption, e.g., in US households socio-economic status tends to differentiate adopters and their use patterns for personal and network computing (Fischer 1977; Dutton et al. 1987; Rogers 1995; Cooper 2000). That is, early adopters are generally better educated and receive more media exposure (Rogers 1995). For less affluent populations new technology adoption patterns are similar. Early adopters are usually community leaders and diffusion is via social networks (Hudson 1984; Wresch 1996; Kavanaugh et al. 2002). However, less affluent socio-economic group leaders have fewer social ties to computer users and accordingly acquire less computing skill. While public access to information technology (IT) is available, disadvantaged individuals are not typically comfortable in these social contexts. For example, less educated and affluent individuals often have negative associations with public schools and are not comfortable gaining Internet access in this manner. Similarly, individuals with poor reading skill are less likely to frequent public libraries. Clearly, IT diffusion is slower within disadvantaged populations. However, when IT is integrated in community social settings it is used. IT diffusion problems are exacerbated in developing countries with large uneducated populations that receive low incomes and are not computer literate.

Most individuals, regardless of socio-economic status, are not expert in computing. While most US citizens access the Internet, typically users conduct routine tasks, such as sending e-mail and browsing. Changes to software, Web capability and e-mail functions require users to continuously augment their knowledge stocks. As such, computer literacy is enhanced by experience and learning arrangements within social contexts. Learning is a social activity, i.e., individuals exchange information about programming, configuring and using computer technology in work and community settings. While formal contexts are important for learning about new technology much knowledge is gained informally (Carroll et al. 2002). Learning-by-doing occurs through Web software exploration, obtaining product technical support and so on. Such activity is often of a problem solving nature, and supported by family and modeling on Web sites. For example, Kavanaugh et al. (2003) find teenage adopters instruct household residents on their computer use. As such, computer literacy accrues more rapidly when informal (and authentic) situations motivate problem solving. Less affluent communities, in developing and developed countries, are less commonly involved in computer literate social networks from which informal learning occurs.

Received analysis show an individual's knowledge stock is augmented by formal and informal training, and through experience (National Commission on Excellence in Education 1983; Soloway 1995; Norman and Spohrer 1996). That is, effective learning occurs when individuals have a model (or analog), based on prior knowledge and experience, with which to absorb information. For example, manual typewriter keyboards evolved into the computer keyboard. However, disadvantaged groups have no model for either the keyboard or computer. Computer interface requires basic literacy and application instruction, e.g., e-mail and Web browsers. Without computer literacy computing skills must be gained from 'small media' (e.g., radios) use (Schramm 1964). A mobile telephony model—fixed telephone service—is widespread, even for the computer illiterate. That is, the interface of mobile telephony is similar to that for fixed telephony, viz., numbers, letters and buttons in familiar format. Moreover, telephony operations require minimal literacy. Schramm documents small media that are more effective than 'big media' (such as TV) in educating the developing country poor as these populations possess radios but few have TVs. Further, the computer illiterate in developing countries do not view telephones as computers or themselves as computer users. However, many learn to perform basic computer tasks while operating mobile telephones, e.g., speed dialing, storing frequently used numbers and sending text messages.

The World Wide Web is a global information repository with most information produced by personnel trained in design, editing and content management. However, beyond basic text editing capability—available in systems based on object oriented multi-user domain (MOO) technology—the authoring of original material is generally not well supported, e.g., in community organizations, public sector and small business (Bruckman 1998; Haynes and Holmevick 1998).[1] Appropriate Web server infrastructure and page editing software makes Web page production of similar difficulty to that for desktop publishing or word processing. However, such Web pages are static and difficult to customize without encountering document version, overwriting and logistical problems (Carroll et al. 2000a, b). Tools

---

[1] MOOs are Web sites that offer combined synchronous and asynchronous communication mechanisms. Namely, MOO users create, modify and manipulate objects, thus changing the MOO for later users. MOOs are spatial in that content is organized into 'rooms' and users navigate the information structure via directional commands.

inspired by 'wiki' software increase non-expert participation and information production.[2] For instance, wiki tools enabled the development of the Java-based toolkit, BRIDGE, which produces Web-viewable documents, annotated images and interactive collaborative environments (Isenhour et al. 2001). Further, these tools allow synchronous or asynchronous content collaboration and manipulation such as images, geographical system maps and spreadsheets within a Web browser. This function enables the editing of spreadsheets within a Web page. That is, users are not required to exit a browser, access proprietary software, create a document, transfer the file to a directory and create a link from the html file to a spreadsheet. This integration of editing functions into standard Web tools removes information production and maintenance barriers. BRIDGE relies on underlying object replication mechanisms that synchronize distributed replicas of objects when edited (Isenhour et al. 2001). Many developing countries wish to increase their national Web content by developing Web pages that contain government, cultural, historical, business and educational information. However, a paucity of qualified experts is available to construct, edit and manage desired content. For instance, while Tunisia has a national committee charged with the enhancement of national Internet content production, the project is constrained by a shortage of IT engineers and technicians. Clearly, wiki tools make the construction and editing of Web-based content feasible for nations lacking widespread computer literacy by substantially reducing training required to gain proficiency in constructing and customizing Web content. Wiki tools, based on open source software, are more affordable and compatible across computer platforms than is proprietary software.

Mobile telephony is included among small media along with personal digital assistants and similar hand held devices that provide a foundation for the building of computing skills, especially in developing countries. Because of the scalability of mobile telephone technology, subscribers require few skills to operate telephone handsets for basic voice communication purposes. However, subscribers are able to learn more complex telephone functions, such as saving frequently dialed numbers in a directory and programming voice-activated numbers, incrementally. Such knowledge accumulation is important as mobile telephones are essentially networked computers. Accordingly, as subscribers become more familiar with basic system features their functionality is extended to encompass network-based applications. From mobile telephone text messaging, novice subscribers learn to send and receive e-mail on Web-enabled telephones. From checking train schedules, subscribers gain fundamental Web browsing concepts. Local regulatory and service provisions may restrict network access via mobile telephones. However, in countries where mobile telephone networks are connected to IP based infrastructure, it is possible for subscribers to extend their computing skills from basic voice communication to more complex Internet messaging and Web browsing service within a supportive social context using a familiar tool.

---

[2] See, Georgia Tech [http://www.cc.gatech.edu/fac/mark.gudzial/cseet/tsld019.htm], University of Hawaii [http://c2.com/cgi-bin/wiki] and Virginia Tech [http://hci.cs.vt.edu].

Demography and Economic Indicators

Algeria, Egypt, Libya, Mauritania, Morocco and Tunisia comprise the North African region. The region's population is mostly Arabic speaking and Muslim. Moreover, Algeria, Libya, Mauritania, Morocco and Tunisia are Maghreb states.[3] Algeria, Morocco and Tunisia have French colonial history, indigenous Berber populations, and share Mediterranean, Atlas Mountain and Saharan desert boundaries in common. Ninety-five percent of Algeria's 31 million inhabitants reside in the northern coastal strip between the Atlas Mountains and Mediterranean Sea. Morocco has a population of 30 million inhabitants but less natural mineral wealth, while an ongoing conflict with Western Sahara depletes national saving. Egypt has twice the Moroccan and Algerian population, but similar income per capita. Tunisia has a relatively small population by regional standards of 10 million inhabitants. Algeria, Egypt, Morocco and Tunisia are classified by the World Bank as lower-middle income, with per capita incomes from US$ 1,200 to US$ 2,000 (World Bank 2003). Algeria has oil and natural gas reserves that contribute to gross domestic product (GDP), while Egypt depends on foreign aid and US investment. Morocco is reliant on phosphate mineral exports, and Tunisia's main source of foreign exchange is agriculture, textiles and tourism. GDP growth is highest for Tunisia and Egypt at 5.5% and 5.3%, respectively. Adult literacy is highest in Tunisia (72%) and Algeria (68%). By contrast, only half the Moroccan population is literate. North African states have capitalist, private-public sector mix and socialist political regimes. However, whatever the doctrine, there is a strict governance throughout, whether by revolutionary council or authoritarian rule. Political and leadership styles affect development strategy and telecommunications sector reform.

Table 16.1. Demographic and Economic Indicators, 2001

Country	Population (million)	GDP per capita (US$)	Adult Literacy (%)	GDP CAGR (%)
Algeria	31	1,639	68	3.0
Egypt	65	1,530	56	5.3
Morocco	29	1,180	49	3.3
Tunisia	10	2,070	72	5.5

Source: The World Bank (2003), United Nations Development Programme (2003)

---

[3] Libya and Mauritania are excluded from this study as their communications markets are less accessible and developed than those of other Maghreb states. Moreover, Libya does not allow public access to the Internet via mobile telephony.

Information Sector Reform

North African states commenced telecommunications sector reform in the 1990s (Kavanaugh 1998; Arab Advisors Group 2001, 2002, 2003; Lahouel 2002). In February 1998, the Moroccan state corporation, the National Office of Post and Telecommunications (ONPT), split into the Itisalat Al Maghrib (IAM, telecommunications development including provision of an international Internet hub), Barid Al-Maghrib (postal service) and the National Regulatory Agency for Telecommunications (interconnection and litigation settlement oversight) entities (ESIS 2000). Morocco converted ONPT to IAM through a share float on the Casablanca Stock Exchange. Also in 1998, Egypt converted Telecom Egypt from a telecommunications authority to a joint stock company (Law 19). However, the potential for conflict is apparent with the Egyptian Minister of Communication and Information Technology the Head of the Telecommunications Regulatory Authority (Lahouel 2002). The Algerian operating entity remains a department within the Ministry of Post, Telephone and Telegraph (PTT). State corporation conversion plans are stagnant because of political unrest. Algeria, Egypt, Morocco and Tunisia opened their terminal equipment (computer, facsimile, modems and satellite antennae) markets to competition. In particular, private ownership in certain telecommunications markets (equipment manufacturing, retail services and software engineering) is encouraged so as to stimulate both equipment industry competition and value-added technology and service (databases, Internet and videotex) market growth. Moroccan telecommunications sector reform is associated with improved growth, relative to that for the sluggish Algerian and Tunisian markets. Egypt and Morocco granted most mobile telephony licenses to foreign corporations. Additionally, Egypt and Morocco opened their Internet service markets to competition. Morocco introduced wireless application protocol service (WAP) in July 2000. That market provides e-mail, Web browsing, *e-souk* (e-commerce), financial information, horoscope, sport, telephone directory, train schedule, TV schedule and weather services. Mobile Internet is priced at US$ 7.50 per half minute. Finally, Tunisia allowed competition in mobile communication markets in 2002 with the awarding of a second GSM license. Despite Tunisia's relatively high adult literacy and economic growth, government control of telecommunications continues to hamper the sectors growth.

Information Networks

Fixed-line service consists of wire lines and is measured by direct exchange lines (DELs).[4] Fixed-line telephone service is typically originated by a PTT. Mobile telephony infrastructure primarily consists of wireless telephone towers that receive

---

[4] A DEL is a telephone line connected to buildings that has a unique telephone number. DELs are distinguished from many handsets connected to a line and sharing a common telephone number.

and hand-off signals. Developing country governments often allow private sector corporations to provide mobile telephony service via licensing agreements. In the North African region DEL density is low. In particular, Table 16.2 shows that while cumulative annual growth rate (CAGR) for fixed-line telephony is rising steadily in Algeria, Egypt and Tunisia from 1995 to 2002, it is declining in Morocco due to rising mobile telephone subscription.

Table 16.2. DEL Indicators

Country	DEL / 100 Inhabitants		CAGR (%)
	1995	2002	
Algeria	4.19	6.10*	6.4
Egypt	4.67	11.83*	14.2
Morocco	4.24	3.80	−1.5
Tunisia	5.82	12.23	11.2

Note: * Estimate based on 2001 ITU data. CAGR is for 1995 to 2002

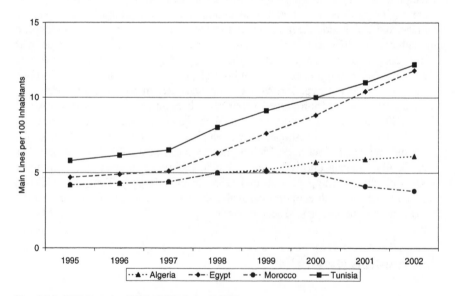

Fig. 16.1. DEL Density 1995-2002. Source: ITU

Table 16.3 shows that North African fixed-line telephone service is characterized by frequent equipment breakdown and prohibitively long subscription application waiting lists. More importantly, regional fixed-line telephone networks are heavily congested. Such congestion is to some extent caused by PTTs making politically convenient decisions and providing fixed-line access, thus reducing DEL waiting lists, without any corresponding network upgrade or increase in trunk line capacity (Saunders and Wellenius 1993; Kavanaugh 1998). Accordingly, subscribers receive busy signals for unacceptably long time intervals, especially during peak calling periods.

Table 16.3. Quality of Service

Indicator	Algeria		Egypt		Morrocco		Tunisia	
	1995	2001	1995	2001	1995	2001	1995	2001
Wait list DEL (000s)	676	727	1,300	583	93	5	129	109
Wait time (days)	n.a.	1,972	n.a.	694	n.a.	37	n.a.	33
Faults / DEL (%)	73	6	9	1	49	25	79	n.a.

Note: n.a. is not available.

In 2002, Mobile telephony subscription exceeded that for fixed-line network subscription in many developing countries, including Jordan, Lebanon, Morocco, Paraguay, Saudi Arabia and Venezuela. Further, Table 16.4 shows approximately 85% of Moroccan telephony customers subscribe to mobile telephone service. Furthermore, Morocco has the fastest growing mobile telecommunications market in the North African region, with a reported CAGR of 149% from 1995 to 2002. Egypt and Tunisia report CAGRs of 115% and 123%, respectively, for mobile telephone subscription. Algeria recorded a relatively modest 81% CAGR for the same period. Table 16.4 also shows that Morocco (21%) has the highest North African region mobile telephone subscription density, while not surprisingly Algeria (less than 1%) has least penetration.[5] Further, combined Moroccan mobile and fixed-line (DEL) telephone density is approximately 25%. Corresponding Egyptian and Tunisian telephone densities are 14.7 and 14.9, respectively. For Algeria, composite telephone density is unaffected by mobile service penetration, viz., national penetration is 6.4 for fixed-line service and 6.42 for combined service.

Table 16.4. Mobile Telephony Subscription

Country	Subscription (000s)		CAGR	Subscription / 100		Mobile penetration
	1995	2002	(%)	1995	2002	2002
Algeria	5	300	81	0.02	0.9	13.8
Egypt	7	4,412	115	0.01	6.7	39.7
Morocco	30	6,199	149	0.11	20.9	84.6
Tunisia	3	389	123	0.04	4.1	26.9

Note: CAGR for 1995-2002.

---

[5] Density is the number of subscribers per 100 inhabitants.

In Morocco, mobile telephone subscription grew from a base of 1,800 subscribers in 1991 to approximately 40,000 subscribers at 1996 (30,000 of whom subscribed to digital service). At 2000, more than 2 million subscribers accessed service in the urban areas of Casablanca, Fes, Kenitra, Marrakech, Meknes and Rabat. Mobile telephone traffic increased rapidly in the late-1990s and early-2000s. However, the composition of traffic broadened to encompass e-mail, while Internet access is gained through WAP. By 2002, the Moroccan system had in excess of 6 million subscribers.

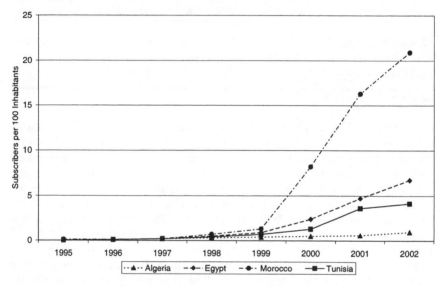

Fig. 16.2. Mobile Telephone Density 1995–2002. Source: ITU

Analog mobile telephone service, introduced to Tunisia in early-1985, grew more slowly than the Moroccan market because of strict government controls. Until 2001, the state corporation Sotetel managed the market while working closely with the Ministry of Communications. In response to unanticipated latent demand, the Ministry quickly doubled the size of the network. By 1995, network subscription had risen from a base of 50 subscribers to approximately 4500 subscribers. In March 1998, Tunisia introduced digital mobile telephone service with a European transmission standard (*groupes de systemes mobiles* or GSM) and further boosted subscriptions to 8,000. An absence of competition had restricted mobile and fixed-line telephony market growth. Finally, on 15 January 2001, a new telecommunications law Number 1-2001 moved toward the liberalization of Tunisian telecommunications markets (Arab Advisors Group 2003). This law established an independent regulatory oversight and opened the market to private company entry. Further, the law introduced a licensing regime for the supply to telecommunication services and networks. Although Tunisian per capita GDP and adult literacy are substantially higher than those of Algeria, Egypt and Morocco, GSM penetration remains low at 3.6% in 2001. Indeed this penetration is among the lowest in Arab nations and higher only than Algeria, the Sudan and Syria. After much delay, Tunisia awarded a second GSM license to Orascom Telecom (based in Egypt) in consortia with Kuwaiti National Mobile Telecommunications for US$ 454 million. Orascom Telecom Tunisie launched service in December 2002. The Arab Advisors Group (2003) forecast mobile telephony subscriptions to reach 4.4 million by 2006. While it is possible that the Tunisian government could introduce competition to the fixed-line telephony market post-2004, it is more likely to remain a state-owned monopoly operator for the foreseeable future. Arabian state-owned fixed-line operators have a mandated monopoly over the operation of international gateways, with the exception of Morocco. GSM private operator Medi Telecom offers subscribers international long-distance service.[6] GSM license prices in North African states are high. For example, in Morocco Medi Telecom paid US$ 1 billion, while in 1998 Egypt sold two GSM licenses at US$ 0.5 billion. Also, Algeria sold a second GSM license in 2001 for US$ 737 million.[7]

Developing country Internet market growth is typically constrained by poor quality fixed-line telephony and data networks. Already congested fixed-line networks are usually not able to accommodate any increase in fixed-line subscription or traffic demand. Most subscribers during the 1990s coped with chronically congested dial-up connections. This situation changed markedly with mobile communication and value-added service markets opened to competition, in particular markets for mobile telephony that connects to the Internet. That is, private companies deploy necessary infrastructure, e.g., routers, switches and towers to provide mobile telephony. Where local regulations allow Internet connectivity through mobile devices, Internet access experiences a growth spurt. Also, Internet connection via mobile networks avoids fixed-line network bottlenecks. Morocco, after the Republic of South Africa, is the most networked African nation, with 39 ISPs (ONPT 1997). At 2002, Egypt had more Internet hosts that other North African nations, however, the PC and Internet user density of Egypt and Morocco is less than that for Tunisia with a population of less than 10 million inhabitants (see Table 16.5).[8] Finally, North Africa, excluding Tunisia, has lower PC density than World Bank classified lower-middle income nations, and less than the Middle East and North Africa (MENA) region average. Tunisia has proportionately more Internet users but fewer Internet hosts than other North African states. This imbalance results from allowing relatively few computers to have Internet connection. For the period 1996 to 2002, Figure 16.3 shows host numbers, while hosts per

---

[6] Medi Telecom consortium is comprised of Telefonica of Spain, Portugal Telecom, BMCE and AFRIQUIA.

[7] License terms differ by country, however, so such comparisons are not exact. For example, the Egyptian government collects annual license fees based on network subscription, while the Algerian, Moroccan and Tunisian governments collect license fees at the time of granting the license and offer tax concessions to operators (Arab Advisors Group 2002).

[8] Internet hosts are computers connected to the Internet.

10,000 inhabitants or host density are depicted in Fig. 16.4. While Egypt has most host computers, Morocco has the higher host density at almost one per 10,000 inhabitants at 2002. For North African states, Internet user numbers grew rapidly after 2000, following the liberalization of value added services markets and price reduction from competition among ISPs (Fig. 16.5).

Table 16.5. Personal Computer and Internet Indicators, 2002

Country / Region	PC / 100	Internet users / 100	Internet hosts / 100
Algeria	0.71	1.60	0.0026
Egypt	1.55	1.22	0.0047
Morocco	1.37	1.69	0.0090
Tunisia	2.63	5.15	0.0035
Lower Middle	2.23		
MENA	3.12		

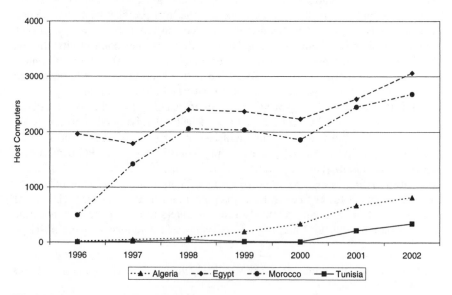

Fig. 16.3. Internet Hosts. Source: ITU

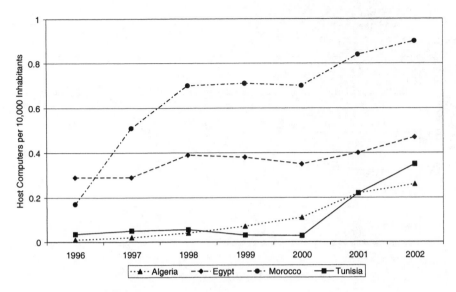

Fig. 16.4. Internet Host Density. Source: ITU

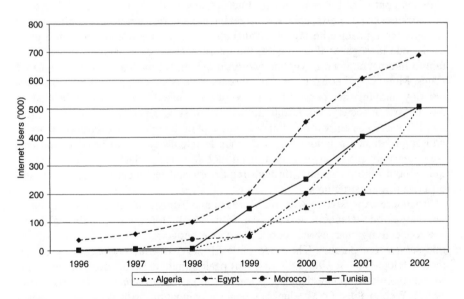

Fig. 16.5. Internet Users. Source: ITU

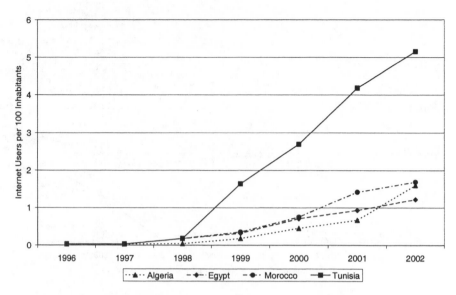

Fig. 16.6. Internet User Density. Source: ITU

Despite Tunisia having comprehensive laws and regulations concerning censorship and control of Internet content, Internet density is relatively high. In particular, laws governing newspaper and magazine content restrictions are applied to the Internet through separate laws (Kavanaugh 1998). In other North African states, press code laws generally extend to the Internet, although separate laws also pertain to Internet content. Egypt and Morocco are among the least Internet regulated states. Morocco is known for its liberal press code laws. Comparison of mobile telephony and Internet penetration shows mobile telephony deeply embedded in North African society. While Internet use is growing, especially among the affluent, the market is embryonic. Importantly, mobile telephone subscription is more widespread than for computing, viz., mobile telephony has diffused to the less well educated. However, transition from mobile voice to data communication requires both the opportunity, through regulatory and technical infrastructure, and efficient pricing structures.

It is more expensive in Algeria, Morocco and Tunisia to subscribe to the Internet than mobile telephone service (see Table 16.6). Nonetheless, additional Internet connectivity charges via mobile telephone networks make both forms of on-line service comparable. The price of WAP service—Internet access by digital mobile telephone—is US\$ 0.80 for a half peak minute or an off-peak minute. Moroccan mobile telephone handset price at 2000 is approximately US\$ 80. However, a Web-enabled GSM telephone handset compatible with WAP service, such as Nokia R320, Eriksson 7110, is priced at US\$ 265. Accordingly, with Moroccan GDP per capita US\$ 1,180 at 2001, Web-enabled GSM service is typically unaffordable. However, mobile telephony equipment is less expensive than an Apple iMac 350 MHz with Mac OS 9 operating system (US\$ 1,130) or Hewlett Packard

Vectra Vei89 PC with Intel Pentium III 550 MHz processor (US$ 1,365). At 1996, Tunisian mobile telephony equipment prices were US$ 3,850, with calling charges three times that for local fixed-line calling. By 2002, mobile telephone charges and equipment prices, though falling considerably, remain expensive. Conversely, PC prices are similar to US prices as Tunisian equipment markets are competitive. Further, the unlimited dial-up Internet access price is approximately US$ 26 per month plus US$ 50 for installation. The *Agence Tunisienne d'Internet* (ATI), established in 1996, provides Internet gateway access (via Sprint to the US and via Telecom Italia to Italy) to private and public agency ISPs. ATI operates the Tunisian Internet backbone, services and top-level domain from the *Institute de Recherché Scientifique en Informatique et Telecommunication*. In 2002, several public agencies served as ISPs to the public sector, while five licensed commercial ISPs offer service to the public and private companies. Additionally, there are approximately 300 public Internet access centers across the country. The Moroccan incumbent operator IAM offered unlimited dial-up Internet access for US$ 10 per month, since competition from private providers led to initial price reductions in 1999. Menara offers a package containing multi-media computer, Internet subscription, Microsoft software and technical assistance for US$ 28 per month. Private ISPs offer similarly structured packages. Egyptian prices are similar to those for Tunisian and Moroccan Internet access markets. Not surprisingly, Algeria lags in Internet infrastructure and subscription. The CERIST research center provided Internet service for several years. However, dial-up connection is slow and expensive at US$ 70 per month. The few Internet cafes are strictly controlled by the government. At 2000, Algeria allowed private application for ISP licenses.[9] An inter-ministerial commission licensed six companies. Dial-up access tariffs are from US$ 13 per month. With access charges falling subscription rapidly increased.

Table 16.6. Monthly Internet and Mobile Subscription Price (US$), 2001

Country / Region	Internet	Mobile Telephony
Algeria	27	17
Egypt	9	16
Morocco	26	13
Tunisia	25	14
Lower-Middle Income	18	
MENA	27	

---

[9] An Executive Decree issued in 1998 ended CERIST's monopoly.

Conclusions

This chapter argues that mobile telephony, especially in developing countries, is an appropriate transition technology for teaching computer networking skills to mostly illiterate populations, as technical support is provided through social networks. North African data show poor quality fixed-line telephone service, rapid mobile telephone and Internet subscription growth when service is provided via competitive markets. The Internet subscription market, however, is populated only by the well educated and affluent. It is reasonable to expect that this market will reach saturation in the North African region within five years. That is, to effectively use the Internet requires reading and computer literacy skills, and an on-line social network. Social networks are important in motivating individuals to access the Internet. Mobile telephone technology enables voice communication however, even with minimal literacy, mobile telephony subscribers can also extend their mobile network use to store and retrieve data. Further, through Web-enabled telephony, lay persons are able to learn simple networking tasks informally with technical support provided by their social networks. Simply, the Internet will diffuse more broadly in North Africa and developing country populations, more generally, via mobile communication, and not through desktop computers attached to fixed-line telephone network connections. Policy should reflect this reality so that the economic benefits of information networks can be more widely appropriated by developing country populations.

References

Algeria (1995-1997) Ministry of Post and Telecommunications. Documents and statistics on telecommunications, Algiers

Arab Advisors Group (2001) Morocco Internet and Datacomm landscape report. Available at http://www.arabadvisors.com

Arab Advisors Group (2002) Orascom Telecom snatches Tunisia's second GSM license. Available at http://www.arabadvisors.com

Arab Advisors Group (2003) Tunisia Internet and Datacomm landscape report. Available at http://www.arabadvisors.com

Bruckman A (1998) Community support for constructionist learning. Computer Supported Cooperative Work 7: 47-86

Carroll JM, Rosson MB, Isenhour P, Kavanaugh A (2002) Mutually leveraging computer literacy and community networks. National Science Foundation (IIS-0307806)

Carroll JM, Rosson MB, Isenhour PL, Van Metre C, Schaefer WA, Ganoe CH (2000a) MOOsburg: Multi-user domain support for a community network. Internet Research 11: 65-73

Carroll JM, Rosson MB, Isenhour PL, Van Metre C, Schaefer WA, Ganoe CH (2000b) MOOsburg: Supplementing a real community with a virtual community. In: Furnell S (ed) Proceedings of the Second International Network Conference. University of Plymouth, Plymouth, pp 307-16

Coleman J, Katz E, Menzel H (1957) The diffusion of an innovation among physicians. Sociometry 20: 253-70

Cooper M (2000) Disconnected, disadvantaged and disenfranchised: Exploration in the digital divide. Consumer Federation of America, Washington

Dutton WH, Rogers EM, Jun SH (1987) The diffusion and impacts of information technology in households. In: Zorkoczy PI (ed) Oxford surveys of information technology. Oxford University Press, New York

ESIS (2000) Regulatory developments: Morocco. European Survey of Information Society Master Report April. Available at http://europa.eu.int/esis

Fischer C (1977) Networks and places. Free Press, New York

Gartner (2000) The digital divide and American society. Gartner Group, Washington

Haynes C, Holmevik JR (eds) (1998) High wired: On the design, use, and theory of educational MOOs. University of Michigan Press, Ann Arbor

Hudson H (1984) When telephones reach the village: The role of telecommunications in rural development. Ablex Publishing, Norwood

International Telecommunication Union (2002) Internet for a mobile generation. ITU, Geneva

International Telecommunication Union (2003a) Data available at http://www.itu.int/ITU-D/ict/publications/world/world.html and http://www.itu.int/ITU-D/statistics

International Telecommunication Union (2003b) World telecommunications development report. ITU, Geneva

International Telecommunication Union (various) Yearbook of telecommunications statistics. ITU, Geneva

Internet Software Consortium (2003) Data available at http://www.isc.com

Isenhour PL, Rosson MB, Carroll JM (2001) Supporting interactive collaboration on the Web with CORK. Interacting with Computers 13: 655-76

Katz E, Lazarsfeld P (1955) Personal influence. Free Press, New York

Kavanaugh A (1998) The social control of technology in North Africa: Information in the global economy. Praeger, Westport

Kavanaugh A, Carroll JM, Rosson MB, Reese DD (2003) Using community computer networks for civic and social participation. Transactions of Computer Human Interaction Special Issue, in press

Kavanaugh A, Patterson S (2001) The impact of community computer networks on social capital and community involvement. American Behavioral Scientist 45: 496-509

Kavanaugh A, Schmitz J, Patterson S (2002) The use and social impact of telecommunications and information infrastructure assistance on local public and nonprofit sectors. Final Grant Report for Broad Agency Announcement. (# 50-SBN-TOC-1033) US Department of Commerce

Lahouel MH (2002) Telecommunication services in the MENA region: Country case analysis of markets, liberalization and regulatory regimes in Egypt, Morocco and Tunisia. International Affairs Institute, Rome

Markus ML (1987) Toward a critical mass theory of interactive media: Universal access, interdependence and diffusion. Communication Research 14: 491-511

National Commission on Excellence in Education (1983) A nation at risk. Washington

Norman DA, Spohrer JC (1996) Learner centered education. IEEE Computer 39: 24-7

Morocco Ministry of Communications (2003) Telecommunications documentation and statistics. Rabat

Morocco National Office of Post and Telecommunications (1997) Le maroc, 2eme pays branche d'Afrique. Recherche, Internet Maroc. At http://www.maghrebnet.net.ma

Rogers E (1995) Diffusion of innovation. Free Press, New York

Saunders J, Wellenius B (1993) Telecommunications and economic development. Johns Hopkins University Press, Baltimore

Schramm W (1964) Big media, little media. Academy for Educational Development, Washington

Soloway E (1995) Beware techies bearing gifts. Communications of the ACM 38: 17-24

Tunisia Ministry of Communications (2003) Telecommunications documentation and statistics, Tunis

Valente T (1995) Network models of the diffusion of innovations. Hampton Press, New York

United Nations Development Programme (2003) Available at http://www.undp.org/rbas/

Wellman B, Salaff J, Dimitrova D, Garton L, Gulia M, Haythornthwaite C (1996) Computer networks as social networks: Collaborative work, telework, and virtual community. Annual Review of Sociology 22: 213-39

World Bank (2003) Data available at http://www.worldbank.org/data/wdi2003/index.htm and http://www.worldbank.org/data/countrydata/ictglance.htm

Wresch W (1996) Disconnected: Haves and have-nots in the information age. Rutgers University Press, New Brunswick

17 OECD Broadband Market Developments

Dimitri Ypsilanti and Sam Paltridge

Introduction

A recent Organization for Co-operation and Development (OECD) report to Ministers, concerning information and communications technology (ICT) and the digital economy, concluded that, "... considering the ongoing spread of ICT and its continued importance for growth, policy makers should foster an environment that helps firms seize the benefits of ICT" (OECD 2003a: 7). The deployment of high-speed (or broadband) Internet networks is integral for ICT diffusion generally, and in particular the availability broadband service.[1] Broadband networks, as a general purpose technology, have the capability to deliver benefit to many economic sectors. That is, broadband networks can provide a vehicle to deliver, e.g., government, education, cultural and health services. Given the potential social dividend obtained from broadband technology use, it is important to ensure that this technology is widely diffused within nations—and not concentrated at the main, typically urban, national economic hubs. With broadband technology penetration increasing in urban areas, several OECD Member Country governments are addressing potential national broadband digital divide issues, viz., that national divides may develop should broadband service not be made readily available to rural and remotely located populations. Accordingly, this study provides an overview of broadband network developments within OECD Member Countries, and analyzes recent Member Country national broadband policy. Subsequent to this review, the potential for intra-Member Country national digital divides to emerge is considered, as are national initiatives to provide high-speed access to rural and remote populations. The chapter concludes with recommendations for policy makers.

Broadband and the OECD

A fundamental, and often overlooked question, is why broadband technology availability matters. Broadband technology is important because it is a unique

---

[1] There is no universally accepted definition of broadband. While broadband is commonly viewed as Internet service delivery at a 250 kbps minimum speed, broadband is often described by access speeds greater than ISDN capability, i.e., 128 kbps. Much of this confusion arises from the technology in use, viz., digital subscriber lines employ fixed-line telecommunication copper loop; cable modem use cable TV networks; local area network Ethernets; satellites; broadband fixed wireless and mobile broadband wireless.

conduit capable of providing high-speed, two-way—downstream and upstream—communication and information good transmission, and it also has an 'always-on' connection capability. For example, broadband infrastructure is able to stream audio and video via Internet at high quality. The economic costs of broadband connectivity are far lower than those for traditional telecommunication products that deliver equivalent service, e.g., leased lines. Thus for many users, including small and medium enterprise (SME), fast always-on connection is affordable. Accordingly, business subscribers through broadband technology are able to increase productivity and substantially reduce their networking costs. Broadband technology is also a platform for the delivery of consumer electronic commerce, education, health, entertainment and e-government service. As such, broadband is viewed as an important mechanism for economic growth and development in both central and 'peripheral' economic regions. For rural and remote regions, broadband is viewed as an important means to ensure improved access to economic markets, core government service—including health and education—and access to cultural products. Further, broadband technology is viewed as means through which remote and rural populations are better able, in an inexpensive and effective way, to telework. The importance placed on making broadband available to rural and remote populations is apparent from existing national government support programs. Such programs are policy responses intended to stimulate broadband network infrastructure deployment and service delivery to less populated regions when the private sector views such activity unprofitable. Not surprisingly, geographically larger nations are more active in this regard. For example, the Australian Commonwealth Government developed a National Broadband Strategy, with a funding base of US$ 92 million, to provide affordable broadband service access to regional Australia (see Alston 2003). Similarly, the Government of Canada recently allocated US$ 75 million for communities to obtain broadband for rural and Northern development. Also, the US Department of Agriculture (USDA) provides grants and low interest loans for rural broadband network development.[2]

Broadband Subscription

Broadband subscription is growing rapidly within the OECD. At end-1999, there were more than 3 million broadband subscribers within Member Countries. By end-2002, this OECD subscription had risen to 56 million subscribers. Further, in the first quarter of 2003, broadband networks grew by 6.6 million subscribers. Based on this trend, the OECD predicts more than 80 million broadband subscribers within the OECD by end-2003. While such subscription growth is among the highest for new communication services, OECD broadband access subscriber penetration remains low at 5 subscribers per 100 inhabitants, especially when compared to a telephony penetration of 54 access channels per 100 inhabitants. At

---

[2] USDA allocates US$ 1.4 billion for broadband adoption in rural US, and augments the Rural Development Standard High-speed Telecommunications Loan program.

present the main technology for broadband access are digital subscriber line (DSL) and cable modem. Cable networks initially provided the main platform for high speed Internet access however, since 2001 the role has been taken by DSL technology for the OECD.[3] Nevertheless, in 12 of 30 OECD Member Countries cable modem subscribers out number those for DSL (see OECD 2003b). Fig. 17.1 shows that OECD Member Country broadband diffusions vary considerably. Some countries are in the early development stage (e.g., Greece, Ireland, and Turkey), while Canada and Korea are relatively advanced in their penetration of 12 and 22 broadband access subscribers per 100 inhabitants, respectively. Among the main reasons for such differences are variations in competition—in particular facilities based competition—and in regulatory frameworks that facilitate access to incumbents networks through local loop unbundling policy.

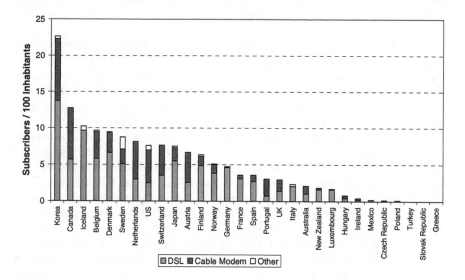

Fig. 17.1. OECD Broadband Diffusion at March 2003

Between OECD Member Country Digital Divide

From data reported in Fig. 17.1 it is apparent that a digital divide is emerging between OECD Member Countries. Some countries are experiencing rapid growth in broadband subscription, however, despite this growth it will require several years for laggard Member Countries to achieve the current broadband penetration of OECD leaders. However, as leading countries continue to expand their national

---

[3] At present asymmetric digital subscriber line (ADSL) technology is mainly in use.

broadband penetration the target moves. Rather than penetration per se, another issue that should be of concern to OECD policy makers is the emergence of substantial differences among countries in the standard of basic broadband service provided. In Korea and Japan residential minimum broadband Internet bandwidth access offered is from 4 Mbps to 8 Mbps. This is at variance with most other OECD Member Countries using cable or DSL that offer minimum bandwidth of between 250 kbps to 512 kbps. This minimum speed offering is largely determined by the competitiveness of given markets. For example, in the US, Time Warner offers residential cable modem service at 2 Mbps, and a commercial service at between 2 Mbps to 4 Mbps.[4] In response Verizon supply residential service from 1.5 Mbps and business service to 7.1 Mbps. Within Europe differences in minimum available bandwidth are substantial and often depend on the nature of the competitive environment facing incumbents. Residential subscribers in Belgium have broadband access at 3 Mbps for DSL service and 4 Mbps for cable modem service.

This bandwidth significantly exceeds highest speeds available to business users in many European countries.[5] Cable networks in Belgium are widely available and this means that the incumbent public switched telecommunications network (PSTN) operator faces considerable broadband market competition. While significant differences in communication service penetrations by OECD Member Countries are readily apparent, this inequality is not as substantial as those for broadband bandwidth. The performance of DSL or cable modem service offered at 256 kbps is significantly different to that of an 8 Mbps service. The potential to use such higher speeds to provide superior or more sophisticated services offering is also clearly different. OECD Member Countries that lead in the broadband market penetration are forging ahead with high-speed access through platforms such as very high speed DSL (VDSL) and residential fiber. In Japan, residential fiber service is available at 100 Mbps from a cable company for US$ 45 per month. Moreover, in Korea and Japan, roaming through a growing number of Wireless-LAN hot-spots is marketed as an extremely inexpensive option for fixed network broadband subscribers, e.g. less than US$ 20 for unlimited monthly service. In Europe, VDSL is available in Sweden with residential bandwidth service offered to 26 Mbps, depending the customer distance from the exchange. Such differences in service offering warrant close attention from policy makers.

Within OECD Member Country Digital Divide

National broadband penetration relies on the upgrade of PSTN to support DSL technology. This is important for regional areas as PSTN networks in most OECD countries are universally served. Table 17.1 provides data on DSL coverage.

---

[4] For example, see the New York Road Runner service at http://www3.twcnyc.com.
[5] This phenomenon is highlighted by a Swedish market entrant that introduced a 10 Mbps service for business formerly only available to residential users in apartments.

Table 17.1. OECD Member Country DSL Coverage

Country	Launch	Indicator	2000	2001	2002	2003
			Coverage (%)			
Australia	August 2000	Lines	50	72	75	
Austria	November 1999	Lines	72	77	80	
Belgium	October 1999	Population	75	93	98	98
Canada	1996	Households	69	70		
Czech Rep.	Trial only		0	0	0	44
Denmark	July 1999	Lines	65	90	95	95
Finland	May 2000	Sonera lines only	50	60	75	85
France	November 1999	Lines	32	76	86	91
Germany	August 1999	Lines	60	80	90	
Greece	May 2003		0	0	0	
Hungary	December 2000	Lines		20	38	45
Iceland	Start 2000	Lines	33	51	78	
Ireland	May 02	Lines	0	0	25	50
Italy	December 1999	Lines	45	67.5		
Japan	September 2000	Households		73.5	80	90
Korea	April 1999	Lines		70		
Luxembourg	2001		0	65	89	
Mexico	September 2001		0	0		
Netherlands	June 2000	Lines	40	67	85	
New Zealand	June 1999	Customers	60	69	83	84
Norway	December 2000	Lines	20	50	52	54
Poland	2001	Lines	0	3.5	10	
Portugal	December 2000	Population	50		60	
Slovak Rep.	Not launched		0	0		
Spain	1999	Lines	62.2	81.3	89.3	
Sweden	October 2000	Lines		70	75	
Switzerland	Mid 2001	Lines	0	85	90	
Turkey	February 2001	Lines	0	0.01	2.5	5
UK	July 2000	Lines	50	60	66	
US	1997	Lines	36	50	62	65

Note. US data is NTCA citing Morgan Stanley Dean Witter. Some carriers have higher availability in their service area, e.g., Bell South at 70%. Cable modems to reach 71% of households by 2002. BT indicates DSL is available to 90% of UK by 2005. 2003 data are projections. Cable networks provision of broadband access is limited by the cable TV market size and the economics of building small cable networks. The connectivity problem is to gain backhaul to high speed Internet backbone networks, e.g., in Canada many community cable TV networks find though backbone networks pass close by, backhaul connection coats are prohibitive. Cable networks pass 60% of OECD households, so their reach is not as extensive as PSTN. Also, a third of OECD cable networks are owned by incumbent telecommunication carriers with little incentive to extend their network to provide cable modem service, and prefer to offer DSL service. Source: OECD

A third of OECD Member Countries are able to readily provide DSL service to 70% of their telephone subscriber populations. Belgium leads with DSL available to 98% of the population—a corollary to highest national cable TV penetration for an OECD country. DSL service is also available to 96% of the Danish population. When it became apparent that cable TV networks would provide high-speed Internet access via cable modems, incumbent telecommunication carriers quickly acted to upgrade the PSTN to provide DSL. In Denmark, deployment of fixed wireless broadband stimulated an incumbent response. In particular, Danish fixed wireless networks cover 96% of the geographical land mass and 99% of the population—even higher than DSL. Not surprisingly the provision of broadband access to rural and remote regions is not a concern for these countries.

In other OECD Member Countries, PSTN subscribers without DSL service availability include communities served by exchanges that are not upgraded, and regional populations located too far from an upgraded exchange to receive service. Table 17.1 shows that, as business premises are typically located closer to urban centers than residential PSTN subscribers, the proportion of business users with access DSL is slightly higher than national penetration. For example, at end-2001 Telecom Italia provided DSL service to 68% of Italian Internet subscribers, but penetration to SME is 74%. Similarly, Telecom New Zealand provides 70% SME DSL coverage, at a time when national coverage has not reached 60%. The challenge of national broadband coverage is evident from Table 17.2 that shows a low DSL coverage to New Zealand farms. At 2002, 16% of New Zealand farms are covered by DSL. This coverage is expected to rise an additional 18% should Telecom New Zealand expect latent demand is sufficient to justify network investment. However, it appears unlikely that more than 50% of New Zealand farms, given Telecom New Zealand projections, will ultimately receive broadband via DSL. Finally, 30% of New Zealand schools are unlikely to be served by DSL.

Table 17.2. Actual and Projected New Zealand DSL Coverage

Customer reach	Proportion (%)				
	June 2002	Demand Required	Difficult	Unlikely	Residual
Residential	58	19	5	5	13
Corporate	67	16	4	5	8
SME	73	14	3	3	7
Farm	16	18	12	12	42
School	43	20	7	5	25
Health Provider	75	15	2	2	6
Total	**58**	**22**	**5**	**5**	**10**

Source: Telecom New Zealand

Bridging the Divide

Telecommunication policy has the potential to play an important role in reducing the magnitude of the digital divide, and in doing so ensure rural and remote regions obtain desired broadband access. However, deployment of new technology networks and upgrading existing networks require time to implement. A further complicating factor is that commercial telecommunications network investment is costly, and so is deployed initially in geographic locations where the expected returns are both high and, more or less, immediate. Network investment is also made in markets where competitor entry is easier. Such markets are usually densely populated and comprised of high income urban residents. Conversely, initial broadband access subscription in areas covered by alternative platforms such as Wireless Fidelity (WiFi), are sometimes higher than that for markets served by DSL or cable modem. For instance, in the US start-up wireless Internet service providers (ISPs) have reported higher broadband service subscription than the national average.[6] In such areas latent demand is at least as great as that for urban areas—as long as price and service quality are similar. For some new technology networks there is not so much a divide, but a delay in deployment. It is important that this possibility is clarified. In developing policy, there is a need to treat separately geographic areas that are not economically viable to serve and those that will eventually be served. In this regard, only through the proper working of the market and competition will not profitable service areas be clearly identified.[7]

Subscriber Access Thresholds

A frequent complaint made by rural populations is a lack of information concerning the potential for broadband access provision. In particular, information regarding the scheduling of exchange upgrades, and threshold population or demand magnitudes required for broadband access are often sought. Transparent processes assist interested groups. When rural populations are informed that a carrier is intending to upgrade an exchange or that some communities will not be served post-upgrade, this allows community agents to search for alternative solutions and alerts policy makers to not served populations based on incumbent telecommunication carrier commercial determinations. In several OECD Member Countries, telecommunication operators establish thresholds for DSL provision. Among the lowest thresholds is that for Iceland where their goal is to provide DSL access to

---

[6] In Pocahontas Iowa, Evertek delivers up to 0.5 Mb/s downloads for US$ 35 a month. Within Pocahontas, over 200 of its 900 households subscribe by mid-2003, i.e., service penetration exceeds 22%—well above the US average. Refer WCAI Awards 2003. http://www.wcai.com.

[7] Telecommunication incumbents generally exaggerate the extent to which certain markets are not profitable. This is the case, e.g., for universal service where estimates of universal service costs by incumbents are high when compared to regulator estimates.

towns with a population of at least 50 persons. More commonly, thresholds are identified by subscribers required to trigger a network upgrade. For example, France Telecom intends to upgrade networks with 100 potential subscribers. Such thresholds enable communities to organize the required subscriber base. This approach is also a feature of the UK approach, where the incumbent BT sets thresholds for providing DSL to communities with 200 to 700 potential subscribers. Triggers depend on the cost to upgrade telephone exchanges to support DSL—the more costly an upgrade, the greater is the required subscriber base. Core cost components are attributed to equipment, accommodation and backhaul. Both equipment costs required for exchange upgrade and associated with accommodation upgrade are relatively negligible. The main variable cost is typically associated with providing a backhaul link to an exchange, i.e., the provision of a link between the local exchange and a backbone network. However, other UK providers have substantially lower DSL broadband access benchmarks. For example, Pipemedia, operating in Lesctershire, set a benchmark of 45 subscribers to provide asymmetrical DSL and 20 subscribers for symmetrical DSL. Symmetrical DSL is mostly subscribed to by business and has high use charges. Therefore, a combination of subscriber types assists communities in reaching trigger levels faster, rather than only gaining the required asymmetrical DSL subscriber base. Further, benchmarks provide rural communities with a better understanding of alternative broadband access modes. However, such information needs to be readily available. Web sites with running totals are an effective form for doing so.

Technical Solutions

Several technical modes are available for broadband subscription within rural and remote areas. For example, in 2001 a commercial two-way interactive satellite service was launched in the US. Prior to this initiative satellite subscribers relied on the PSTN to provide the upstream link, with the downstream link provided by the satellite. Two-way satellite services are currently available in other OECD Member Countries, although they may rely on the PSTN to provide the upstream link. Satellite technology offers a wide footprint but is highly priced when compared to terrestrial technology prices. In Australia, e.g., to obtain broadband access via satellite requires subscribers pay three times the dial-up Internet access price and twice the baseline DSL price. In addition to price differences, satellite connection has a relatively inferior performance due to inherent transmission latency. Weather also reduces and distorts signals. Satellite service also entails higher equipment and installation charges. Clearly, when satellite broadband is available, subscription is much lower than for terrestrial options.

Apart from providing direct service when alternative modes of broadband delivery are not available, satellites have an important role in providing backhaul to rural areas. An emerging technology to provide broadband access is WiFi, e.g., 802.11b or 802.11g. Indeed, the principal barrier to rural area broadband service delivery may ultimately not be the last 'country' mile, due to wireless ISPs providing service to areas not covered by fixed networks. However, when local access is by fixed wireless networks the backhaul provision problem remains. In such circumstances satellites can be combined with fixed wireless to provide broadband access. Satellites, however, are not the only backhaul option. Some ISPs employ fixed wireless to provide broadband access and use leased lines for their backhaul. Eventually, fixed wireless networks may also provide solutions. Further, the emergence of wireless ISPs and WiFi is leading to prices charged by wireless ISPs to occasionally be less than those for urban DSL and cable modem subscription. Moreover, advertised access speeds are higher than those for DSL and cable. In the UK, e.g., a wireless ISP recently advertised a price of US$ 15 per month for 54 Mbps service. While this service has a Gbyte download limit, such limits are familiar to DSL and cable modem subscribers.[8] Finally, subscribers have an option to pay US$ 28 for an increased 16 Gbyte download cap. To sum, this service offers higher performance at a lower price than many traditional broadband services.

A policy message from such activity is to encourage innovation and not design programs around single platforms or an incumbent operator. Incumbents can derive benefit from new technology however they must react more quickly in competitive situations. Additionally, governments should review spectrum policy to ensure underutilized resources are made available to facilitate rural area broadband growth. In Japan, e.g., the Federal government intends to reallocate, by March 2004, 18 gigahertz to local governments to create broadband networks and connect public facilities and households (Japan Today 2003; Yomiuri Shimbun 2003).

Addressing the Backhaul Problem

While broadband access technology solutions exist, the backhaul to high speed backbone networks problem remains. Recent BT threshold experience shows that backhaul substantially impacts on DSL availability. Also, even when communities obtain broadband access, e.g., through WiFi, connection to backbone networks is still required. Entrants have begun to specialize in rural WiFi backhaul, e.g., Invisible Network's in the UK. Once enough of the local population expresses interest in broadband subscription, Invisible Network's install a 2 Mbps leased line linked to a backbone network and establish access points with an omni-directional antenna covering a 200 meter range. Prices are usually cheaper than those offered in urban areas by the incumbent telecommunication operators. Further, an Australia Parliamentary Committee recently recommended that a mechanism be developed to solve the backhaul problem by allowing small community wireless ISPs to negotiate wholesale prices for Internet backbone connection (Standing Committee on Communications, Information Technology and the Arts 2002: 92). Additionally, that Committee noted another approach is to include certain conditions in the universal service obligation. They recommended eligibility be automatic in those

---

[8] See WRBB pricing is reported at: http://www.wrbb.net/servicePlans.html.

areas where DSL is not available. Alternatively, technical developments may continue to provide solutions. Several equipment firms are, e.g., working on WiMAX or the 802.16a standard. This fixed wireless technology promises point-to-point connectivity, as required for backhaul of 74 Mbps at distances of 50 kilometers.

Policy and Conclusions

Any government assessing the deployment of broadband networks and their subscription must gather relevant data. Mapping broadband availability and allowing firms to define their threshold for the providing service is an integral part of the process. However, making that information public is also important to ensure the transparency of the process. Further, government should strive to strengthen competition in local access markets. In rural areas, entrants may provide innovative solutions for areas that are not profitable to serve via DSL or cable modem. A corollary proposal is to support competing platforms and open existing platforms to competition. Initial evidence indicates that unbundling has implications for small exchanges, with some entrants setting lower exchange upgrade thresholds than incumbents. Government should also review spectrum allocations to facilitate entry and experimentation. Further, government can act to aggregate broadband demand so that it is profitable for efficient firms to upgrade their networks—both access and backhaul. Finally, government should deliver, e.g., health and education services via broadband networks to stimulate use, provide more efficient delivery and increase rural access. The underlying principle in implementing such broadband policy, with regard to the supply side, is to ensure there are no distortions to competition and that policy is consistent with increasing broadband access to rural areas.

Acknowledgement

The authors are communication policy analysts at the OECD. The views expressed in the paper are those of the authors and do not necessarily reflect those of the OECD.

References

Alston R (2003) Government response to regional telecommunications inquiry. Media Release, Canberra, 25 June at http://www.dcita.gov.au/Article/0,,0_4-2_4008-4_115488,00.html
Japan Today (2003) Government to bridge urban-rural 'digital divide'. 19 July
OECD (2003a) Seizing the benefits of ICT in a digital economy. Meeting of the OECD Council at Ministerial Level, OECD, Paris, p 7

OECD (2003b) Broadband and telephony services over cable television networks. DSTI/ICCP/TISP(2003)1/Final, OECD, Paris at http://www.oecd.org/sti/telecom

House of Representatives Standing Committee on Communications, Information Technology and the Arts (2002) Connecting Australia! Wireless broadband. p 92, para. 7.48, November

Yomiuri Shimbun (2003) Wireless access planned to put rural areas online. 18 July

18 Understanding the Evolving Digital Divide

Russel Cooper and Gary Madden

Introduction

Much activity of international aid and development agencies focuses on the North-South digital divide.[1] Available evidence shows a staggering gap between developed and developing country information and communication technology (ICT) access, e.g., at the end of the 20th century the average OECD Member Country had an eleven times greater per capita income than that for a typical South Asian country. Further, it possessed 40 times more computers, 146 times the mobile telephones and 1,036 times Internet hosts (Rodríguez and Wilson 2000).[2] Most developing countries are not catching up with developed countries, and so the gap continues to grow. As Temple (1999) concludes:

> Poor countries are not catching up with the rich, and to some extent the income distribution is becoming polarized. Countries do converge to their own steady states, but at an uncertain rate. One reason for this uncertainty is that countries catch up by adopting technologies from abroad, as well as by investing in physical capital and education. It is easy to envisage a hypothetical long-run equilibrium in which countries grow at the same rate, but over the last thirty years, rates of efficiency growth have almost certainly varied widely (Temple 1999: 151).

Further, through the last decade the speed of technical change within the ICT sector has increased with the Internet, mobile telephony advance, and the convergence of information technology and communication technology. To the extent that an economy needs to be properly positioned to take advantage of technology change, more rapid change may hinder the catch-up process. Conversely, rapid technological change allows developing nations to avoid making old technology investment, e.g., by establishing mobile telephony networks rather than deploying costly fixed-line networks. The existence of substantial ICT access gaps between developed and developing countries does not of itself necessitate that a reduction in this disparity should be a priority in development agendas. After all, developing countries also have fewer doctors, factories and lower calorie intake than developed nations. Among reasons why this growing gap is of concern is its size—much larger than explained solely by national income differences—and the impli-

---

[1] Major international organizations in the policy-making process include ICANN, ITU, UNCITRAL, UNCSTD, World Bank, W3C and WTO (Bridges.org 2001).
[2] The definition of ICT adopted by the OECD ICCP panel of statistical experts is that it is the set of activities that facilitate by electronic means the processing, transmission and display of information.

cation that many developing regions could be forced into an ICT-related poverty trap. That is, the issue is not connectivity per se. Rather, connectivity for economic growth and the broader agenda of sustainable development. Whether connectivity leads to increased growth to an extent sufficient to begin to close the welfare divide depends, in a market economy, on the ability of consumers to recognize benefits from increased connectivity and their willingness to modify consumption behavior, both to achieve saving necessary to finance the growth and to create the market for products based on new regime technology.

These considerations are important for an ICT sector subjected to substantial recent technology change and the convergence of mobile telephony and the Internet. While mobile telephony is currently playing a major role in developing country communication markets—mobile telephony has overtaken fixed-line telephony network reach and subscription in many developing nations—the Internet is less important.[3] This chapter explores links between investment opportunity, production possibility and consumer willingness to adopt new ICT in a representative North (developed) and South (developing) vertically integrated firm general equilibrium model.[4] Vertical integration is modeled as occurring in the North, South and across the divide. The model as such allows examination of the implications of network ICT investment for growth in the context of a dynamic North-South model that explicitly incorporates uncertainty. Uncertainty is introduced through returns arising from the vertical integration of telecommunications carrier activity with application sector developments such as operations software. This aspect of uncertainty is selected for examination as it recognizes a key to determining the ease of bridging the North-South ICT divide lies in understanding factors that influence application sector adaptation of North ICT to regional contexts (Bresnahan and Greenstein 2002). For example, opportunity potentially available to the South application sectors from North ICT advance may be outweighed by local human capital shortage—obtaining adequate training is a lengthy process—or inadequate business infrastructure (Jalava and Pohjola 2002).[5] That is, the more alike are North-South industry circumstances, the more likely are benefits realized from investment in North invention, e.g., the greater is national banking activity the more likely are gains to accrue from new generic banking software. Empirical research shows such benefits arrive slowly and unevenly (Bresnahan and Greenstein 2002).

---

[3] Among the reasons for this situation include cost and pricing, and the readiness of business and consumers to adopt Internet technology.

[4] A vertically integrated firm is considered here as it is common incumbent strategy to expand activity into new but related operations such as mobile communications and Internet access. More ambitious extensions are into Internet portals, transaction platforms, video content and hardware production. Advantages sought include extension of market power and economies of scope in production (Noam 2003).

[5] The unevenness of gains cross country is not typically considered in analyses of the impact of ICT. For example, Röller and Waverman (2001) argue that, non-OECD countries might only realize the growth effects through telecommunications investment like their OECD counterparts, if a critical significant improvement in the telecommunications infrastructure is achieved.

The following section provides informal evidence as to the existence, composition, magnitude and growth pattern of the North-South digital divide. Next, the production possibilities facing a vertically integrated firm that produces and uses general-purpose network ICT is specified. North and South firms are differentiated by their productivity and by allowing the South downstream application sectors to invest in North ICT. Asymmetry in the impact of uncertainty is considered explicitly. Responses to 'good' and 'bad' news are allowed to be different in the North and South. The implications of these features of the model on the digital divide are analyzed. Further, the structure of capital stock transition equations obtained from North-South firm investment decisions made in an uncertain environment are derived. The model is illustrated using isoelastic consumer preferences and a simplified technology specification. Implications of the optimal investment results on the digital divide are presented. The discussion highlights the complexity of North and South firm investment and growth relationships, and suggests that hoped for technology transfer 'solutions' to necessitate growth and convergence are too simplistic.[6] In particular, study findings show that asymmetric uncertainty can exacerbate the digital divide.

Digital Divide Statistics

A fundamental indicator of the international digital divide is the number of access lines per 100 inhabitants. Table 18.1 indicates an impressive recent growth of fixed-line and mobile telephony penetration in developing countries. For the decade 1991 through 2000, developing country (with the exception of sub-Saharan Africa) telephone line and mobile subscription per capita growth outstripped that for North America, Scandinavia and Western Europe. However, this growth comes from a substantially lower base, and disguises wide disparity in terms of absolute levels and relative penetration. The story is much bleaker for Internet host penetration. Table 18.2 provides data on Internet host penetration by region at 2000. North America, Scandinavia and Western Europe account for 90% of the world total, and this rises to 98% when the Asia-Pacific region is included. The situation does not substantially improve when the estimated number of users is considered, with 94% of users located in North American, Scandinavian, Western European and the Asia-Pacific.

---

[6] Crucially, the impact of asymmetric uncertainty on business decision making is missing from such approaches.

Table 18.1. Regional Telecommunications Access, 1991-2000

	1991	1992	1993	1994	1995	1996	1997	1998	1999	2000
Telephone mainlines per 100 persons										
North America	55.5	56.3	57.4	58.9	60.6	62.0	64.2	66.3	67.1	69.7
Scandinavia and W. Europe	44.4	45.3	46.5	47.8	49.5	50.9	52.2	53.3	54.4	55.3
Eastern Europe	15.2	15.8	16.6	17.5	18.5	20.0	21.7	23.0	24.5	20.8
Asia-Pacific	4.0	4.3	4.6	5.2	5.8	6.6	7.3	8.1	9.0	10.4
Middle East and N. Africa	7.8	8.5	9.4	10.1	10.6	11.3	12.2	13.1	14.1	14.5
Sub-Saharan Africa	2.5	2.5	2.6	2.7	2.8	3.0	3.2	3.5	3.8	3.5
- excluding RSA	0.6	0.7	0.7	0.7	0.8	1.0	1.1	1.2	1.3	1.5
Mobile telephones per 1000 persons										
North America	29.8	42.5	60.4	89.8	124.4	161.1	200.0	248.2	306.6	386.5
Scandinavia and W. Europe	12.8	16.6	24.3	38.9	60.6	93.5	145.3	244.6	403.8	636.4
Eastern Europe	0.0	0.2	0.5	1.5	2.9	6.4	14.3	27.7	52.6	96.4
Asia-Pacific	1.0	1.4	2.1	3.9	8.3	16.6	25.1	35.4	52.1	75.5
Middle East and N. Africa	0.9	1.2	1.6	2.5	5.9	11.5	20.0	32.7	59.0	112.0
Sub-Saharan Africa	0.1	0.1	0.3	1.9	3.0	5.3	8.8	14.8	30.7	49.4
- excluding RSA	0.0	0.0	0.0	0.1	0.2	0.4	1.0	2.2	6.0	12.8
Internet hosts per 1000 persons										
North America	2.0	3.5	5.4	11.6	22.0	36.3	72.1	105.2	180.9	271.1
Scandinavia and W. Europe	0.3	0.7	1.4	2.6	5.5	8.9	13.7	18.5	24.2	29.4
Eastern Europe	0.0	0.0	0.1	0.3	0.8	1.9	3.2	4.7	6.1	7.7
Asia-Pacific	0.0	0.0	0.1	0.1	0.3	0.5	0.8	1.1	1.8	2.8
Middle East and N. Africa			0.0	0.1	0.2	0.3	0.6	0.8	1.1	1.2
Sub-Saharan Africa	0.0	0.0	0.1	0.1	0.3	0.5	0.6	0.8	0.9	1.0
- excluding RSA				0.0	0.0	0.0	0.0	0.0	0.1	0.1
Internet users per host										
North America	5.8	4.8	3.7	2.7	3.3	3.0	2.1	2.1	1.6	1.3
Scandinavia and W. Europe		4.9	3.6	3.6	3.4	3.7	4.4	5.2	6.7	8.1
Eastern Europe	1.0	8.3	5.2	5.5	3.8	3.0	3.0	3.6	4.0	4.8
Asia-Pacific	6.5	6.0	7.3	6.1	5.5	6.0	8.2	10.6	12.4	13.8
Middle East and N. Africa			4.0	4.6	3.9	4.3	6.6	8.8	13.6	18.5
Sub-Saharan Africa	10.8	13.8	12.8	12.2	9.5	6.3	6.7	9.1	11.7	15.0
- excluding RSA				9.0	13.5	9.2	12.8	16.5	24.5	37.9

Source: ITU (2002)

Table 18.2. Internet Diffusion at 2000

	Population (million)	Internet Users (million)	Internet Hosts (million)	Users / Population (%)	Hosts / Users (%)
Region					
North America	305.9	108.1	82.9	35.3	76.7
Scandinavia and W. Europe	379.1	90.3	11.1	23.8	12.3
Eastern Europe	304.4	11.1	1.2	3.7	11.2
Asia-Pacific	3,173.4	121.7	8.8	3.8	7.2
Middle East and N. Africa	263.2	5.8	0.3	2.2	5.4
Sub-Saharan Africa	168.5	2.4	0.2	1.4	7.8
- excluding RSA	124.8	0.7	0.0	0.0	0.0
Economic development					
Post-industrial	685.0	198.4	94.1	29.0	47.4
OECD	1,116.2	282.1	102.4	25.3	36.3
Developing	3,521.9	58.0	2.2	1.6	3.8

Source: ITU (2002)

Further, the growth rate for Internet hosts in OECD and post-industrialized countries currently, and historically, outstrips that for developing nations. Given this experience, the unequal distribution of ICT will arguably lead to further divergence in economic performance. For example, countries which lack a critical mass of ICT may experience lower growth rates. To make useful policy recommendations to resolve the digital divide, the nature of critical mass needs to be examined. The existence of a critical mass is often discussed in the context of increasing returns in production due to network externalities. Clearly this dynamic is an important contributing factor for network industry growth. However, concentration on network effects that operate through increasing returns in production overlooks another factor capable of creating a critical mass effect, viz., the pervasiveness of ICT. When an economy is sophisticated, and the extent to which ICT permeates is a measure of this, then any positive economic impact from new application success is enhanced. In particular, good news effects are magnified through many sectors due to the pervasiveness of ICT, while bad news effects are more readily absorbed. To examine the issue, this study considers the interrelationship between asymmetric uncertainty and the existence of ICT critical mass. That is, given that technical advance in ICT is globally available with minimal lag, to what extent does co-invention in the form of new products (or applications) development by the South application sectors enhance growth and development prospects as measured by new ICT investment? In the theoretical modeling conducted below, North and South firms differ in the impact of uncertainty associated with co-invention undertaken by application sectors, and also by consumer willingness to modify their consumption plans in the face of uncertainty.

North-South Production and ICT Investment

A generic specification for the production possibilities of a representative vertically integrated firm producing both ICT services and associated application sector output is, for carrier services $g = G(k)$ and applications $y = A(g)$, where k is an index of ICT capital (accumulated investment of producers and service providers, e.g., network ICT and producer equipment), g is ICT sector output (e.g., telecommunications services and user equipment) and an application sector input, and y is application sector output (e.g., software). Vertically integrated production is described by $y = A(G(k)) = Y(k)$. Firms in the North and South are not identical in terms of their production possibilities. That is, when they operate in isolation, representative vertically integrated firms in the North and South are subject to separate technology $y_j = Y^j(k_j)$ for $j = N, S$. Production functions $Y^j(k_j)$ are assumed increasing and concave in k_j for both the North ($j = N$) and South ($j = S$).

Firms in the North typically have at their disposal 'superior' ICT and more 'coherent' application sectors, e.g., more Internet hosts suggest the quality of infrastructure is superior and associated application sectors more 'populous'. The situation for the North firm contrasts to that of the South firm with a relatively less productive ICT stock. This productivity assumption introduces relative inefficiency in ICT use—due to local institutional factors—in addition to an absolute ICT endowment disparity that further inhibits South-firm productivity. In doing so, this added complexity broadens the notion of digital divide to include the use of ICT within application sectors. In this context, the South firm must choose between employing more productive North ICT or its less productive ICT in downstream application sectors. Superior North technology implies $Y^N(k_s) > Y^S(k_s)$.[7] Due to concavity, it does not necessarily follow that, pre-joint venture, actual average and marginal products are higher in the North than the South, since the South is operating from a lower base.[8]

---

[7] To illustrate, let $Y^N(k_N) = A_N k_N^{\beta_N}$ and $Y^S(k_S) = A_S k_S^{\beta_S}$, where relative technological disadvantage (poor quality ICT) by the South is reflected in $\beta_S < \beta_N$. Relatively poor infrastructure or institutional environment suggests $A_S < A_N$. Together these parameter specifications imply a technological and institutional advantage for the North abstracting from absolute size effects, viz., $Y^N(k_S) > Y^S(k_S)$.

[8] Let $r_N = \partial Y^N(k_N)/\partial k_N$ and $r_S = \partial Y^S(k_S)/\partial k_S$. In this illustration $r_N = \beta_N A_N k_N^{\beta_N - 1}$ and $r_S = \beta_S A_S k_S^{\beta_S - 1}$. Clearly, when k_N is sufficiently large relative to k_S it is possible that $(\beta_S A_S)/(\beta_N A_N) > (k_S)^{1-\beta_S}/(k_N)^{1-\beta_N}$, and so $r_S > r_N$. The relationship between the marginal products depends on the degree of institutional and ICT quality disadvantage relative to ICT quantity disadvantage. However, it is shown below that post-joint venture the most likely outcome is $r_S < r_N$.

Joint Ventures and Optimized Current Output

Intuitively, North-South joint ventures must be motivated by firms seeking an advantage. The South firm gains access to superior technology, while the North firm enhances market reach. Accordingly, the outcome of bargaining means both firms gain when agreement is reached.[9] Indeed, r_N and r_S tend to be related as the private sector searches for profit opportunity, however, any realized joint venture returns are not necessarily shared equally.[10] To proceed, represent expected outcomes by the functions $F^N(k_N,k_S,x_N)$ for the North firm and $F^S(k_S,k_N,x_S)$ for the South firm, where x_N and x_S are the financial commitments of North and South firms to the joint venture. These are expected net output functions.[11] Clearly, to reach an agreement, $F^N(k_N,k_S,x_N) > Y^N(k_N)$ and $F^S(k_S,k_N,x_S) > Y^S(k_S)$.

Investment and Applications Sector Volatility

Returns from investment in general purpose network ICT are not certain. Further, ICT output is itself an input into application sector production. The application sectors of North and South vertically integrated firms are different in their geographic concentration, human capital (availability and quality) and product innovation history. This does not mean that risk is absent from North application sectors, but that the 'upside' innovation (success) shock is typically greater in the North, while 'downside' innovation (failure) is the greater in the South. For the capital stock transition equations to be meaningful they must reflect such underlying asymmetric stochastic processes. The capital stock transition equations for North and South firms are modeled as geometric Brownian motions with drift.[12] Accordingly, a Wiener process $d\xi$, with $E(d\xi)=0$ and $E(d\xi)^2 = dt$, is added to

---

9   Because of the macro-orientated approach of this chapter, the specific features of the implied game leading to this equilibrium outcome are not examined. It is implicitly assumed that governments on both sides of the divide provide a supporting environment to ensure the existence of Nash equilibrium.

10 The relationship may be modeled by recognizing that, through joint ventures, firms have an incentive to equalize their adjusted risk premiums, placing the marginal dollar where the greatest return for a given degree of risk is gained.

11 The functions are net in the sense that they represent output post-technology transfer net of bargaining transfers. Consider any joint venture as analogous to a financial investment and there must be an expected return, say μ, that is sufficiently high to compensate for any risk. In the absence of information asymmetry, it is assumed that both sides agree in their expectations, even though bargaining may result in an unequal sharing of benefits (or costs) from news arising from the joint venture.

12 However, the drift and volatility are variable. The drift is variable because of explicit firm decision making, viz., intertemporal optimization. Variation in the volatility coefficient reflect the reality facing the ICT industry.

the deterministic investment identity: Investment $\equiv$ Net Output – Consumption. Since the ICT sector is highly globalized, good news ($d\xi > 0$) is treated as equally applicable to the North and South—conditional on their ICT sectors being of similar size. The impact of bad news ($d\xi < 0$) is also, in principle, global. Since $d\xi$ is a standardized process, it is scaled by a common global (world ICT uncertainty) variance parameter σ. However, recognizing that much ICT investment is sunk, a variable parameter τ is introduced to reduce σ as a base case. In general, the direction of investment shock volatility is treated as asymmetric. Additionally, in recognizing the weaker South application sectors—a weakness that is expected to be correlated with network size—the asymmetric volatility parameter τ is specified as sensitive to network size.[13] With c_j, $j = N, S$, denoting consumption (distribution of output to shareholders) by the North and South firms, the capital stock transition equations are, respectively,

$$dk_N = (Y^N(k_N) + (\mu - r_N)x_N - c_N)dt + \tau_N x_N d\xi \qquad (18.1)$$

and

$$dk_S = (Y^S(k_S) + (\mu - r_S)x_S - c_S)dt + \tau_S x_S d\xi. \qquad (18.2)$$

To illustrate the modeling of network dependent asymmetric volatility, consider the specifications:

$$\tau_j^+ = \sigma(k_N + k_j)/(k_N + \eta_j) \qquad (18.3)$$

and

$$\tau_j^- = \sigma(k_N + \eta_j)/(k_N + k_j) \qquad (18.4)$$

for $j = N, S$, where τ_j^+ applies when $d\xi > 0$ and τ_j^- when $d\xi < 0$.[14]

---

[13] The model treats the number of firms as given and, without loss of generality, is set at unity—effectively subsuming the number of firms into the functional forms. Since the firm is representative, k is interchangeable as an index of firm and network size.

[14] An effect of asymmetric volatility is that $E(\tau_N d\xi) \neq 0$. Implied in Eq. 18.1 is an expected net output function relevant to the North firm in its bargaining defined as $F^N(k_N, k_S, x_N) = Y^N(k_N) + (\mu - r_N)x_N + E(\tau_N x_N d\xi)$. Similarly for the South firm, expected net output is defined as $F^S(k_S, k_N, x_S) = Y^S(k_S) + (\mu - r_S)x_S + E(\tau_S x_S d\xi)$. In this stylized representation, in addition to receiving returns to compensate for risk, asymmetric volatility is used as an indicator the North and South ability to deliver in the joint venture, and so is included in the outcome to bargaining. As a result of asymmetric volatility, $(\mu - r_j)x_j + E(\tau_j x_j d\xi)$ must be positive for region j, $j = N, S$ to accept the risk in-

Specification Eq. 18.3 applies when a technology shock is positive and yields $\tau_N^+ = \sigma k_N/(\frac{1}{2}k_N + \frac{1}{2}\eta_N)$ for the North firm and $\tau_S^+ = \sigma(k_N + k_S)/(k_N + \eta_S)$ for the South firm. The extent of influence of an upside shock is related to how important the ICT sector is to an economy. For the North, η_N is a critical measure, with $k_N^* = \eta_N$ defining the point where $\tau_N^+ = \sigma$ —a normalizing value of volatility independent of the size of the ICT shock.[15] The specification also suggests the North is autonomous regarding the propagation effect of positive shocks, i.e., τ_N^+ does not depend on the South ICT sector. Alternatively, the South depending on the North is built into the specification, with τ_S^+ modeled as a function of North and South ICT stock, and the extent of South ICT sector development relative to the norm η_S. In particular, $k_S < \eta_S \Rightarrow \tau_S^+ < \sigma$, so the expansionary effect of an upside shock is lower in the South than in the North.[16] Conversely, when a technology shock is negative Eq. 18.4 applies, and yields $\tau_N^- = \sigma(\frac{1}{2}k_N + \frac{1}{2}\eta_N)/k_N$ for the North firm and $\tau_S^- = \sigma(k_N + \eta_S)/(k_N + k_S)$ for the South firm. When $\eta_N = k_N$, upside and downside shocks are of equal magnitude in the North, with $\tau_N^+ = \tau_N^- = \sigma$. Thus, $k_N \geq \eta_N$ indicates the North has an ICT stock sufficiently large to ensure that upside shocks are at least as beneficial as downside shocks are detrimental. This specification represents a market whereby bad news is contained and good news permeates through the economy. In this sense, $k_N^* = \eta_N$ is interpretable as a North ICT capital critical mass above which good outweighs bad news. At this critical value volatility is symmetric and equal to σ. Similarly, $k_S^* = \eta_S$ is the critical value for symmetric volatility for the South firm. As is the case for positive shocks, however, the specification allows the extent of influence in the South to be modified by the North network size. This situation represents, in a stylized manner, that technological innovation predominantly originates in the North.

When k_N is large relative to η_N the North firm is insulated against innovative failure to a greater extent, with $k_N > \eta_N$ implying $\tau_N^+ > \sigma > \tau_N^-$. Given $k_S < \eta_S$, the South firm is not as well insulated. Here $\tau_S^+ < \sigma < \tau_S^-$ reflects greater uncertainty of outcome and less support infrastructure for South application sectors. A higher $\eta_j, j = N,S$ value represents greater downside volatility in the North and South respectively. It is possible that η_j is amenable to change, with a lower η_j value sustained by a more ICT-friendly environment, e.g., flexible workplace arrangements that encourage greater 3G technology subscription. A lower η_j value offers improved protection from downside shocks and greater stimulus from upside shocks. In this stylized model, η_S represents an opportunity for government policy inter-

---

volved in the joint venture, and not $(\mu - r_j)x_j$. Clearly, $(\mu - r_j)x_j$ may be negative as long as $E(\tau_j x_j d\xi)$ compensates.

[15] The North is defined by $k_N \geq \eta_N$, implying $\tau_N^+ \geq \sigma$.

[16] Correspondingly, a defining characteristic of the South is $k_S < \eta_S$, implying $\tau_S^+ < \sigma$.

vention.[17] Given Eq. 18.1 and Eq. 18.2, the optimal investment strategy for the representative North and South ICT firms requires optimal choice of control variables c_N, x_N and c_S, x_S, respectively. The mean growth rates in the North and South are influenced by their technology, network dependent volatility and optimal choices. Specifically:[18]

$$\tfrac{1}{dt} E(dk_N) = Y^N(k_N) + (\mu - r_N)x_N - c_N + \tfrac{1}{5}(\tau_N{}^+ - \tau_N{}^-)x_N \tag{18.5}$$

and

$$\tfrac{1}{dt} E(dk_S) = Y^S(k_S) + (\mu - r_S)x_S - c_S + \tfrac{1}{5}(\tau_S{}^+ - \tau_S{}^-)x_S . \tag{18.6}$$

Optimal Investment Strategy

A rational investment strategy is characterized as the outcome of a representative firm dynamic optimization program. The objective of the firm is to maximize the discounted future shareholder utility stream. The optimization problem of the representative North and South ICT firms are interrelated, with an optimal strategy affecting the outcome of the other. The influence of network size on volatility is modeled as an externality, viz., in determining an optimal strategy the firm ignores the fact that its actions influence the size of the effect. Let $\bar{k}_N$ ($\bar{k}_S$) denote the network size in the North (South) when viewed as an external effect.[19] Taking this assumption into account, the North ICT transition equation is:

$$dk_N = (Y^N(k_N) + (\mu - \bar{r}_N)x_N - c_N)dt + \bar{\tau}_N x_N d\xi \tag{18.1a}$$

---

[17] This chapter concerns macro relationships through which opportunity to influence volatility impacts on the digital divide, and not the micro structure underlying endogenous volatility.

[18] τ is conditional on $d\xi$, $d\xi$ is symmetric around 0 and $E(\tau d\xi) = 0.5\tau^+ E(d\xi \mid d\xi > 0)$ $+ 0.5\tau^- E(d\xi \mid d\xi < 0) = 0.5(\tau^+ - \tau^-)E(d\xi \mid d\xi > 0)$. $d\xi \sim N(0,1)$ $dt \Rightarrow E(d\xi \mid d\xi > 0)$ $= 0.398942 \, dt$. Hence $E(\tau d\xi) \approx 0.2 \, (\tau^+ - \tau^-)dt$.

[19] This implies the principal concern is with private optimum. This approach is reasonable as the concern is to model the ability of the private sector to contribute to the digital divide. For similar reasons, the risk free rates of return r_N and r_S are treated as outside the control of North and South firms. The same is assumed for the degree of risk aversion, which is endogenous in the general equilibrium. For similar reasons, where it enters in the volatility functions, k_S (k_N) is outside the control of the North (South) firm. These assumptions are represented by overbars on variables.

where $\bar{\tau}_N = 2\sigma \bar{k}_N /(\bar{k}_N + \eta_N)$ when $d\xi > 0$ and $\bar{\tau}_N = \frac{1}{2}\sigma(\bar{k}_N + \eta_N)/\bar{k}_N$ when $d\xi < 0$. From the point of view of the South firm, the ICT transition equation is:

$$dk_S = (Y^S(k_S) + (\mu - \bar{r}_S)x_S - c_S)dt + \bar{\tau}_S x_S d\xi \qquad (18.2a)$$

where $\bar{\tau}_S = \sigma(\bar{k}_N + \bar{k}_S)/(\bar{k}_N + \eta_S)$ when $d\xi > 0$ and $\bar{\tau}_S = \sigma(\bar{k}_N + \eta_S)/(\bar{k}_N + \bar{k}_S)$ when $d\xi < 0$. Let $\hat{c}_N, \hat{x}_N$ and $\hat{c}_S, \hat{x}_S$ represent optimal choices for c_N, x_N and c_S, x_S in the respective North and South firm optimization problems, subject to the equilibrium conditions $k_N = \bar{k}_N$ and $k_S = \bar{k}_S$. Denote these optimal equilibrium feedback controls by $\hat{c}_N = C^N(\bar{k}_N, \bar{k}_S)$, $\hat{x}_N = X^N(\bar{k}_N, \bar{k}_S)$ and $\hat{c}_S = C^S(\bar{k}_S, \bar{k}_N)$, $\hat{x}_S = X^S(\bar{k}_S, \bar{k}_N)$. For model consistency, transition equations for the network stocks must satisfy:

$$d\bar{k}_N = (Y^N(\bar{k}_N) + (\mu - \bar{r}_N)X^N(\bar{k}_N, \bar{k}_S) - C^N(\bar{k}_N, \bar{k}_S))dt + \bar{\tau}_N X^N(\bar{k}_N, \bar{k}_S)d\xi \qquad (18.1b)$$

and

$$d\bar{k}_S = (Y^S(\bar{k}_S) + (\mu - \bar{r}_S)X^S(\bar{k}_S, \bar{k}_N) - C^S(\bar{k}_S, \bar{k}_N))dt + \bar{\tau}_S X^S(\bar{k}_S, \bar{k}_N)d\xi. \qquad (18.2b)$$

For the North firm, the problem is:[20]

$$\max_{c_N, x_N} E_0 \int_0^\infty e^{-\rho t} U^N(c_N)dt \qquad (18.7)$$

subject to Eq. 18.1a, Eq. 18.1b, Eq. 18.2b with initial conditions $k_N(0) = k_{N0}$, $\bar{k}_N(0) = k_{N0}$ and $\bar{k}_S(0) = k_{S0}$. The corresponding program for the South firm is: [21]

$$\max_{c_S, x_S} E_0 \int_0^\infty e^{-\rho t} U^S(c_S)dt \qquad (18.8)$$

subject to Eq. 18.2a, Eq. 18.1b, Eq. 18.2b with initial conditions $k_S(0) = k_{S0}$, $\bar{k}_N(0) = k_{N0}$ and $\bar{k}_S(0) = k_{S0}$.

To simplify presentation of optimal solution implications for investment differences particular functional forms for pre-joint venture output are employed, viz.[22]

---

[20] Assume a common time preference rate ρ in the North and South.

[21] Here the equilibrium condition $k_j = \bar{k}_j$, $j = N, S$ is imposed on the initial conditions.

[22] The functional forms are too simple for empirical implementation. However, they convey qualitative features of the results which are present in more empirically realistic specifications.

$$Y^N(k_N) = A_N k_N^{\beta_N} \text{ and } Y^S(k_S) = A_S k_S^{\beta_S} . \tag{18.9}$$

Further, define the respective opportunity cost of ICT capital in the North and South by $r_N = \partial Y^N(k_N)/\partial k_N$ and $r_S = \partial Y^S(k_S)/\partial k_S$, and first-order conditions to $\Pi^N(r_N) \equiv \max_{k_N}(Y^N(k_N) - r_N k_N)$ and $\Pi^S(r_S) \equiv \max_{k_S}(Y^S(k_S) - r_S k_S)$, as:[23]

$$r_N = \beta_N A_N k_N^{\beta_N - 1} \text{ and } r_S = \beta_S A_S k_S^{\beta_S - 1} . \tag{18.10}$$

Also, utility is specified as isoelastic to allow different values for the inter-temporal elasticity of substitution in the North and South, ε_N and ε_S respectively, i.e.,

$$U^N(c_N) = (c_N^{1-1/\varepsilon_N} - 1)/(1-1/\varepsilon_N) \text{ and } U^S(c_S) = (c_S^{1-1/\varepsilon_S} - 1)/(1-1/\varepsilon_S) . \tag{18.11}$$

This specification allows different degrees of consumer flexibility.

Let $J^N(k_N, \bar{r}_N)$ denote the optimal value function for Eq. 18.7. The optimal feedback solutions for c_N, x_N and c_S, x_S are derived in terms of J_k^N and J_{kk}^N. After solving for the optimal value function, these solutions are written as:[24,25]

$$\hat{x}_N = \varepsilon_N(\mu - \bar{r}_N + \tfrac{1}{5}(\bar{\tau}_N^+ - \bar{\tau}_N^-))(\Pi(\bar{r}_N)/\bar{r}_N + k_N)/(\tfrac{1}{4}(\bar{\tau}_N^+)^2 + \tfrac{1}{4}(\bar{\tau}_N^-)^2) \tag{18.12}$$

and

---

[23] Consequently, viewed as external to the decision makers, $\bar{r}_N = \beta_N A_N \bar{k}_N^{\beta_N - 1}$ and $\bar{r}_S = \beta_S A_S \bar{k}_S^{\beta_S - 1}$. In equilibrium $r_N = \bar{r}_N$ and $r_S = \bar{r}_S$.

[24] For problem Eq. 18.7, $\rho J^N = \max_{c,x}\{U^N(c_N) + J_k^N \tfrac{1}{dt}E(dk_N) + \tfrac{1}{2}J_{kk}^N \tfrac{1}{dt}E(dk_N)^2\} + \Omega^N$ defines the Hamilton-Jacobi-Bellman (HJB) equation, where Ω^N collects terms, such as the effects of the evolution of r_N, considered external to the private firm. Further, Eq. 18.5, and Eq. 18.9 imply $\tfrac{1}{dt}E(dk_N) = A_N k_N^{\beta_N} + (\mu - \bar{r}_N)x_N - c_N + \tfrac{1}{5}(\bar{\tau}_N^+ - \bar{\tau}_N^-)x_N$ and $\tfrac{1}{dt}E(dk_N)^2 = \tfrac{1}{4}((\bar{\tau}_N^+)^2 + (\bar{\tau}_N^-)^2)x_N^2$. That is, the HJB equation is directly optimized and it is straightforward to obtain the solutions in terms of J_k^N and J_{kk}^N as $\hat{x}_N = -(\mu - \bar{r}_N + \tfrac{1}{5}(\bar{\tau}_N^+ - \bar{\tau}_N^-))J_k^N/(\tfrac{1}{4}((\bar{\tau}_N^+)^2 + (\bar{\tau}_N^-)^2)J_{kk}^N)$ and $\hat{c}_N = (J_k^N)^{-\varepsilon_N}$.

[25] It can be verified that the optimal value function for Eq. 18.7 is given by the expression: $J^N = (\varepsilon_N/(\varepsilon_N - 1))((\varepsilon_N \rho + (1 - \varepsilon_N)(\bar{r}_N + \alpha_N \varepsilon_N))^{-1/\varepsilon_N}(\Pi(\bar{r}_N)/\bar{r}_N + k_N)^{1-1/\varepsilon_N} - 1/\rho) + \Omega^N/\rho$ with $\Pi^N(\bar{r}_n) = (A_N)^{1/(1-\beta_N)}(\beta_N^{\beta_N/(1-\beta_N)} - \beta_N^{1/(1-\beta_N)})(\bar{r}_N)^{-\beta_N/(1-\beta_N)}$ and $\alpha_N = 2(\mu - \bar{r}_N + \tfrac{1}{5}(\bar{\tau}_N^+ - \bar{\tau}_N^-))^2/((\bar{\tau}_N^+)^2 + (\bar{\tau}_N^-)^2)$. By direct differentiation of J^N with respect to k_N it follows that $J_k^N = (\varepsilon_N \rho + (1 - \varepsilon_N)(\bar{r}_N + \alpha_N \varepsilon_N))^{-1/\varepsilon_N}(\Pi(\bar{r}_N)/\bar{r}_N + k_N)^{-1/\varepsilon_N}$ and $J_{kk}^N = -(1/\varepsilon_N)(\varepsilon_N \rho + (1 - \varepsilon_N)(\bar{r}_N + \alpha_N \varepsilon_N))^{-1/\varepsilon_N}(\Pi(\bar{r}_N)/\bar{r}_N + k_N)^{-1/\varepsilon_N - 1}$. Combining these results with those of Footnote 22, the HJB equation holds identically, and the candidate J^N is verified.

$$\hat{c}_N = (\varepsilon_N \rho + (1 - \varepsilon_N)(\bar{r}_N + \alpha_N \varepsilon_N))(\Pi(\bar{r}_N)/\bar{r}_N + k_N).$$ (18.13)

Next, define planned investment terms $g_j \equiv \frac{1}{dt} E(dk_j)/k_j$ for $j = N, S$. So, the optimal planned investment strategy for the North firm is summarized as:[26]

$$g_N = g_{N1} + g_{N2}$$ (18.14)

where $g_{N1} = (r_N - \rho)\varepsilon_N / \beta_N$, $g_{N2} = \alpha_N(1 + \varepsilon_N)\varepsilon_N / \beta_N$, $\theta_N = (\tau_N^+)^2 + (\tau_N^-)^2$ and $\alpha_N = 2(\mu - r_N + \frac{1}{5}(\tau_N^+ - \tau_N^-))^2 / \theta_N$.

When $r_N > \rho$, then marginal productivity outweighs impatience and $g_{N1} > 0$ —which is regarded here as the typical case. However, even when this is not so, there is a tendency for planned ICT investment to be positive because of the component $g_{N2} > 0$.[27,28] However, for the purpose of later comparison with the situation in the South it is convenient to break α_N (and hence g_{N2}) into the component parts, viz.,

$$\alpha_N = \alpha_{N1} + \alpha_{N2} + \alpha_{N3}$$ (18.15)

with $\alpha_{N1} = 2(\mu - r_N)^2 / \theta_N$, $\alpha_{N2} = \frac{4}{5}(\mu - r_N)(\tau_N^+ - \tau_N^-)/\theta_N$ and $\alpha_{N3} = \frac{2}{25}(\tau_N^+ - \tau_N^-)^2 / \theta_N$. Component α_{N1} is unambiguously non-negative. In general, the critical mass of ICT stock is identified as that for which volatility is symmetric.[29] Since it is reasonable to assume that k_N is at least equal to the North critical mass then upside shocks are not less volatile than downside shocks in the North, and α_{N3} is non-negative. However, α_{N2} can be non-positive when $\mu < r_N$. The role of ICT critical mass is evident in α_{N2}, but the direction of influence switches depending on expectation for the joint venture relative to marginal productivity in the North. For comparison with prospects in the South it is helpful to consider a base case in the

---

[26] Result Eq. 18.14 follows directly from the calculations in Footnote 22 and Footnote 23. Having found the solutions for a competitive representative firm the overbars in what follows are dropped. Similar derivations give Eq. 18.16 as the solution to Eq. 18.8.

[27] Even in the recent downturn aggregate ICT investment remained positive, though substantially reduced in most cases. Accordingly, the pertinent issue is the relative size of positive ICT investment in the North and South.

[28] The solution for x_N in Eq. 18.12 takes as given that $\mu - r_N + \frac{1}{5}(\tau_N^+ - \tau_N^-) > 0$. Otherwise, no portfolio investment is undertaken as the risk free return $\bar{r}_N$ exceeds the asymmetric variance adjusted expected return $\mu + \frac{1}{5}(\tau_N^+ - \tau_N^-)$. This also implies that it is not necessary for μ to exceed r_N for the North firm to profitably engage in a joint venture. This is because the North firm also benefits from asymmetric volatility when its ICT stock is greater than the critical value, giving $\tau_N^+ - \tau_N^- > 0$.

[29] In the example given in Eq. 18.3 and Eq. 18.4, this was denoted as η_N for the North.

North with symmetric volatility. This threshold case represents the minimum extent to which the North has an advantage over the South. It implies $\alpha_{N2} = \alpha_{N3} = 0$ and only α_{N1} is non-negative (and is unambiguously positive except for the knife edge $\mu = r_N$).

The investment strategy of the South firm is analogous. The results are:

$$g_S = g_{S1} + g_{S2} \tag{18.16}$$

where $g_{S1} = (r_S - \rho)\varepsilon_S / \beta_S$ and $g_{S2} = \alpha_S(1 + \varepsilon_S)\varepsilon_S / \beta_S$, and where

$$\alpha_S = \alpha_{S1} + \alpha_{S2} + \alpha_{S3} \tag{18.17}$$

with $\alpha_{S1} = 2(\mu - r_S)^2 / \theta_S$, $\alpha_{S2} = \frac{4}{5}(\mu - r_S)(\tau_S^+ - \tau_S^-)/\theta_S$, $\alpha_{S3} = \frac{2}{25}(\tau_S^+ - \tau_S^-)^2 / \theta_S$ and $\theta_S = (\tau_S^+)^2 + (\tau_S^-)^2$. For the South firm, it is reasonable to assume k_S is below the South critical mass, so that $\tau_S^+ < \tau_S^-$. Furthermore, it is necessary to have $\mu > r_S$ when the South firm commits positive resources (x_S) to a joint venture.[30] Therefore, α_{S1} and α_{S3} are positive, but α_{S2} is negative. This contrasts to the most likely situation for α_{N2} . Crucially, if the South is to achieve convergence with the North, some of the corresponding positive terms in the South must exceed their corresponding North values.

Implications for ICT Investment and the Digital Divide

Eq. 18.14 through Eq. 18.17 show that the planned rate of investment in ICT depends on an interrelationship between risk premiums and volatility related network size effects. The components also depend on the degree of consumer flexibility relative to productivity— ε_N relative to β_N in the North and ε_S relative to β_S in the South. The role of the intertemporal elasticity of substitution in influencing investment plans is evident from the component growth terms. This influence is overlooked by a logarithmic utility specification, since $\varepsilon_N = \varepsilon_S = 1$. Similarly, the role of the capital-output elasticity— β_N in the North and β_S in the South—is ignored for linear homogeneous production function specifications for which this elasticity is unity.[31] Although logarithmic utility and linearly homoge-

---

[30] The requirement for positive x_S is that $\mu - r_S + \frac{1}{5}(\tau_S^+ - \tau_S^-) > 0$.

[31] In the modeling context, the capital-output elasticity is better termed capital-profit elasticity. While variable inputs are not explicitly considered here, an interpretation of the output functions $Y^N(k_N)$ and $Y^S(k_S)$ is that variable inputs are optimized out and that they are variable profit functions. With this interpretation, an underlying linearly homogeneous technology implies the variable profit function is linear in quasi-fixed stock.

neous production are common theoretical specifications, they are almost certainly empirically unrealistic. Given greater flexibility in the North, ε_N should typically be greater than ε_S. Econometric estimates from similar models suggest the intertemporal elasticity of substitution increases with wealth. In a study of general investment in South Africa which distinguishes 'capital poor' and 'capital rich' behavior, Cooper and Donaghy (2000) find ε to be low (at about 0.79) for the capital poor, whereas it was relatively large (at about 4.85) for the capital rich. In a more specialized cross country study of wealthier OECD Member Countries (which exclude Turkey and Mexico and concentrates on investment in Internet hosts), Cooper and Madden (2002), find a variable parameter specification of ε to be increasing in wealth (using Internet hosts as the wealth measure), ranging from a low of 3.32 (for average OECD wealth at 1996) to an estimated asymptotic upper bound of 16.66 (the estimated theoretical maximum value of the elasticity for saturation numbers of Internet hosts). Based on these considerations, it is argued that $\varepsilon > 1$ for the North, but is lower for the South and possibly below unity. Hence, for sake of discussion assume that $\varepsilon_S < 1 < \varepsilon_N$. Additionally, superior productivity in the North suggests $\beta_N > \beta_S$. However, since externalities in production are not considered here, let $\beta_S < \beta_N < 1$. Consequently, a reasonable scenario is $\varepsilon_N / \beta_N > \varepsilon_S / \beta_S$. These considerations also imply that $(1 + \varepsilon_N)\varepsilon_N / \beta_N > (1 + \varepsilon_S)\varepsilon_S / \beta_S$. Further, superior supporting institutional infrastructure in the North ($A_N > A_S$) together with technological superiority ($\beta_N > \beta_S$) suggest the marginal product of capital is higher in the North, even though there is the prospect that a lower base of ICT in the South may counter this to some extent due to decreasing returns.[32]

Given this result, a comparison of growth rates is made under a base scenario where volatility is specified by Eq. 18.3 and Eq. 18.4 with $k_N = \eta_N$ and $k_S < \eta_S$, i.e., volatility is symmetric in the North, so that $\tau_N^+ = \tau_N^- = \sigma$, but there is asymmetric volatility in the South with $\tau_S^+ < \tau_S^-$. In this case, $\theta_N = 2\sigma^2$ and $\theta_S = 2(1 + \gamma)\sigma^2$ where:

Further, $\beta_j < 1$, $j = N, S$ is a more reasonable technology specification when capital input k_j in $A_j k_j^{\beta_j}$ represents privately owned capital. External effects can be present and represented through A_j.

[32] As is argued above, the private sector is expected to operate to the point of equalization of risk premiums across the divide for a given expected return μ and same degree of fundamental risk σ. The tendency to equalize $\mu - r_N + \frac{1}{5}(\tau_N^+ - \tau_N^-)$ with $\mu - r_S + \frac{1}{5}(\tau_S^+ - \tau_S^-)$ implies a relationship between the returns r_N and r_S after adjustment for asymmetric volatility. Further, the equality of risk premiums imply that $r_N = r_S + \frac{1}{5}((\tau_N^+ - \tau_S^+) - (\tau_N^- - \tau_S^-))$. Juxtaposed with larger upside shocks in the North and larger downside shocks in the South, this implies $r_N > r_S$. This effect holds when volatility in the North is symmetric.

$$\gamma = \tfrac{1}{2}((k_N + k_S)^2 - (k_N + \eta_S)^2)^2 / (k_N + k_S)^2 (k_N + \eta_S)^2 .$$ (18.18)

By construction $\gamma > 0$, so $\theta_S > \theta_N$. Moreover, the components α_{N2} and α_{N3} in Eq. 18.15 are zero under the scenario of symmetric volatility in the North. Under these conditions comparison of the planned investment results Eq. 18.14 through Eq. 18.15 for the North with Eq. 18.16 through Eq. 18.17 for the South imply the planned growth differential:

$$g_N - g_S = \kappa_1 + \kappa_2 + \kappa_3 + \kappa_4$$ (18.19)

where:

$$\kappa_1 = (r_N - \rho)\varepsilon_N / \beta_N - (r_S - \rho)\varepsilon_S / \beta_S \qquad \text{Positive}$$

$$\kappa_2 = ((\mu - r_N)^2 (1 + \varepsilon_N)\varepsilon_N / \beta_N - (1 + \gamma)(\mu - r_S)^2 (1 + \varepsilon_S)\varepsilon_S / \beta_S) / \sigma^2 \qquad \text{Ambiguous}$$

$$\kappa_3 = -\tfrac{4}{5}(\mu - r_S + \tfrac{1}{5}(\tau_S^+ - \tau_S^-))((\tau_S^+ - \tau_S^-)/\theta_S)(1 + \varepsilon_S)\varepsilon_S / \beta_S \qquad \text{Positive}$$

$$\kappa_4 = \tfrac{2}{25}((\tau_S^+ - \tau_S^-)^2 / \theta_S)(1 + \varepsilon_S)\varepsilon_S / \beta_S . \qquad \text{Positive}$$

Although the result depends on actual parameter values, Eq. 18.19 indicates that there are more factors acting to increase the digital divide than to narrow it. The divide may be narrowed, but only when κ_2, which is in general ambiguous, is sufficiently negative to outweigh the positive terms. Crucial to the sign of κ_2 is the size of γ. From Eq. 18.18, γ rises as η_S falls. That is, a lower critical mass helps.

Conclusion

This study provides the structure of a model developed to address issues associated with the digital divide between developed (North) and developing (South) nations. Building on insights in growth theory, the analysis posits firms in the North and South with characteristics appropriate to firms in the ICT sector that have an incentive to joint venture to capture additional network benefit. The question arises, in an era of significant technological evolution, whether the benefits are spread or are capable of being spread, in such a fashion so as to ameliorate the substantial economic imbalance between the North and South. It is evident from the reported results that hoped for technology transfer solutions to necessitate growth and convergence, are too simplistic. A particularly important factor that is highlighted is the influence of substantial asymmetry in the impact of uncertainty. Since a high speed of technology change can exacerbate asymmetric uncertainty, it is particularly important to investigate mechanisms through which this effect influences private sector decision making and identify available policy parameters to ameliorate the impact of the evolving digital divide.

When the South ICT capital stock is low relative to the critical mass required for symmetric volatility, asymmetric volatility tends to favor a divergence in growth rates across the digital divide, even with optimal technology transfer from the perspective of the South firm. This process increases the degree of asymmetry, exacerbating the problem. Although this is a pessimistic conclusion, an advantage of the formulation is that it offers a cogent description of the nature of the evolving digital divide, raising the possibility that estimating forms based on this specification are capable of contributing to a better understanding of how and why, from the private sector's perspective, such divergence may be the result of optimal decision making. Further, the analysis raises the prospect of the application of policy, represented here in a stylized fashion by attention to parameters such as those describing the ICT critical mass, to influence private sector incentives and so enhance the prospects for private sector investment decisions being compatible with the bridging of the digital divide.

References

Bresnahan TF, Greenstein S (2002) The economic contribution of information technology: Value indicators in international perspective. Stanford University, mimeo

Bridges.org (2001) Spanning the digital divide: Understanding and tackling the issues. http://www.bridges.org

Cooper RJ, Donaghy KP (2000) Risk and growth: Theoretical relationships and preliminary estimates for South Africa. 8th World Congress of the Econometric Society, Seattle, US

Cooper RJ, Madden G (2002) Network externalities and the Internet. Presented at 14th Biennial Conference of the International Telecommunications Society, Seoul, Korea

ITU (2002) World telecommunication indicators database. CD-ROM, International Telecommunication Union, Geneva

Jalava J, Pohjola M (2002) Economic growth in the new economy: Evidence from advanced economies. Information Economics and Policy 14(2): 189–210

Noam E (2003) Corporate and regulatory strategy for the new network century. In: Madden G (ed) The international handbook of telecommunications economics, vol III: World telecommunications markets. Edward Elgar, Cheltenham, pp 1–11

Rodríguez F, Wilson EJ (2000) Are poor countries losing the information revolution? infoDev Working Paper, The World Bank, Washington

Röller LH, Waverman L (2001) Telecommunications infrastructure and economic development: A simultaneous approach. American Economic Review 91(4): 909–23

Temple J (1999) The new growth evidence. Journal of Economic Literature 37(1): 112–56

List of Contributors

Jiwoon Ahn, PhD Candidate
   Techno-Economics and Policy Program, Seoul National University
   San 56-1, Shilim-dong, Kwanak-Gu, Seoul, 151-742
   Korea
   E-mail: indra4@freechal.com

James H. Alleman, Associate Professor
   College of Engineering & Applied Science, University of Colorado
   CB 530, Boulder, CO 80309-0530
   USA
   E-mail: james.alleman@Colorado.edu

Aniruddha Banerjee, PhD, Vice President
   NERA Economic Consulting
   One Main Street, 5th Floor, Cambridge, MA 02142
   USA
   E-mail: andy.banerjee@nera.com

Johannes M. Bauer, Professor
   Department of Telecommunication, Information Studies and Media
   Michigan State University
   Room 409, Communication Arts Building, East Lansing, MI 48824-1212
   USA
   E-mail: bauerj@msu.edu

Jeffery I. Bernstein, Chancellor's Professor
   Department of Economics, Faculty of Public Affairs and Management
   Carleton University
   1125 Colonel By Drive, Ottawa Ontario, K1S 5B6
   Canada
   E-mail: jeff_bernstein@carleton.ca

Sang-Kyu Byun, PhD, Senior Researcher
   Electronics and Telecommunications Research Institute, ITMRG
   161 Gajeong-dong, Yuseong-gu, Daejeon, 305-350
   Korea
   E-mail: skbyun@etri.re.kr

Grant Coble-Neal, PhD Candidate
Communication Economics and Electronic Markets Research Centre
Curtin University of Technology
GPO Box U1987, Perth, WA 6845
Australia
E-mail: maddeng@cbs.curtin.edu.au

Russel Cooper, Professor
Australian Expert Group in Industry Studies, School of Economics and Finance
University of Western Sydney, Nepean
Level 8, 263 Clarence Street, Sydney, NSW 1230
Australia
E-mail: r.cooper@uws.edu.au

Jerry Hausman, John and Jennie S. MacDonald Professor
Department of Economics, Massachusetts Institute of Technology
E52-271a, 50 Memorial Drive, Cambridge, MA 02142-1347
USA
E-mail: jhausman@mit.edu

Andrea L. Kavanaugh, PhD, Assistant Director
Center for Human Computer Interaction, Department of Computer Science
Virginia Tech
660 McBryde Hall (0106), Blacksburg, VA 24061
USA
E-mail: kavan@vt.edu

Hak Ju Kim, PhD Candidate
Department of Information Science and Telecommunications
School of Information Sciences, University of Pittsburgh
Information Sciences Building, 135 N. Bellefield Ave, Pittsburgh, PA 15206
USA
E-mail: hjkim@mail.sis.pitt.edu

Jeong-Dong Lee, Professor
Techno-Economics and Policy Program, Seoul National University
San 56-1, Shilim-dong, Kwanak-Gu, Seoul, 151-742
Korea
E-mail: leejd@snu.ac.kr

Jongsu Lee, PhD
Techno-Economics and Policy Program, Seoul National University
San 56-1, Shilim-dong, Kwanak-Gu, Seoul, 151-742
Korea
E-mail: jxlee@snu.ac.kr

Gary Madden, Professor
   Communication Economics and Electronic Markets Research Centre
   Curtin University of Technology
   GPO Box U1987, Perth, WA 6845
   Australia
   E-mail: maddeng@cbs.curtin.edu.au

Phil Malone, Manager, E-commerce
   National Office for the Information Economy
   Department of Communications, Information Technology and the Arts
   GPO Box 2154, Canberra ACT 2601
   Australia
   E-mail: phil.malone@noie.gov.au

M. Ishaq Nadiri, Jay Gould Professor of Economics
   Department of Economics, New York University
   269 Mercer Street, 7th Floor, New York, NY 10003
   USA
   E-mail: m.ishaq.nadiri@nyu.edu

Banani Nandi, PhD, Principal Technical Staff
   AT&T Shannon Laboratories
   180 Park Avenue, Building 103, Florham Park, NJ 07932
   USA
   E-mail: ban@homer.att.com

Sam Paltridge, PhD, Communication Analyst
   Information, Computer and Communications Policy Division
   Directorate for Science, Technology and Industry
   Organisation for Economic Co-operation and Development
   2, rue André-Pascal, 75775 Paris Cedex 16
   France
   E-mail: sam.paltridge@oecd.org

Paul N. Rappoport, Associate Professor
   Economics Department, School of Business and Management
   Temple University
   Philadelphia, PA 19122
   USA
   E-mail: prapp@sbm.temple.edu

Yong-Yeap Sohn, Associate Professor
    Department of Economics, College of Business Administration
    Chonnam National University
    300 Yongbong-dong, Buk-gu, Gwangju, 500-757
    Korea
    E-mail: yysohn@chonnam.kr

Charles Steinfield, Professor
    Department of Telecommunication, Information Studies and Media
    Michigan State University
    Room 436, Communication Arts Building, East Lansing, MI 48824-1212
    USA
    Email: steinfie@msu.edu

Lester D. Taylor, Professor
    College of Business and Public Administration, University of Arizona
    Room ECON 322, 322 Economics Building, Tucson, AZ 85721
    USA
    E-mail: ltaylor@bpa.arizona.edu

James H. Tiessen, Associate Professor
    Michael G. DeGroote School of Business, McMaster University
    1280 Main Street, Hamilton, Ontario L8S 4M4
    Canada
    Email: tiessenj@mcmaster.ca

Glenn A. Woroch, Professor
    Department of Economics, University of California Berkeley
    549 Evans Hall #3880, Berkeley, CA 94720-3880
    USA
    E-mail: glenn@econ.berkeley.edu

Dimitri Ypsilanti, Head, Telecommunication and Information Policy Section
    Information, Computer and Communications Policy Division
    Directorate for Science, Technology and Industry
    Organisation for Economic Co-operation and Development
    2, rue André-Pascal, 75775 Paris Cedex 16
    France
    E-mail: dimitri.ypsilanti@oecd.org

Charles Zarkadas, PhD, Special Consultant
    NERA Economic Consulting
    One Main Street, 5th Floor, Cambridge, MA 02142, Cambridge
    USA
    E-mail: charles.zarkadas@nera.com

Index